ESO ASTROPHYSICS SYMPOSIA
European Southern Observatory

Series Editor: Bruno Leibundgut

Peter J. Quinn Krzysztof M. Górski (Eds.)

Toward an International Virtual Observatory

Proceedings
of the ESO/ESA/NASA/NSF Conference
Held at Garching, Germany, 10-14 June 2002

Springer

Volume Editors

Peter J. Quinn

European Southern Observatory
Karl-Schwarzschild-Strasse 2
85748 Garching, Germany

Krzysztof M. Górski

Jet Propulsion Laboratory
M/S 169/327
4800 Oak Grove Drive
Pasadena, CA, 91109, USA

Series Editor

Bruno Leibundgut

European Southern Observatory
Karl-Schwarzschild-Strasse 2
85748 Garching, Germany

ISBN 978-3-662-30817-2 ISBN 978-3-540-39908-7 (eBook)
DOI 10.1007/978-3-540-39908-7

Library of Congress Control Number:2004107508

springeronline.com

Originally published by Springer-Verlag Berlin Heidelberg New York in 2004
Softcover reprint of the hardcover 1st edition 2004

Typesetting: Camera-ready by the authors/editors
Cover design: Erich Kirchner, Heidelberg

Printed on acid-free paper 55/3141/du - 5 4 3 2 1 0

Preface

Astronomy faces a data avalanche. Breakthroughs in telescope, detector, and computer technology allow astronomical instruments to produce terabytes of images and catalogs. These datasets will cover the sky in different wavebands, from gamma- and X-rays, optical, infrared, through to radio. In a few years it will be easier to "dial-up" a part of the sky than wait many months to access a telescope. With the advent of inexpensive storage technologies and the availability of high-speed networks, the concept of multi-terabyte on-line databases interoperating seamlessly is no longer outlandish. More and more catalogs will be interlinked, query engines will become more and more sophisticated, and the research results from on-line data will be just as rich as that from "real" telescopes. Moore's law is driving astronomy even further: new survey telescopes now being planned will image the entire sky every few days and yield data volumes measured in petabytes. These technological developments will fundamentally change the way astronomy is done. These changes will have dramatic effects on the sociology of astronomy itself.

Over the past two years the concept of the Virtual Observatory has emerged rapidly to address the data management, analysis, distribution and interoperability challenges. Astronomers world-wide are seeking to build a Virtual Observatory (VO) with international connectivity and capabilities. The Virtual Observatory will be a system that allows astronomers to interrogate multiple data centres in a seamless and transparent way, which provides new powerful analysis and visualisation tools within that system, and which gives data centres a standard framework for publishing and delivering services using their data. This is made possible by standardisation of data and metadata, by standardisation of data exchange methods, and by the use of a registry which lists available services and what can be done with them. The long term vision is not one of a fixed specific software package, but rather one of a framework which enables data centres to provide competing and co-operating data services, and which enables software providers to offer a variety of compatible analysis and visualisation tools and user interfaces.

The first step for the VO projects worldwide, is to develop the standardised framework which will allow such creative diversity. Once in place, the framework must be taken up by data providers to allow them to publish their holdings and make their services and facilities available. The framework will then empower scientists and developers to provide new tools for research and to undertake new research programs to tackle complex astronomical and astrophysical problems. There are, therefore, three fundamental tasks to be undertaken to make the VO a successful research infrastructure.

1. The completion and deployment of a VO technical infrastructure
2. The uptake of this infrastructure by data providers
3. The support of the research and development community to utilize this infrastructure and data content to discover new knowledge and build new capabilities

A vital first step along the road to address these tasks was taken at our meeting in Munich. During the course of the meeting, the leaders of the AVO, NVO and ASTROGRID projects proposed the formation of an International Virtual Observatory Alliance (IVOA, `http://www.ivoa.net`). This alliance is providing the essential international collaboration and coordination necessary to achieve a unified and universal technical and interoperability infrastructure for the IVO. At this time 14 international projects have joined the IVOA and are working together on a jointly agreed development roadmap with priorities set for the demonstration of new VO capabilities over the next 2–3 years. Astronomy now stands at the edge of a new frontier for discovery, enabled by modern information technology (such as the Grid) and by political and technical collaboration among international partners.

Munich, *Peter Quinn*
November 2003

Contents

Poster Papers

The conference participants (photo courtesy of R. Fosbury)

List of Participants

Allen, Mark
CDS Strasbourg
allen@astro.u-strasbg.fr

Ansari, Salim
ESA - ESTEC
Salim.Ansari@esa.int

Arviset, Christophe
ESA RSSD, Madrid
Christophe.Arviset@esa.int

Babu, G. Jogesh
Pennsylvania State University
babu@stat.psu.edu

Banday, Anthony
MPA Garching
banday@MPA-Garching.MPG.de

Baruffolo, Andrea
INAF - Observatory of Padova
baruffolo@pd.astro.it

Benvenuti, Piero
ST-ECF Garching
pbenvenu@eso.org

Bergier, Alex
Columbia Astrophysics Lab.,
New York
alex@astro.columbia.edu

Blanchard, Alain
LAS Toulouse
alain.blanchard@ast.obs-mip.fr

Boch, Thomas
CDS Strasbourg
boch@vizier.u-strasbg.fr

Bonnarel, François
CDS Strasbourg
bonnarel@astro.u-strasbg.fr

Brusberg, Göran
Solac, Bromma
gbrusb@algonet.se

Bykov, Oleg
Pulkovo Observatory, St. Petersburg
oleg@OB3876.spb.edu

Caux, Emmanuel
CESR Toulouse
caux@cesr.fr

Cesarsky, Catherine
ESO Garching
ccesarsk@eso.org

Cimatti, Andrea
Osservatorio Astrofisico di Arcetri
cimatti@arcetri.astro.it

Corbin, Michael
STScI Baltimore
corbin@stsci.edu

Costa, Marco
Critical Software S.A., Coimbra
jcar@criticalsoftware.com

Cresitello-Dittmar, Mark
Smithsonian Astrophysical Observatory
mdittmar@cfa.harvard.edu

Cseresnjes, Patrick
Columbia Astrophysics Lab., New York
patrick@astro.columbia.edu

Csillaghy, Andre
University of Applied Sciences, Windisch
csillsg@fh-aargau.ch

da Costa, Luiz
ESO Garching
ldacosta@eso.org

Davis, Lindsey
NOAO Tucson
davis@noao.edu

de Grijs, Richard
IoA Cambridge
grijs@ast.cam.ac.uk

De Young, David
NOAO Tucson
deyoung@noao.edu

Debray, Bernard
Observatoire de Besançon
debray@obs-besancon.fr

Dekel, Avishai
Hebrew University, Jerusalem
dekel@phys.huji.ac.il

Delmotte, Nausicaa
CDS Strasbourg / ESO Garching
delmotte@astro.u-strasbg.fr

Derriere, Sébastien
CDS Strasbourg
derriere@astro.u-strasbg.fr

Diamond, Philip
Jodrell Bank Observatory
pdiamond@jb.man.ac.uk

Djorgovski, George
Caltech, Pasadena
george@astro.caltech.edu

Dolensky, Markus
ST-ECF Garching
Markus.Dolensky@eso.org

Doroshkevich, Andrei
Theoretical Astrophysics Center, Copenhagen
dorr@tac.dk

Durand, Daniel
CADC Victoria
Daniel.Durand@nrc.ca

Egret, Daniel
CDS Strasbourg
Daniel.Egret@astro.u-strasbg.fr

Eichhorn, Günther
CfA Cambridge
gei@cfa.harvard.edu

Engels, Dieter
Hamburger Sternwarte
dengels@hs.uni-hamburg.de

Fabbiano, Giuseppina
Smithsonian Astrophysical Observatory
pepi@cfa.harvard.edu

Fernique, Pierre
CDS Strasbourg
fernique@simbad.u-strasbg.fr

Fosbury, Robert
ST-ECF Garching
rfosbury@eso.org

Friel, Eileen
National Science Foundation, Arlington
efriel@nsf.gov

Gal, Roy
Johns Hopkins University
rrg@pha.jhu.edu

Gallegos, Julio
Instituto de Fisica de Cantabria, Santander
jgallego@ifca.unican.es

Garrington, Simon
MERLIN/VLBI National Facility
stg@jb.man.ac.uk

Genova, Françoise
CDS Strasbourg
genova@astro.u-strasbg.fr

Giaretta, David
Rutherford Appleton Laboratory
d.giaretta@rl.ac.uk

Gilmore, Gerry
IoA Cambridge
gil@ast.cam.ac.uk

Giménez Cañete, Álvaro
ESA – ESTEC
Alvaro.Gimenez@esa.int

Godlowski, Wlodzimierz
Krakow Astronomical Observatory
godlows@oa.uj.edu.pl

Gómez de Castro, Ana I.
Instituto de Astronomia y Geodesia
aig@mat.ucm.es

Górski, Krzysztof
ESO Garching
kgorski@eso.org

Gray, Jim
Microsoft, San Francisco
Gray@Microsoft.com

Grosbøl, Preben
ESO Garching
pgrosbol@eso.org

Grupe, Dirk
MPE Garching
dgrupe@mpe.mpg.de

Gutierrez, Raul
LAEFF Madrid
raul@laeff.esa.es

Haase, Jonas
ST-ECF Garching
jhaase@eso.org

Hanisch, Robert
STScI Baltimore
hanisch@stsci.edu

Hatziminaoglou, Evanthia
ESO Garching
ehatzimi@eso.org

Helfand, David
Columbia Astrophysics Lab., New York
djh@astro.columbia.edu

Hessman, Frederic
Universitäts-Sternwarte Göttingen
hessman@uni-sw.gwdg.de

Hudec, Rene
Astronomical Institute Ondrejov
rhudec@asu.cas.cz

Ibata, Rodrigo
CDS Strasbourg
ibata@astro.u-strasbg.fr

Kalinina, Nadejda
Special Astrophysical Observatory
kana@sao.ru

Kembhavi, Ajit
IUCAA, Pune
akk@iucaa.ernet.in

Kerber, Florian
ST-ECF Garching
fkerber@eso.org

Kontizas, Mary
National & Kapodistrian University of Athens
mkontiza@cc.uoa.gr

Kroll, Peter
Sonneberg Observatory
pk@stw.tu-ilmenau.de

Kudryavtsev, Sergey
Maris Technologies, Moscow
ksm@tsinet.ru

Kuijken, Konrad
Kapteyn Institute, Groningen
kuijken@astro.rug.nl

Kunze, Marcel
Forschungszentrum Karlsruhe
Marcel.Kunze@hik.fzk.de

Lawrence, Andrew
Royal Observatory Edinburgh
al@roe.ac.uk

Leahy, Patrick
Jodrell Bank Observatory
jpl@jb.man.ac.uk

Lewis, James
IoA Cambridge
jrl@ast.cam.ac.uk

Linde, Tony
University of Leicester
tol@star.le.ac.uk

Linsolas, Romain
IAS – CNRS, Orsay
romain.linsolas
@medoc-ias.u-psud.fr

Liu, Charles
American Museum of Natural History, New York
cliu@astro.amnh.org

Longo, Giuseppe
Università Federico II di Napoli
longo@na.infn.it

Lorenzen, Dirk
PRESS – German National Radio
DLorenzen@compuserve.com

Louys, Mireille
CDS Strasbourg
louys@astro.u-strasbg.fr

Maccacaro, Tommaso
Brera Astronomical Observatory
tommaso@brera.mi.astro.it

Magagna, Carlo
INAF – Observatory of Padova
magagna@pd.astro.it

Malkov, Oleg
Institute of Astronomy, Moscow
malkov@inasan.ru

Mann, Robert
Royal Observatory Edinburgh
rgm@roe.ac.uk

McDowell, Jonathan
Smithsonian Astrophysical Observatory
jcm@cfa.harvard.edu

McGlynn, Thomas
Goddard Space Flight Center
tam@lheapop.gsfc.nasa.gov

McLean, Brian
STScI Baltimore
mclean@stsci.edu

McMahon, Richard
IoA Cambridge
rgm@ast.cam.ac.uk

Mellier, Yannick
Institut d'Astrophysique de Paris
mellier@iap.fr

Micol, Alberto
ST-ECF Garching
amicol@eso.org

Moore, Reagan
University of California, La Jolla
moore@sdsc.edu

Nakanishi, Koichiro
National Astronomical Observatory, Tokyo
nakanisi@nro.nao.ac.jp

Naylor, Tim
School of Physics, University of Exeter
timn@astro.ex.ac.uk

Norris, Ray
CSIRO ATNF, Epping
ray.norris@csiro.au

Ochsenbein, François
CDS Strasbourg
francois@astro.u-strasbg.fr

Olnon, Friso
Joint Institute for VLBI in Europe, Dwingeloo
olnon@jive.nl

Omont, Alain
Institut d'Astrophysique de Paris
omont@iap.fr

Oosterloo, Tom
ASTRON, Dwingeloo
oosterloo@astron.nl

Ortiz, Patricio
Osservatorio Astronomico di Capodimonte
portiz@na.astro.it

Ostriker, Jeremiah
IoA Cambridge
jpo@ast.cam.ac.uk

Paczynski, Bohdan
Princeton University
bp@astro.princeton.edu

Page, Clive
University of Leicester
cgp@star.le.ac.uk

Pasian, Fabio
INAF – Observatory of Trieste
pasian@ts.astro.it

Peebles, P. James
Princeton University
pjep@pupgg.princeton.edu

Pennypacker, Carlton
University of California, Berkeley
pennypacker@lbl.gov

Peron, Michele
ESO Garching
mperon@eso.org

Pierfederici, Francesco
ST-ECF Garching
fpierfed@eso.org

Pirenne, Benoit
ESO Garching
bpirenne@eso.org

Pirzkal, Norbert
ST-ECF Garching
npirzkal@eso.org

Plante, Raymond
University of Illinois
Urbana-Champaign
rplante@ncsa.uiuc.edu

Pope, Adrian
Johns Hopkins University
pope@pha.jhu.edu

Pottschmidt, Katja
INTEGRAL Science Data Centre,
Versoix
katja.pottschmidt@obs.unige.ch

Prugniel, Philippe
CRAL, Observatoire de Lyon
prugniel@obs.univ-lyon1.fr

Quinn, Peter
ESO Garching
pquinn@eso.org

Reardon, Kevin
Osservatorio Astrofisico di Arcetri
kreardon@arcetri.astro.it

Renzini, Alvio
ESO Garching
arenzini@eso.org

Richards, Anita M.S.
Jodrell Bank Observatory
amsr@jb.man.ac.uk

Rite, Charles
ESO Garching
crite@eso.org

Rixon, Guy
IoA Cambridge
guy_rixon@hotmail.com

Rodríguez Fernández, Alejandra
University of La Coruña
alex@mail2.udc.es

Rosa, Michael
ST-ECF Garching
mrosa@eso.org

Rots, Arnold
CfA/ULP Cambridge
arots@head-cfa.harvard.edu

Schaaff, Andre
CDS Strasbourg
schaaff@astro.u-strasbg.fr

Schade, David
CADC Victoria
David.Schade@nrc.ca

Scholl, Isabelle
IAS - CNRS, Orsay
isabelle.scholl
@medoc-ias.u-psud.fr

Schreier, Ethan
Associated Universities, Inc.,
Washington
ejs@aui.edu

Shaw, Richard
NOAO Tucson
shaw@noao.edu

Smareglia, Riccardo
INAF – Observatory of Trieste
smareglia@ts.astro.it

Solano, Enrique
LAEFF Madrid
esm@laeff.esa.es

Staiano, Antonio
University of Salerno
astaiano@unisa.it

Sygnet, Jean-François
IAS – CNRS, Orsay
Jean-Francois.Sygnet
@ias.u-psud.fr

Szalay, Alexander
Johns Hopkins University
szalay@jhu.edu

Szokoly, Gyula Pal
MPE Garching
gpszokoly@aip.de

Takata, Tadafumi
National Astronomical Observatory,
Tokyo
takata@naoj.org

Taylor, Philip
Observatory Sciences Ltd., Eastbourne
pbt@observatorysciences.co.uk

Tody, Doug
NOAO Tucson
tody@noao.edu

Valentijn, Edwin
Kapteyn Institute, Groningen
valentyn@astro.rug.nl

Vandame, Benoit
ESO Garching
bvandame@eso.org

Verkhodanov, Oleg
Special Astrophysical Observatory
vo@sao.ru

Verkhodanova, Natalia
Special Astrophysical Observatory
nat@sed.sao.ru

Vettolani, Giampaolo
Istituto di Radioastronomia CNR,
Bologna
vettolani@ira.cnr.it

Vijayaraman, T.M.
Persistent Systems, Pune
vijay@persistent.co.in

Vitkovskij, Vladimir
Special Astrophysical Observatory
vvv@sao.ru

Voges, Wolfgang
MPE Garching
wvoges@mpe.mpg.de

Voit, G. Mark
STScI Baltimore
voit@stsci.edu

Walton, Nicolas
IoA Cambridge
naw@ast.cam.ac.uk

Watson, Mike
Astrogrid, University of Leicester
mgw@star.le.ac.uk

Way, Michael
NASA Ames Research Center,
Moffett Field
mway@mail.arc.nasa.gov

Weghorn, Hans
University of Cooperative Education,
Stuttgart
weghorn@ba-stuttgart.de

Wenger, Marc
CDS Strasbourg
wenger@astro.u-strasbg.fr

White, Simon
MPA Garching
swhite@mpa-garching.mpg.de

Wicenec, Andreas
ESO Garching
awicenec@eso.org

Williams, Roy
Caltech Pasadena
roy@cacr.caltech.edu

Yadav, J.S.
Tata Institute of Fundamental Research, Mumbai
jsyadav@tifr.res.in

Yamamoto, Kazuhiro
MPA Garching
kyama@mpa-garching.mpg.de

Yasuda, Naoki
National Astronomical Observatory, Tokyo
naoki.yasuda@nao.ac.jp

Zarate, Nelson
NOAO Tucson
zarate@noao.edu

Zhao, Yongheng
National Astronomical Observatories, Beijing
yzhao@lamost.org

Zhelenkova, Olga
Special Astrophysical Observatory
zhe@sao.ru

Zinnecker, Hans
AIP Potsdam
hzinnecker@aip.de

Zurek, David
American Museum of Natural History, New York
dzurek@amnh.org

Nineteenth and Twentieth Century Clouds Over the Twenty-First Century Virtual Observatory

P.J.E. Peebles

Joseph Henry Laboratories, Jadwin Hall,
Princeton University, Princeton NJ 08544, USA

Abstract. Physical science has changed in the century since Lord Kelvin's celebrated essay on *Nineteenth Century Clouds over the Dynamical Theory of Heat and Light*, but some things are the same. Analogs in what was happening in physics then and what is happening in astronomy today serve to remind us why we can be confident the Virtual Observatory of the twenty-first century will have a rich list of challenges to explore.

1 Introduction

Astronomy has enjoyed a very good century. Have the basic problems now been solved, leaving for the astronomers of the 21st century the task of working out the pesky details? The question is little discussed – astronomers are too busy with ongoing research – but worth considering from time to time. I shall argue that we have a useful guide to the long-term prospects for research in astronomy from analogs to the present situation in was happening in physics 100 years ago. In both cases there is a basis of fundamental concepts that are strikingly successful, apart from some stubborn clouds, or, as we would now say, challenges for research. The clouds over electromagnetism and thermal physics at the start of the 20th century foreshadowed relativity and quantum physics. We can't say what will be learned from the clouds over present-day astronomy – I shall mention aspects of the dark sector, strong space curvature, and the meaning of life – but we can be sure they will continue to drive difficult but fascinating research in astronomy for quite some time to come.

2 Physics at the Start of the 20th Century

The elements of the situation in physics a century ago have been retold to generations of students, and rightly so; these are golden moments in the history of physical science. And I think they are an edifying example for our assessment of the present state of research in astronomy.

At the start of the 20th century physicists had good reason to believe they had securely established laws of electromagnetism and thermal physics, well tested in the laboratory and applied in rapidly growing power and communications industries; Lord Kelvin's fortune came from his contributions to the design of the transatlantic telegraph cable. But he and others were well aware of flaws, or clouds, in the physics, as famously summarized in Kelvin's [1] essay in 1901.

Kelvin's Cloud No. I is the luminiferous ether. The experimental situation is also discussed in Lecture VIII, *The Ether*, in Michelson's *Light Waves and Their Uses* [2]. You can read about the familiar experiments – Michelson's discovery of the isotropy of the velocity of light, and the Fizeau measurement of the addition of velocities of light and fluid in a moving fluid – and others that are less celebrated but as remarkable. My favorite among the latter is the measurement of annual aberration in a telescope that is filled with water so as to reduce the velocity of light. The results are no surprise to us, but a real problem for Kelvin and Michelson. Kelvin mentions with approval the contraction idea of Fitzgerald and Lorentz, but concludes "I am afraid we must still regard Cloud No. I as very dense" [1]. Einstein's brilliant insight cleared the cloud, and gave us special relativity theory. If that had not happened I have to believe people would soon have pieced together the full theory from these remarkable measurements.

Kelvin's Cloud II is the inconsistency of the law of partition of energy at thermal equilibrium with the measured ratios C_p/C_v of heat capacities of gases at constant pressure and volume.[1] Kelvin [1] quotes Rayleigh's assessment [3]: "The difficulties connected with the application of the law of equal partition of energy to actual gases have long been felt. In the case of argon and helium and mercury vapour the ratio of specific heats (1.67) limits the degrees of freedom of each molecule to the three required for translatory motion. The value (1.4) applicable to the principal diatomic gases gives room for three kinds of translation and for two kinds of rotation. Nothing is left for rotation round the line joining the atoms, nor for relative motion of the atoms in this line. Even if we regard the atoms as mere points, whose rotation means nothing, there must still exist energy of the last-mentioned kind, and its amount (according to the law) should not be inferior." Something certainly is wrong. Kelvin accepted the mechanics and questioned the assumption of strict statistical equilibrium. Planck (1900) hit on the fix, to the mechanics, in the model for blackbody radiation, and Einstein (1907) applied the fix to heat capacities, in early steps to quantum physics.

It is often said, at least in introductory remarks in courses on modern physics, that people were a lot more impressed by the successes of physics in 1900 than by the clouds, that the feeling was that physics is essentially complete, apart from fixing a few problems and adding decimal places. The famous example is Michelson's statement (in [2], p.23), that the "more important fundamental laws and facts of physical science have all been discovered, and these are now so firmly

[1] This memorable story has been taught to generations of students in the introduction to quantum mechanics, some of whom I hope actually appreciated it. Equipartition in classical mechanics says that at thermal equilibrium at temperature T the mean energy belonging to each quadratic term in the Lagrangian is $kT/2$. It follows that if each atom or molecule in a gas has ν quadratic terms associated with its internal structure then, taking account of the pdV work at constant p, the ratio of heat capacities is $C_p/C_v = (5+\nu)/(3+\nu)$. Thus classical physics predicts that a gas of point-like particles has $C_p/C_v = 5/3$, and a gas of atoms with a rich internal structure, so ν is large, has C_p/C_v close to unity.

established that the possibility of their ever being supplanted in consequence of new discoveries is exceedingly remote." This is clear enough, and Badash [4] shows Michelson repeated these sentiments elsewhere, so at the time he must have meant it. But consider Michelson's summary statement in the same book, at the end of the chapter on the ether ([2], p.163): "The phenomenon of the aberration of the fixed stars can be accounted for on the hypothesis that the ether does not partake of the Earth's motion in its revolution about the sun. All experiments for testing this hypothesis have, however, given negative results, so that the theory may still be said to be in an unsatisfactory condition." And earlier in the summary he says "Little as we know about it [the ether], we may say that our ignorance of ordinary matter is still greater." Here Michelson sounds like someone who sees very real challenges.

These challenges drove hard work, as in Fizeau's remarkable waterworks and Michelson's [2] massive arrangements to suppress vibrations: "the apparatus was mounted on a stone support, about four feet square and one foot thick, and this stone was mounted on a circular disc of wood which floated in a tank of mercury." I see no evidence of complacency in Kelvin's ([1], p.17) struggle to visualize mercury vapor atoms, which are capable of producing a rich line spectrum but at thermal equilibrium in laboratory conditions seem to be incapable even of rotating, or in Rowland's 1899 presidential address to the American Physical Society [5]: "What is matter; what is gravitation; what is ether and the radiation through it; what is electricity and magnetism; how are these connected together and what is their relation to heat? These are the greater problems of the universe. But many infinitely smaller problems we must attack and solve before we can even guess at the solution of the greater ones."

Badash [4] gives a valuable survey of opinions across a broader range of the academic community, and concludes that at the end of the 19th century the idea that science is reaching completeness 'was more a "low-grade infection," *but nevertheless very real.*' This sounds right, but my impression is that the infection had little effect on the research of leading physicists, including Michelson.

The confidence in the established parts of physics at the start of the 20th century was well placed: we still use and teach this electromagnetism and thermodynamics – though we now think of it as part of a hierarchy of approximations that for all we will ever know may run arbitrarily deep. Concerns about the 19th clouds could not have anticipated the vast enlargement of physics and our worldview in the 20th century, but the point for our purpose is that the clouds were recognized and driving research.

I offer some parallels to the situation in present-day astronomy. We know how stars like the Sun shine, but there are big gaps in our understanding of how stars form, at high redshift and even in our own galaxy. I classify star formation as a Rowland-type "smaller problem:" it is fiendishly difficult but approachable by well-motivated lines of research involving standard physics (as far as we know). Such Rowland-type problems are the key to a healthy science, and astronomy has them in abundance. We know the universe is evolving, and the evidence is that general relativity theory gives a good description of the dynamics. But we don't know what the universe is made of – apart from the five percent or so in the visible sector – or what happens when spacetime curvature gets large, as was the case in the very early universe and happens now in the centers of galaxies. These are Kelvin-level clouds: critical issues whose resolution would greatly advance our understanding of the material world. We don't know what the present-day clouds are hiding, but we can be sure they will continue to be a good focus for research.

3 Astronomy at the Start of the 21st Century

The situation in astronomy in 1900 was close to the academic myth about physics. Badash [4] gives a good quote from Newcomb [6]: "we do appear to be fast approaching the limits of our knowledge ... one comet is so much like another that we cannot regard one as adding in any important degree to our knowledge. The result is that the work which really occupies the attention of the astronomer is less the discovery of new things than the elaboration of those already known, and the entire systemization of our knowledge." The main systemization was the cataloging of angular positions, apparent magnitudes, and spectral classifications of literally hundreds of thousands of stars. But this dreary labor led to wonderful new things; consider these two examples of research trajectories.[2]

Eddington's (1924) gas spheres gave Bethe (1938) the physical conditions for nuclear reactions in stars, and a way out of the discrepancy between the Helmholz–Kelvin (1860) Solar cooling time and the much greater geological times from radioactive decay ages. A beautiful recent development is the demonstration that the Solar neutrino luminosity really is in satisfactory agreement with the theory of the Solar nuclear reaction rates, to be understood with the help of the demonstration of nonzero neutrino masses.

Kapteyn (1901) set the distance scale for star counts in our island universe, Shapley (1918) enlarged the island, and Hubble (1925) placed it in the near homogeneous realm of the nebulae. Hubble's linear relation between his distances to the nebulae and Slipher's (1914) redshifts led Lemaître (1927) to the now standard model for the expanding universe. The most direct evidence that our universe actually is evolving – expanding and cooling – was completed with the demonstration by the USA COBE and Canadian UBC experiments (1990) that the 3 K cosmic background radiation spectrum is very close to thermal. In the 1930s Hubble commenced the great program of cosmological tests to check the relativistic Friedmann–Lemaître model for the expanding universe. Now, seven

[2] I have taken the liberty of indicating contributions by several people, and even groups, under the name of a representative leading figure, with an approximate year for developments that in some cases occurred over many years. I hope it is understood that another reviewer could choose very different representative examples of what happened in 20th century astronomy. Harwit [7] presents a well-documented and much more complete analysis of discoveries in astronomy and the prospects for discoveries of new astronomical phenomena.

decades later, we are approaching a satisfactory application of the tests, which the relativistic cosmology passes so far.

A byproduct of the cosmological tests is evidence that structure grew out of a mass distribution at high redshift that is specified by one function of one variable, the near scale-invariant power spectrum of a random Gaussian process. There are problems with details, as will be discussed, but the evidence pretty strongly indicates this is a good approximation to the way it is. The Rowland-type problem, of breathtaking scope and complexity, is to demonstrate that standard physics actually can account for the origin of the worlds and their spectacular variety of phenomena out of this simple initial condition.

I have mentioned stories with some happy endings, in reasonably conclusive resolutions of lines of research that have occupied generations of astronomers. We cannot say whether more happy endings to big puzzles are in store, but we get some feeling for the prospects by considering present-day clouds over astronomy. I shall comment on two from the 20th century and one from the 19th century.

3.1 Cloud No. I: the Dark Sector

The dark sector includes the nonbaryonic matter that is thought to dominate the outer parts of galaxies and clusters of galaxies; Einstein's cosmological constant, Λ, or dark energy that acts like it; and the vacuum energy density. The darkest part of the cloud is over the vacuum energy. I draw these comments from a review of the issues in [8] and the executive summary in [9].

Nernst [10] seems to be the first to have discussed the energy of the quantum vacuum, in 1916. His zero-point energy for each mode of oscillation of the electromagnetic field is off by a factor of two, remarkably good considering this was before Heisenberg and Schrödinger. Nernst showed that the sum over zero-point energies of the modes with laboratory wavelengths is on the order of 1 g cm^{-2}. Pauli (in [11], p.250) was quite aware that this mass density would be ruinous for relativistic cosmology; he advised that we just ignore the zero-point energy of the electromagnetic field. This is a prescription, of course, and not even a rational one. Pauli certainly knew that one must take account of zero-point energies to get the right binding energies in nonrelativistic particle mechanics. We now know the same applies to gravitational masses. And in standard physics the zero-point energies of fields are just as real. The problem with the vacuum energy density has persisted – if anything grown more puzzling – through all the spectacular advances in physics in the 20th century. I like Wilczek's phrase: this aspect of our physics is "profoundly incomplete" [12]. It is a Kelvin-level cloud: within physics that is wonderfully well tested and successful in a broad range of applications there is a distinct glitch.

We have observational probes that might be helpful. If the vacuum presents the same properties to any inertial observer, its effect on spacetime curvature is the same as Einstein's cosmological constant, Λ. The evidence from the cosmological tests is that the expansion of the universe actually is dominated a term that acts like Λ – though the absolute value is ridiculously small compared to what is suggested by current ideas in particle physics. The case for detection is serious, but since it depends on difficult observations and insecure models I am inclined to limit the odds to maybe five to one. But work in progress should convincingly show us whether a term that acts like Λ really is present.

Until recently the tendency in the astronomy community has been to hope that it could get by with Pauli's prescription, or at worse the phenomenological description of the vacuum by the numerical value of one constant, Λ, leaving the dispersal of this cloud to the physicists. But current ideas are that Λ is only an approximation to a dynamical entity, dark energy, whose mass density varies with time on the scale of cosmic evolution, and varies with position in response to the large-scale irregularities in the matter distribution. Detection of these effects would not solve the vacuum energy density problem, but it would be a spectacularly stimulating clue. We know how it might be done, and I have been hearing ambitious plans to make the astronomical measurements. You may be sure the physicists will be hanging on every word of progress; they are desperate for something to knock them off dead center.

In the standard cosmology the dark sector also contains nonbaryonic matter that dominates the mass in the outer dark halos of galaxies and the mass in clusters of galaxies. I am in sympathy with those who ask for more evidence this nonbaryonic matter really exists, but I think the case already is close to compelling. The clearest exhibition of dark matter is the giant luminous arcs – the gravitationally lensed images of background galaxies produced by the gravitational deflection of light by the masses in clusters of galaxies. No force law I can imagine could produce these smooth arcs out of gravitating matter with the clumpy distribution of the starlight in clusters. There has to be cluster dark matter, and if it were baryonic it would cause ugly problems [8].

We have little empirical guidance to the physics of the dark sector: we are working in the dark. We accordingly adopt the simplest physics we can get away with, which is good strategy, but certainly need not be the whole story: consider that polytropic ideal gas spheres were good enough for Eddington's analysis of the structure of the Sun, but helioseismology reveals a host of new details. If our model for the dark sector is missing details that matter it will be revealed by problems in fitting the observations. And there are hints of problems, from observations of the structure and formation of galaxies. My list is headed by the prediction that elliptical galaxies form by mergers at modest redshifts, which seems to be at odds with the observation of massive quasars at $z \sim 6$; the prediction of appreciable debris in the voids defined by L_* galaxies, which seems to be at odds with the observation that dwarf, irregular, and L_* galaxies share quite similar distributions; and the prediction of cusp-like dark matter cores in low surface brightness galaxies, which is at odds with what is observed. These are Rowland-type problems that draw on the rich phenomenology of astronomy, from the latest observations by the Hubble Space Telescope to the vast accumulation of lore from decades past. Sorting through all this takes time, but I expect will show us whether the problems with the standard picture for the dark sector will be resolved by better understanding of the observations and theory, or will be promoted to a Kelvin-level cloud.

3.2 Cloud No. II: Strong Spacetime Curvature

Cloud II is the singularities of general relativity, where the theory becomes meaningless. It took some time for people to sort out the physical singularities from singular coordinate labels, and to face up to the phenomenological importance of the former. I remember as a graduate student in the late 1950s reading a distinguished physicist's elegant picture of the bounce in an oscillating universe: like turning a glove inside out, one finger at a time. In the mid 1960s Penrose's [13] pioneering approach to singularity theorems forced us to accept that we need deeper physics to see past the formal singularity at infinite redshift in the relativistic Friedmann–Lemaître cosmological model. At about the same time, the discovery of quasars, and the broader recognition of active galactic nuclei, offered an example of strong spacetime curvature in compact objects closer to hand. Now, a half century later, we have rich phenomenologies of compact objects and cosmology, and we still have the singularities.

Analyses of the astrophysics of massive compact objects – those observed at the centers of large galaxies, and star remnants more massive than a white dwarf – usually take as given a Schwarzschild or Kerr black hole geometry with a truly black inside, in discussions of what have grown to be quite detailed observations. There are no problems with this approach, a sign of the remarkable predictive power of general relativity theory. But good science demands that we seek positive evidence in support of the black hole picture, and watch for credible evidence that the standard picture may not be quite right. Maybe advances in fundamental physics will show us what really is happening in the centers of galaxies, or maybe the dispersal of this cloud will be guided by the phenomenology.

Analyses of observations in cosmology finesse the formal singularity of the Friedmann–Lemaître model, and the unknown physics at the Planck scale, by stipulating initial conditions at a more modest redshift, let us say $z = 10^{15}$. Nowadays the initial conditions often are given a pedigree, from the inflation model, and the observational constraints on the initial conditions are used to infer conditions on what was happening during inflation. But, since the inflation scenario can fit a considerable range and variety of initial conditions, we don't know whether these measures of the very early universe amount to anything more than a "just so" story. Three assignments may help.

We look to observational astronomers and cosmologists for tighter constraints on the initial condition at redshift $z = 10^{15}$. And it behooves us to watch for hints that there is more to learn about cosmic evolution than is encoded in this initial condition within the present standard cosmology. The successes of the extrapolation of standard physics to the length and time scales of cosmology are impressive, but the enormous extrapolation certainly allows room for surprises. I am watching for them in the problems with galaxy formation I mentioned in connection with the dark sector.

We look to those exploring ideas about the early universe to try to find alternatives to inflation. If all due diligence yielded none we would have an argument by default that inflation really happened, a dismal closure but better than nothing. Alternatives are under discussion; it will be of great interest to know whether some variant of the ekpyrotic universe [14] has a physical basis comparable to that of inflation, which is not asking all that much.

We look to the physics community to build a firmer basis for cosmology at high redshift. If fundamental physics converged on a complete theory that predicts a definite version of inflation, or some other picture for the early universe, which agrees with the astronomical constraints, it will convincingly complete cosmology. The prediction's the thing, of course.

3.3 Cloud No. III: the Meaning of Life

This is a cloud over a much broader community. We can leave to the experts in other fields the philosophical issues, and the analysis of the molecular basis for life. The task for astronomy and its Virtual Observatory is to search for evidence of extraterrestrial life. This is a Kelvin-level cloud: a powerful driver of research whose outcome could profoundly affect our worldview.

Maybe life on Earth came from primitive extraterrestrial seeds; Hoyle and Wickramasinghe [15] survey the history and present state of ideas. Maybe there are advanced forms of life on other worlds, seeded or evolved out of spontaneously created life. The familiar 19th century example of the search for organized life is Lowell's study of possible signs on Mars; the search continues in the SETI and OSETI projects. The Terrestrial Planet Finder (TPF) will search for Earth-like worlds where life might flourish in a primitive or organized state.

I read that the search for extraterrestrial life is the part of astronomy that most interests most people. I offer four observations of how the big ideas and activities in society have influenced the directions of this research.

First, Charles Darwin's deeply influential arguments for evolution by natural selection forced debate on what the first step in the evolution of life might have been. At about the same time, people were coming to the conclusion that spontaneous generation is an exceedingly rare event, if it happens at all, and maybe contrary to Darwin's principle that life evolves out of life [16,17]. It was natural therefore that people turned to the idea of extraterrestrial seeds. Helmholtz (1874), a most influential physicist and physiologist, argued for the idea, as did an important chemist, Arrhenius (1908). Kelvin (1871) endorsed the general idea, but not natural selection: he argued for "intelligent and benevolent design" [18].

Second, the end of the 19th century was a time of large-scale civil engineering, including completion of the modern Suez Canal in 1869. It is perhaps not so surprising that Lowell looked for signs of big engineering on Mars.

Third, this is an age of computers and information transfer. I think it's not surprising that people are searching for extraterrestrial bar codes. I don't mean to mock serious and important science: a source of bar codes would signify self-aware life by any definition. Imagine the effect on our society of the demonstration that there actually is extraterrestrial self-aware life, that might even have something to say to us.

Fourth, this is an age of big science, that is supported by the wealth of nations. A logical consequence is that research in science is influenced by big government. The TPF is a recent example: this is pure curiosity-driven big science that originated within government funding agencies, rather than being forced by intense pressure from a scientific community.

I offer two lessons from these observations. First, the fascination with the idea of life on other worlds has a long history, back through the 19th century, and, I expect, it has a long future. But societies evolve, and it is natural to expect the focus of the search for extraterrestrial life will evolve too.

Second, the means of support of the scientific enterprise are evolving; the TPF is leading the curve. The TPF certainly may yield wonderful results; we have the inspiring precedent of Slipher's discovery of the cosmological redshift, at the observatory Lowell built with a goal paralleling that of the TPF. But there is the difference that funding agencies have to tend to many masters; they can't have the compulsive attention span of curiosity-driven people like Lowell. The Virtual Observatory is not leading this curve: a community is fighting for it, in the style of what gave us the space telescope, and what happened in physics in the last half century. These are generally happy examples – apart from such glitches as sunset clauses – of what I suppose is an inevitable development: the directions of research in astronomy are increasingly influenced by government as well as society, and astronomers must continue learning how to deal with it.

4 Concluding Remarks

Our ability to explore the physical universe is limited by resources and intellectual energy: the scientific enterprise must eventually reach completion by exhaustion. But we can be sure this will not happen any time soon to astronomy and its Virtual Observatory, because the subject has a rich list of Rowland-type problems to address, and, as I have discussed, a key role to play in the exploration of clear and present Kelvin-level gaps in our understanding of the fundamental basis for physical science. There was no guarantee in 1900 that the clouds over physics would clear, with a wonderful expansion of our knowledge. It would be foolish to try to guess what the present clouds might foreshadow, but we can list the general possibilities. Maybe the clouds will resist all efforts at resolution. If so, convincing people of this certainly will generate a lot of work for astronomers. Maybe the clouds will be cleared and at last leave astronomers to tidy up the pesky details. Or maybe clearing the clouds will reveal a new set, as has happened before.

I have avoided until now commenting on a serious issue under debate in the astronomy community: is this an appropriate time to commit limited resources to an International Virtual Observatory? I respect the arguments against, but am persuaded by personal experience that the growth of the Virtual Observatory is *inevitable* and *would* benefit from intelligent design. Two years ago the walls of my office were covered by about 25 meters of journal rows, dating back to 1965. I loved the convenience of reaching for a copy of the wanted article. But I've discarded the journals; I love even more the much greater convenience and power of ADS, arXiv, and JSTOR. I notice many colleagues feel the same: we have become addicted to these Virtual Libraries. Present-day Virtual Observatories are a useful but limited counterpart. Their further development seems to me to be an inevitable part of what we see happening around us, and surely calls for the proactive community response I have observed at this meeting.

Acknowledgements

I have benefitted from advice from Larry Badash, Neta Bahcall, Jeremy Bernstein, Masataka Fukugita, Rich Gott, Martin Harwit, Gerald Holton, Stacey McGaugh, Bharat Ratra, Paul Schechter, Max Tegmark, and Ed Turner. This work was supported in part by the USA National Science Foundation.

References

1. Lord Kelvin: Phil. Mag. **ii** – sixth series, 1 (1901)
2. A.A. Michelson: *Light Waves and Their Uses* (University of Chicago Press, Chicago 1903)
3. Lord Rayleigh: Phil Mag. **xlix** – fifth series, 98 (1900)
4. L. Badash: Isis **63**, 48 (1972)
5. H.A. Rowland: Bulletin of the American Physical Society **1**, 4 (1899)
6. S. Newcomb: Sidereal Messenger **7**, 65 (1888)
7. M. Harwit: *Cosmic Discovery: the Search, Scope, and Heritage of Astronomy* (Basic Books, New York 1981)
8. P.J.E. Peebles, B. Ratra: astro-ph/0207347
9. P.J.E. Peebles: astro-ph/0208037
10. W. Nernst: Verhandlungen der Deutschen Physikalischen Gesellschaft **18**, 83 (1916)
11. W. Pauli: 'Die allgemeinen Prinzipien der Wellenmechanik'. In: *Handbuch der Physik, Quantentheorie XXIV/1* ed. by H. Geiger and K. Scheel (Springer, Berlin 1933), p. 83
12. F. Wilczek: Physics Today **55**, August, p. 10 (2002)
13. R. Penrose: Phys. Rev. Lett. **14**, 57 (1965)
14. J. Khoury, B.A. Ovrut, P.J. Steinhardt, N. Turok: Phys. Rev. D **64**, 123522 (2001)
15. F. Hoyle, N.C. Wickramasinghe: Astrophys. Space Sci. **268**, pp. vii - vii and 1 - 17 (1999)
16. J. Farley: *The Spontaneous Generation Controversy from Descartes to Oparin* (The Johns Hopkins University Press, Baltimore 1974)
17. M.J. Crowe: *The Extraterrestrial Life Debate, 1750-1900: the Idea of a Plurality of Worlds from Kant to Lowell* (Cambridge University Press, Cambridge 1986)
18. Lord Kelvin: Presidential Address to the British Association, Edinburgh (1871)

Building the Infrastructure for the National Virtual Observatory: An Information Technology Research Initiative of the National Science Foundation

Robert J. Hanisch

Space Telescope Science Institute, 3700 San Martin Drive, Baltimore, MD 21218, USA

Abstract. The U.S. National Science Foundation is sponsoring the development of the infrastructure for the National Virtual Observatory via its Information Technology Research Program. This initiative combines expertise from astronomical observatories, data archive centers, university astronomy departments, and computer science and information technology groups at seventeen different organizations. This paper describes the nature of the project, our approach to managing and coordinating work across such a large collaboration, and the progress made thus far in the initial development activities (metadata standards, systems architecture, and science requirements definition).

1 Introduction

The scientific goals and technical challenges of the U.S. National Virtual Observatory (NVO) initiative have been described previously [1,5]. The joint NASA/ NSF Science Definition Team report provides this concise overview:

> "The NVO will federate the currently disjoint set of digital sky surveys, observatory and mission archives, astronomy data and literature services, and it will greatly facilitate the inclusion of future ones. It will provide powerful tools for the effective and rapid scientific exploration of massive data sets. It will be technology enabled, but science driven. It will generate more efficient and cost-effective science, and even completely new science, by making practical those studies that today are too costly in terms of the efforts and resources required. It will empower scientists and students everywhere to do first-rate astronomy, and it will become an unprecedented venue for science and technology education and for public outreach."

In the U.S., a collaboration of some 20 organizations was formed and submitted a proposal entitled "Building the Framework for the National Virtual Observatory" to the National Science Foundation's Information Technology Research (ITR) program (April 2001). The proposal was selected for funding ($10M over five years) and work began in November 2001.

2 Project Team and Organization

Our project team is composed of a combination of active research scientists, astronomy data management experts, computer scientists, information technology experts, and education and outreach specialists and partners. This diverse team was assembled in order to achieve a balance between science and technology, and between information technology research and pragmatic software development and implementation. The project is led by principal investigator Dr. Alexander Szalay (The Johns Hopkins University) and co-principal investigator Dr. Roy Williams (California Institute of Technology). Project priorities and resource allocations are decided by an Executive Committee comprising the PI and CoPI, the project manager (Dr. Robert Hanisch, Space Telescope Science Institute), the project scientist (Dr. David De Young, National Optical Astronomy Observatories), the system architect (Dr. Reagan Moore, San Diego Supercomputer Center, and two other senior personnel (Dr. George Helou, Infrared Processing and Analysis Center, and Dr. Ethan Schreier, Associated Universities, Inc.). The project leadership also includes an Education and Outreach Coordinator (Dr. Mark Voit, Space Telescope Science Institute). The project organization chart is shown in Fig. 1. Table 1 shows participating organizations.

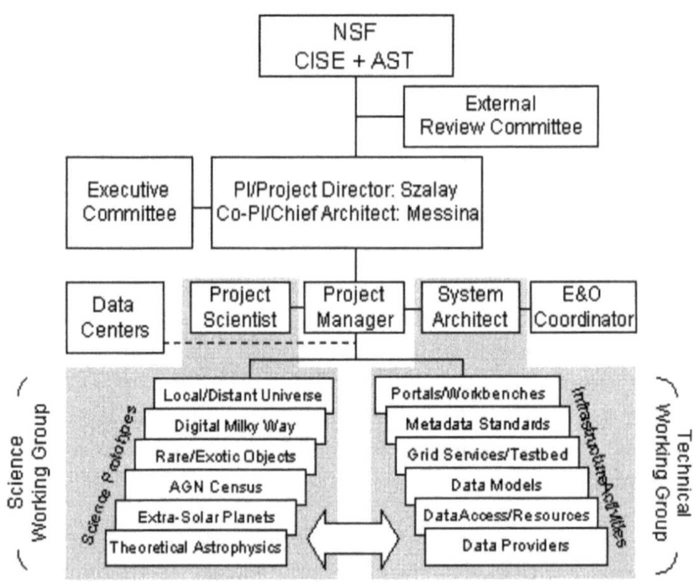

Fig. 1. Organizational structure of the NVO infrastructure development project

Table 1. NVO Project core and collaborating organizations.

Core Project Team
California Institute of Technology
– Astronomy Department
– Center for Advanced Computational Research (CACR)
– Infrared Processing and Analysis Center (IPAC)
– NASA Extragalactic Database (NED)
Carnegie Mellon University (CMU)
Fermi National Accelerator Laboratory (FNAL)
Herzberg Institute of Astrophysics, Canadian Astronomy Data Centre (HIA/CADC)
Johns Hopkins University (JHU)
Microsoft Research
NASA Goddard Spaceflight Center
– Astronomical Data Center (ADC)
– High Energy Astrophysics Science Archive Research Center (HEASARC)
National Optical Astronomy Observatories (NOAO)
National Radio Astronomy Observatory (NRAO)
Smithsonian Astrophysical Observatory, Chandra X-Ray Center (SAO/CXC)
Space Telescope Science Institute (STScI)
United States Naval Observatory (USNO)
University of California San Diego, San Diego Supercomputer Center (UCSD/SDSC)
University of Illinois Urbana-Champaign, National Center for Supercomputer Applications (UIUC/NCSA)
University of Pennsylvania
University of Pittsburgh
University of Southern California, Information Sciences Institute (USC/ISI)
University of Wisconsin

Collaborators
Argonne National Laboratory
Jet Propulsion Laboratory (JPL)
Lawrence Livermore National Laboratory (LLNL)

Carrying out this project – with development efforts spread over so many organizations – requires major efforts in project management, coordination, and communication. To this end we have written an overall project management plan that defines a standard work breakdown structure (WBS), which in our case has eleven principal activities:

1. Project Management
2. Data Models
3. Metadata Standards
4. Systems Architecture
5. Data Access/Resource Layer
6. NVO Services

7. Service/Data Provider Implementation and Integration
8. Portals and Workbenches
9. Testbed
10. Science Prototypes
11. Education and Outreach

The WBS is subdivided to three levels, and every second-level WBS element has a designated lead who is responsible for task assignments, schedule, and deliverables within that area. Progress is monitored through biweekly project review telecons, informal monthly status reports, formal quarterly status reports (which are required deliverables to NSF), and a formal annual report (also an NSF deliverable). The project executive committee holds weekly telecons. The entire team meets for two days three times per year to discuss technical, scientific, and management issues. One of these meetings is scheduled back-to-back with an annual meeting of the external review committee.

Several working groups have been constituted in the first year of the project:

- Metadata Working Group
- Models Working Group
- Architecture Working Group
- Working Group

These working groups hold meetings or telecons of varying frequency; thus far the Metadata Working Group has been the most active, holding weekly telecons. Minutes are kept of telecons, working groups have dedicated e-mail distribution lists, and all electronic communications are archived on the project team web site, `http://us-vo.org/`.

3 Initial Technical Developments

Considerable effort is being spent in the area of metadata standards and data models, two closely related topics. The first major development project, the specification of the VOTable XML formatting standard for astronomical tables, was a joint effort with the Astrophysical Virtual Observatory and AstroGrid projects in Europe [7,11,14]. VOTable builds on experience in the international astronomy community, with the FITS data format standard6 and the immediate predecessor to VOTable, astrores [6,15]. Moreover, the VOTable development effort was exemplary of how dedicated individuals can bring propose, discuss, and agree upon fundamental standards in a relatively short period of time. Already at least four software libraries have been developed for parsing VOTable documents, and work continues on refining the VOTable specification. Like the FITS format definition, we expect to make revisions and enhancements to VOTable as our experience grows.

Within the NVO project we followed up the VOTable specification with a challenge to team members to create the most basic VO web service, a cone search, and provide a response formatted as a VOTable. A cone search is a request for data (catalog entries, names of data sets from an observation log, etc.) located within a specified radius about a particular sky location (i.e., one is describing a very long cone whose vertex is at the observer and base outlines a circular region on the sky). Although many on-line astronomical databases already provide such services (indeed, with much more sophisticated interfaces) we wanted to see if we could all satisfy this minimal specification, and thereby exercise the VOTable generation and parsing software. Within two weeks time seven different groups had provided more than 50 cone search services, and the VOTable outputs were validated against a standard XML parser. We created a cone search registration service, and basic cross-match service was implemented on top of the cone search services [13]. This experiment has shown that simple NVO-compliant services can be built with only marginal extra effort for data and information providers.

Other metadata related activities include the development of a framework for metadata management [8], a draft specification of metadata standards for spatial and temporal information [10], and a draft metadata definition for describing VO resources and services [3] that is based on the Dublin Corei standards [12]. The relationships between data objects and their associated metadata form the basis for the NVO data model [4].

4 Science Prototypes

The NVO Project is "technology enabled, but science driven" [5]. In this spirit, our Science Working Group generated a list of some 40 possible science demonstration projects, from which three were selected for our first-round implementations. These demonstrations have several goals:

- Show the science community the potential benefits of the NVO, through real hands-on tools.
- Guide technical developments so that they are responsive to science requirements.
- Show early, interesting results from NSF's investment in the project.

Our project plan includes a series of science demonstrations, progressing from relatively simple data comparison services to more complex and computationally challenging applications. The initial demonstrations were chosen based on a number of criteria, including availability of necessary data, feasibility of completion by January 2003 (for the winter meeting of the American Astronomical Society), and ability to show results in a matter of a few minutes (i.e., the time one can typically hold the attention of an astronomer passing by a display booth at an AAS meeting). Our initial science demonstrations are described below.

Brown dwarf candidate search: We will deploy web-based cross-matching applications and demonstrate their science value by identifying T-type brown dwarf candidates through cross-matching the 2MASS (infrared) and several optical catalogs that have been published in an NVO-compliant manner. A major target will be the Sloan (optical) source catalogs to identify objects with extreme colors. We will also investigate the use of the newly-released USNOB catalog in determining the completeness of such searches for brown dwarfs having less extreme colors: while the USNOB catalog is not as deep as the Sloan survey and will likely not contain any of the extreme T-dwarfs seen there, it does contain proper motion information and has much greater sky coverage. We will also consider a few other selection criteria, like high redshift QSOs, or high-redshift galaxies (dropouts) as part of our demo suite.

Gamma-ray burst follow-up service: We will develop a quick response service that provides a summary of all known astronomical information regarding a specified region in the sky (typical a few arcminutes to a degree in size). This should include at least:

- IR, radio and X-ray images
- Known stars and galaxies
- Special objects
- Observations by major telescopes including links to appropriate archival data
- Bibliography links

This service shall be integrable with notification services for rapid response events (e.g., BACODYNE GRB notifications).

Galaxy morphology measurement and analysis: This demonstration looks for relationships in galaxy clusters between morphology and the intergalactic environment, specifically the X-ray emitting gas. Given a user-specified list of clusters, this application will create a sample of candidate member galaxies. For each galaxy that has been observed in the X-ray, we will calculate the X-ray flux emitted from the surrounding environment. In addition, we calculate three morphological parameters (mean surface brightness, concentration index, asymmetry index) from image cutouts of the galaxies from a chosen image survey. These calculated values will be combined with other relevant data mined from existing catalogs to create correlation plots.

The demonstration projects are described further on the NVO team web site (`http://www.us-vo.org/projects.html`).

5 Education and Public Outreach

The NSF's competitive research and development programs are judged in part by their broader impacts on education and their benefits to society. From the onset it has been recognized that the NVO has tremendous potential as a vehicle for education and public outreach (EPO), both in astronomy and information technology. Thus, our project incorporates an EPO program that focuses on building partnerships with formal and informal education programs and the news media. Our proposal team includes EPO partners (Table 2) and is open to further collaborators.

On 11-12 July 2002 we will hold an NVO EPO Workshop, bringing together representatives of our established EPO partners plus others with experience in

Table 2. NVO Project education and public outreach partners.

American Museum of Natural History
Association of Science-Technology Centers
Gettysburg College (Project CLEA)
International Planetarium Society
Maryland Space Grant Consortium
National Air and Space Museum
Silicon Graphics (Digital Planetarium)
Spitz (Electric Sky)
UC Berkeley (CSE@SSL)

this area. The goal of this workshop is to identify the requirements EPO content developers would be likely to place on the NVO, and thus to understand the implications for technical development. For example, most outreach products available now do not include astronomical coordinate systems or projection information, which is a severe handicap for developers of museum and planetarium exhibits who wish to show accurate comparisons of data from different telescopes and bandpasses. EPO partners also need to be able to easily identify NVO data products suited to their needs; they are rarely interested in the uncalibrated data from a telescope, but rather want to find the JPEG and GIF images prepared for public consumption along with associated descriptive information. These are both requirements on the metadata for the NVO, through which EPO content providers will be able to develop appropriate user interfaces (NVO portals).

6 International Collaboration

There is perhaps no project in modern astronomy so dependent on international collaboration as the Virtual Observatory. Even though a number of national initiatives have already begun, and have somewhat different priorities and drivers, there is a strong sense of collaboration and coordination among these groups. The goal is to create an International Virtual Observatory, a system of federated databases and interoperable services for which international boundaries are of no importance.

Two major VO initiatives have been constituted in Europe: the Astrophysical Virtual Observatory (AVO), funded by the European Commission and led by P. Quinn (ESO), and AstroGrid, funded by the UK e-Science Program and led by A. Lawrence (Edinburgh). To these we can now add the Canadian Virtual Observatory initiative, led by D. Schade (HIA/CADC). With these three projects and the US NVO project, nearly \$25M is now committed to VO development efforts. Smaller national initiatives are also beginning, in Germany, Russia, Australia, Chile, Japan, and India.

The senior personnel in these projects have organized the International Virtual Observatory Alliance, which is constituted to help assure collaboration on both technical and scientific developments [9]. In June 2002 the IVOA was formally initiated, and a joint roadmap for development was agreed upon. We are openly sharing experiences, documents, and software, and we are committed to building a common framework that supports the exchange of astronomical information regardless of its origin. The IVOA is currently chaired by R. Hanisch and co-chaired by P. Quinn.

Acknowledgements

"Building the Framework for the National Virtual Observatory" is supported by the National Science Foundation's Information Technology Research program via Cooperative Agreement AST-0122449 with The Johns Hopkins University.

References

1. R. J. Hanisch, "Building the infrastructure for the virtual observatory," Proc. SPIE 4477, pp. 191-199, 2001
2. R. J. Hanisch, A. Farris, E. W. Greisen, W. D. Pence, B. M. Schlesinger, P. J. Teuben, R. W. Thompson, & A. Warnock, "Definition of the Flexible Image Transport System (FITS)," Astron. Astrophys. 376, pp. 359-380, 2001
3. R. J. Hanisch & NVO Metadata Working Group, "Resource and service metadata for the Virtual Observatory," `http://bill.cacr.caltech.edu/cfdocs/usvo-pubs/files/ResourceServiceMetadataV3.pdf`
4. J. McDowell, M. Cresitello-Dittmar, J. DePonte Evans, I. Evans, M. Louys, T. McGlynn, M. Noble, R. Plante, A. Rots, & D. Tody, "Data models for the VO: version 0.03," `http://bill.cacr.caltech.edu/cfdocs/usvo-pubs/files/vodm003.ps`
5. NVO Science Definition Team, "Towards the National Virtual Observatory," `http://nvosdt.org/sdt-final.pdf`, 2002
6. F. Ochsenbein, M. Albrecht, A. Brighton, P. Fernique, D. Guillaume, R. Hanisch, E. Shaya, & A. Wicenec, "Using XML for accessing resources in astronomy," ASP Conf. Ser. 216, pp. 83-86, 2000
7. F. Ochsenbein, R. Williams, C. Davenhall, D. Durand, P. Fernique, D. Giaretta, R. Hanisch, T. McGlynn, A. Szalay, & A. Wicenec, "VOTable – A proposed XML format for astronomical tables," ESO Astrophysics Symposia, Toward an International Virtual Observatory, 2002, in press
8. R. L. Plante, "A scalable metadata framework for the Virtual Observatory," `http://bill.cacr.caltech.edu/cfdocs/usvo-pubs/files/fw-draft2.pdf`
9. P. Quinn, & R. Hanisch, "The International Virtual Observatory Alliance: A Mission and Roadmap Statement 2002-2005," `http://bill.cacr.caltech.edu/cfdocs/usvo-pubs/files/IVOA-III.doc`
10. A. Rots, "Space-time coordinate specification for VO metadata," `http://bill.cacr.caltech.edu/cfdocs/usvo-pubs/files/SpaceTime.htm`
11. `http://cdsweb.u-strasbg.fr/doc/VOTable/`
12. `http://dublincore.org/documents/dces/`
13. `http://us-vo.org/metadata/conesearch/`
14. `http://us-vo.org/VOTable/`
15. `http://vizier.u-strasbg.fr/doc/astrores.htx`

ASTRO-WISE
An Astronomical Wide-Field Imaging System for Europe

Edwin A. Valentijn[1] and Konrad Kuijken[1,2]

[1] Kapteyn Astronomical Institute, P.O.Box 800, 9700 AV Groningen, The Netherlands
[2] Sterrewacht Leiden, P.O. Box 9513, 2300 RA Leiden, The Netherlands
www.astro-wise.org

Abstract. With the new one square degree high resolution wide field imagers coming on-line in the near future, such as the 268 Mpix OmegaCAM at ESO's VST and the 360 Mpix MegaCAM at the CFHT, a new European-wide approach has been initiated to handle and disseminate the expected very large data volumes. For OmegaCAM both individual programs, including monitoring programs, and large sky survey programs are planned. Strict data taking procedures facilitate pipeline data reduction procedures both for the calibration and the science data. In turn, the strongly procedurized data handling allows European-wide federations of data-products. On-the-fly re-processing of archival data on the request of individual users with their own plugs-ins or newly derived calibrations sets are facilitated in an internationally distributed system. Compared to the classical more static wide-field image archives the newly designed system is characterized by a much more dynamical type of archiving.

1 Introduction

The data volume produced by the new generation of wide field imagers such as OmegaCAM at ESO's VLT Survey Telescope (VST, Paranal) and the Megacam at the CFHT (Hawaii) will be enormous. OmegaCAM, at a rate of 5 dithered exposures on a particular field in 30 minutes and with 300 nights per year of observing time, will produce over 30 Terabyte of raw data per year. This raw data volume contains roughly 10 Terabyte of calibration data and 20 Terabyte of raw science data. Data processing will then produce another 10 Terabyte of reduced science data and may create, with about 100,000 astronomical objects per OmegaCAM field of one square degree, enormous catalogues. Even the astronomical source lists of measured galaxy parameters can easily accumulate to 3-5 Terabyte per year!

Both the archiving of the data volumes and the processing of the image data go beyond the capabilities of personal work stations, which forces the user communities back to an old operational model of centralized nodes which host processors and storage media. The ASTRO-WISE project described below links the data centers set up in several European countries to support the current generation of wide-field survey instruments:

- The Netherlands, lead partner in the construction of the OmegaCAM instrument (NOVA/Groningen)
- France, European partner in the Megacam project (Terapix, Institute Astrophysique, Paris)
- Italy, lead partner in the construction of the VLT Survey Telescope (INAF/Naples) and partner in OmegaCAM (INAF/Padua)
- Germany, partner in OmegaCAM (Munich Observatory)
- ESO, who will operate VST and VISTA
- UK, lead partner in the construction of VISTA

The European Commission contributes 1.5 Meuro to this project through its FP5 RTD programme.

The aim of Astro-Wise is to set up a common system for processing, analysing and disseminating wide field imaging data. Internally, the centers will be connected via a shared database, while externally data products can be served to the AVO/VO networks.

There is an enormous variety of scientific research anticipated with the wide field imagers, ranging from 'fishing' special objects (moving, variable, or extreme in some colour index) out of millions, to statistical studies of large sample of objects, e.g. for cosmological shear research. The variety of scientific analysis of wide field data precludes the definition of single standard data products, and on top of this the enormous data volumes involved make it nearly impossible to re-process the whole data volume to achieve new releases with better code or improved calibrations data.

To provide the necessary flexibility to the astronomical end-user, Astro-Wise will provide an architecture that facilitates reprocessing of data as needed. It will have a peer-to-peer network between the national data centers, which each provide processing power and storage media, with full knowledge of what has been done at the other node. In this approach the processing of the data is viewed as essentially an administrative problem. The eventual goal is that the system administrates itself, so that when an end-user requests a particular data item, the system has full knowledge of how it has been derived, and how to rederive it if it is based on an 'obsolete' version of code or calibration data, with the result immediately being available at all nodes.

Furthermore, the machine handling of the large images and the big data volumes is non-trivial: particularly the pipeline data reduction, image comparisons and combinations, working with very large source lists, and visualization are all demanding tasks, even with modern hardware. The Astro-Wise consortium aims to share experience, build new tools and disseminate software for visualising and handling large image data.

Here, we describe the Astro-Wise survey system in terms of supporting the OmegaCAM project, which will be its first application, but it could as well host MegaCam data, or future infrared survey data.

1.1 Concepts

To face the data volume problems it is crucial to build an environment that provides, in a systematic and controlled manner, access to all raw and all calibration data while keeping track of all processing and data products in a wide area network.

This environment should allow the astronomer to plan, modify and rerun the reduction and calibration pipelines to fit the particular needs that follow from the astronomical questions posed to the data. In addition, the environment provides systematic and controlled ways of running source extraction algorithms such that other astronomers could benefit from the obtained results. The archives should thus store the reduced data and source lists, or allow regeneration of these data dynamically. Because of the large data volumes and the limitations of local data centers, this archive must link different data centers, i.e. it must be a full-fledged federated database. Users at one data center can then profit from activities at other data centers, where new and possibly better calibrations have been built.

This dynamical archive continuously grows as more raw data enters the system and as more data reductions and calibrations take place. It can be used both for 'small' and for large science projects generating and checking calibration data and exchanging methods and scripts.

A key functionality is the link back from derived source data to the original raw pixel data, associated calibration files and *all* other data items that went into the result. This allows the user of the system to:

- verify the processing steps that have led to a certain product, and
- to qualify the product in terms of personal scientific exploration
- rederive the result with up-to-date calibration, thus providing the basic logistics for on-the-fly re-processing.

1.2 Example Applications

In order to appreciate the above concepts in practical terms, we mention some practical applications:

(a) The VST is to be operated in service mode, and deep multi-colour exposures may be built up from data taken over many nights. Because all data will be accessible, and calibrated in a similar fashion, the optimal combination of data can only be done by selecting observations of a particular quality (quality information is a standard attribute to the archived data). Quality can sometimes only be assessed at the final stages of data reduction, so the linking information back to the raw data is necessary to build a homogeneous final survey input dataset.

(b) Facilitating source list production from well documented final survey images allows the astronomer to select sources on a 1 to 1,000,000 basis as true interesting and *not spurious* sources, for the quality of each individual source extraction is an integral part of the source properties. With the capability of extracting, in a homogeneous way, sources automatically from all reduced frames, variability studies (such as proper motions of asteroids or nearby stars, or just flux variations), can be done easily.

(c) The archive system is the best place to monitor the instrument as all calibration files are meticulously administered. The trend analysis of instrumental properties becomes essentially a push-button operation.

(d) A database environment is also the perfect place to plan observations because one can get a convenient overview of the quality of existing data and plan for filling the gaps in the spatial and quality domain. Based on the already available information, the addition of data in other passbands, the increase in exposure time or requirements for better seeing conditions can be identified and translated into an observing plan. For large surveys, the feedback between the data reduction/archive stage and the observations scheduling is an important tool when creating homogeneous raw datasets.

All this is done in the continuously growing archive.

1.3 Philosophy

The system that should provide above functionality will not be geared to a single data product, but should be a flexible tool. In fact many observations done with VST/OmegaCAM will be made for specific projects, not explicitly part of an all-sky survey *per se.* To capitalize on this diversity, flexibility is essential: even while OmegaCAM is performing individual observing programmes with specific science objectives, much of these data can as well be used for other research programmes with different objectives. Per year of operations the camera will survey the equivalent of 1000 square degrees in 5 passbands. So after a few years of operations the archives will contain a considerable fraction of the Southern sky, in effect delivering the data for large area sky surveys. When the quality and sensitivity of these observations are recorded accurately, then these data can be well used for addressing for example statistical studies, in a similar way as radio luminosity functions can be well derived from radio surveys, in spite of the varying sensitivity over the field of view of the primary beam of radio telescopes.

Overall, the environment should optimise the interaction between users and their data, giving the user easy access to all aspects (attributes) and processing (pipelines) of the data. This, ever growing, dynamic archive will be geared to optical (IR) wide-field image data.

2 The ASTRO-WISE Survey System

To efficiently archive and handle the data volume, the OmegaCAM data acquisition, calibrations and pipeline reductions are strictly procedurized. These procedures are integrated in the design of the pipeline data reductions. Thus the design of calibration and scientific data reduction procedures has focussed on developing standard observing scenarios. It uses object-oriented methods to implement the associated data reduction procedures.

2.1 Procedurizing

The two major components of the data taking are the scientific and the calibration observations. Both need to be procedurized and the associated observations should be performed automatically. This can be achieved by carefully defining observing modes and observing strategies that cover all observational conditions both for building a homogeneous survey and for doing arbitrary scientific observations. The ESO operations model on Paranal naturally allows such a strategy to be defined.

The next sections give an overview of these modes and strategies.

Observing Modes The CCDs in a detector mosaic do not fill the focal plane completely. The basic technique to overcome any gaps or artifacts in the CCD pixels is to take more exposures of the same field with slightly shifted field centers (dithering) and to co-add the images off-line in the pipeline process. This same technique can also be used to filter out cosmic ray hits on the detectors.

We distinguish the following observing modes:

Dither has offsets matching the maximum gap between CCDs, ~ 400 pixels. It will be operated with N pointings on the sky, where $N = 5$ is the standard. Although this mode erases all the gaps in the focal plane and maximizes the sky coverage, the context map (which relates each pixel in the combined image to the original exposures) will be very complex. An advantage is that in this mode the photometry among the individual CCDs can be coupled quite robustly.

Jitter has offsets matching the smallest gaps in CCDs ~ 5 pixels. It is the mode that optimises the homogeneity of the context map and will be used during observations for which the wide CCD gaps are not critical. In this mode all the data from a single sky pixel originate from a single chip.

Stare allows reobserving one fixed pointing position multiple times. It is the main workhorse for monitoring the instrument and allows detection of optical transients.

SSO is the mode for observing Solar System objects. It has non-sidereal tracking. Pipeline processing support for this mode is limited to the standard image processing – specialized techniques are required to extract sources from such images.

Observing Strategies An observing strategy employs one or a combination of the basic observing modes. It also defines a number of additional instructions for the scheduling of the observations. The observing strategy will be recorded in the FITS headers of the observations. Optionally, this header information can be used in data reduction pipelines, particularly those operated by the Consortium when addressing the combination (e.g. stacking) of images. It is not expected that the ESO pipeline will recognize strategies, as the standard ESO pipeline will not combine various runs.

We distinguish between the following strategies:

standard which consists of a single observation (observation block),
deep which does deep integrations, possibly taken at selected atmospheric conditions over several nights,
freq which frequently visits (monitors) the same field on timescales ranging from minutes to months and has overriding priority on the telescope schedule,
mosaic which maps areas of the sky larger than 1°.

2.2 Processing

The observing modes and strategies are fully integrated with the data reduction software. Their precise definition and limited number make it possible to design an associated *data model*, in the form of classes, that drives the pipeline design for data reduction and calibration. Figures 1 and 2 represent an overview of the datamodel which connects the datataking at Paranal, various quality control operations, the derivations of the calibration data and the "image-pipeline" which transforms the raw images into photometrically and astrometrically calibrated images.

Once the data operations, types and classes are defined the pipeline design is relatively straightforward. We discriminate between a *calibration pipeline* producing and qualifying calibration files, often involving a trend analysis, and an *image pipeline* that operates as a black box. By passively applying the calibration files (CalFiles) the *image pipeline* transforms the raw data into astrometrically and photometrically calibrated images. At ESO headquarters these pipelines will run under the Data Flow System pipeline infrastructure. At the national data centers these pipelines will run in an integrated environment where all data and data reduction steps are archived. Because algorithms for data reduction in the optical wide-field imaging arena are well established we can concentrate on other aspects of the data reduction scheme. We can view the pipeline as *an administrative problem*, where most attention should be paid to what ancillary information should be available when.

Calibrations The *calibration pipeline* is the collection of tools specifically designed to obtain all required calibration files (Calfiles). The requirements for these tools are specified as baseline requirements on OmegaCAM calibrations (Valentijn et al 2001). The calibration plan for the VST includes a comprehensive overriding photometric program. At the moment we have identified about 35 requirements, ranging from "check the focus" to "determine and monitor the atmospheric extinction". Each of these requirements are fulfilled by dedicated

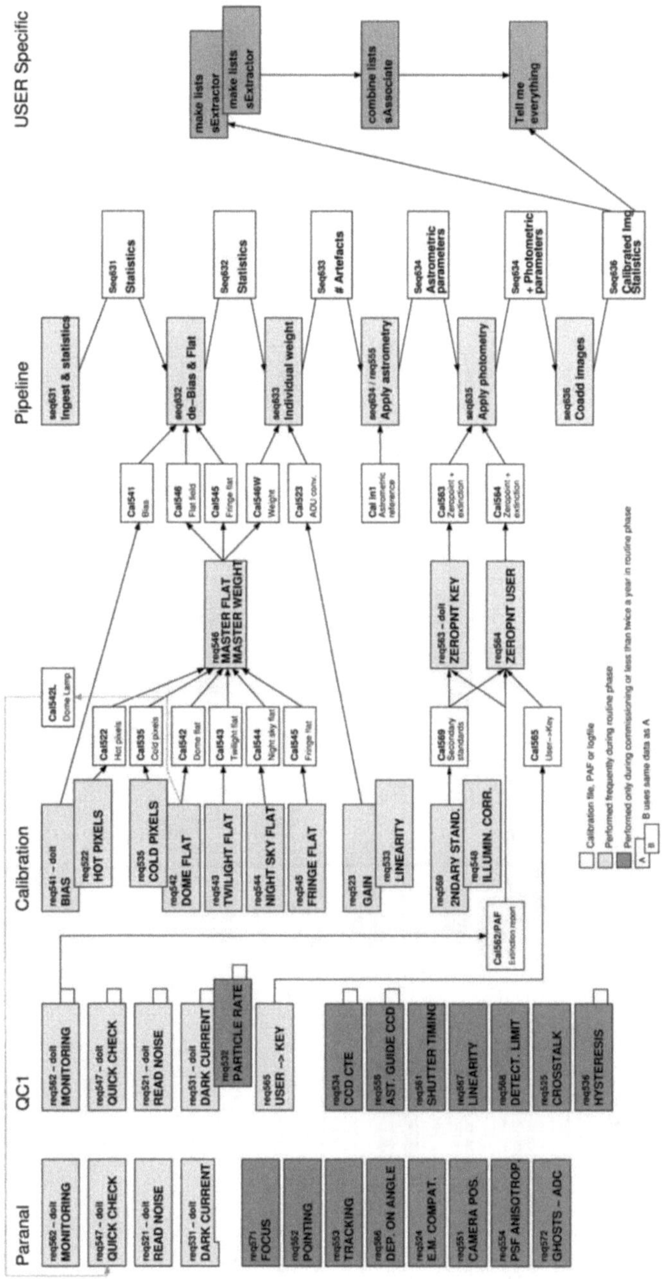

Fig. 1. The OmegaCAM datamodel with local quality control procedures highlighted in dark grey and persistent operations, ie. visible and re-runable by the end-user, highlighted in light grey

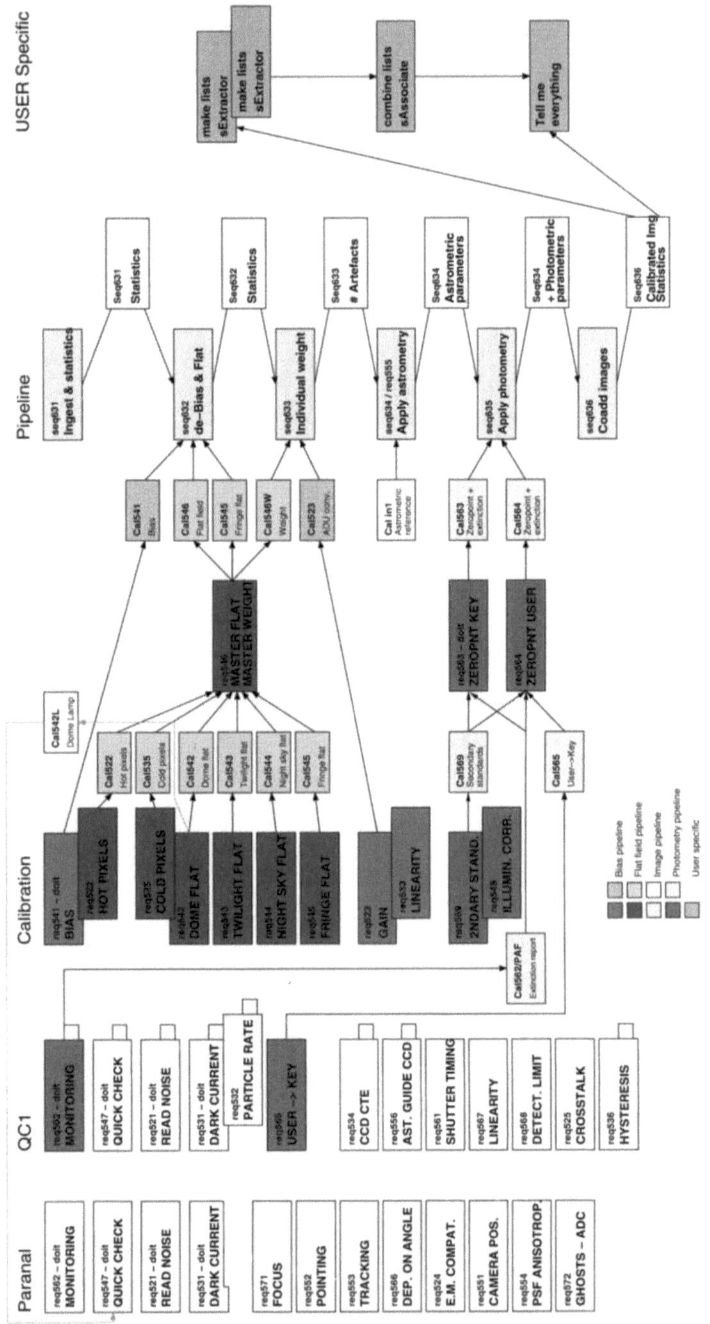

Fig. 2. The OmegaCAM datamodel with reduction pipeline procedures indicated by shades of gray

procedures both for the data acquisition at the telescope and for the calibration pipeline, which produces calibration files. Furthermore, these processes will also result into go-no-go flags. In fact, with the settling of the baseline requirements and the calibration plan all 'classes' in the data reduction have been defined.

The objective is to have a minimal interdependence between these procedures. Thus, the calibration pipeline can run the various derivations of calibration files at various time scales, independent of the derivation of other calibration files. For example, the derivation of the master bias CalFile could be done at a frequency of twice a week, master flat fields once a week, photometric zero point once a night and the cross calibration of filters once a year, with a minimum of interdependence between these processes. The execution frequency of the different procedures of the calibration pipeline is tied to the frequency of the corresponding observations. As a baseline, the various frequencies for different calibration observations are highly standardized. The creation mechanism of the CalFiles includes a time stamping module which, as a result of a trend and/or quality analysis, assigns a time range for which the CalFile is valid. The image pipeline recognizes timestamps.

Science Observations The *image pipeline* transforms the raw science data into calibrated images and passively applies the calibration files (CalFiles) made by the *calibration pipeline* (in fact the image pipeline is used in calibration procedures where required). Thus, the image pipeline produces the calibrated science images, and together with the CalFiles, which were used to derive these images, sets the end product ready for the astronomer for detailed scientific analysis. Unlike the calibration pipeline, the image pipeline does not produce any CalFiles.

The descriptor data of the reduced science images are stored in the database. These descriptors contain a copy of all the FITS header items, but they also contain links to all the data items (i.e. objects) which were used to derive the particular result. The CCD pixel data are not stored in the database; instead, a reference to a frame is added to the descriptor.

The *image pipeline* has many steps. Although it is designed to function as an automated 'streamer', the intermediate results are stored in so-called SeqFiles, again containing FITS-headers, statistics, intermediate results and links to data items. The descriptor will be used to store data of persistent value, and references to the descriptors can be used to track input and output of the various pipeline operations. For example: SeqFile 636 (co-added image) will have a reference to a list of SeqFiles (SeqFile 633 –Astrometrically calibrated image), which were used as input. The descriptors of these SeqFiles can be used to determine, for instance, the distribution of seeings or zero points in the input data, even though the image data for these input images may no longer exist.

Share the Load The huge amount of data that needs to be processed in a limited amount of time necessitates the use of high powered CPUs and large bandwidths. Due to the physical nature of the OmegaCAM camera a natural parallelism is introduced where frames from the 32 CCD's can be processed quite independently through major portions of the data reduction pipelines. The level of parallelization is rather coarse-grained and the implementation of choice is a Linux Beowulf cluster. Having a very large bandwidth for communication between the processing units is essential to allow rapid dissolving of data across the cluster.

Data storage with significant amounts of fast and local disk space (10 - 100 Terabyte), is needed to minimize network traffic and at later stages allow distributed storage of processed data.

The data reduction will go in two stages. First the calibration is derived and CalFiles produced, then the image pipeline is run (at speeds of at least 1 Mpix/s) to produce calibrated science images.

The storage media need to have enough room for the images created throughout the lifetime of the project. This amounts to several 100's Terabyte. The archival storage of source parameters, depending on the use of the system and the total number of users, can easily go beyond the 10 Terabyte level.

In the federated environment the network plays an essential role. In an ideal world there is no need for replication of data, when information stored at a remote data center is needed it is delivered at the time of processing to the processing unit. This requires sustained network connections of 200 Mb/s or better. Such networks are becoming reality in the academic world these days. Even if the network speed is below this critical limit however, a 5 Mb/s network allows full replication of all OmegaCAM data on the 24 hours per day basis.

2.3 Federation

All the I/O of the pipeline processes goes to a federated database. The federated database is *the* archive in Astro-Wise where *all* information regarding the data and the processing of OmegaCAM will be stored. First of all, the raw data are accessible from the archive. The raw data itself will not reside inside the database, but the description of the data, including its location, will be available. This means that the data can only be manipulated through interaction with the database. In fact the methods (or pipelines) for processing of all kinds are also part of the federated database. When these methods are executed, they will interact with the database making sure it correctly describes the state of the OmegaCAM data repository.

Next to the raw data, calibration results, reduced images and source lists (possibly in the form of catalogs) will also be stored in the federated database, either as fully integrated objects or as descriptors.

Concepts of the Federation A federation is a database environment that is spread over different physical locations but maintains a single database in the true sense of the word. The consortium is currently building such a system using Oracle -9i with SQL and Python interface. The choice for Oracle was motivated because of its support for object oriented programming, its scalability up to Terabyte regimes of partitioned tables (needed for source catalogues), its Advanced Replication component supporting our federation and its new component Streams which allows the system to connect to e.g. ESO's Sybase archive and Terapix's MySQL archives. Astro-Wise has been adopted into the company's reference programme, supporting "innovative and leading projects". Also, the availability of these components today (Astro-Wise needs to be able to start operations and receive first data by early 2004) and the size of the company in a very competitive database world has played a role in our decision.

According to Oracle: "A federated database is a logical unification of distinct databases running on independent servers, sharing no resources (including disks), and connected by a LAN."

In this environment full history tracking of all input will be done. To this end we employ object oriented inheritance techniques, links to objects (references in Oracle speak) and the database support for persistent objects. So everything in terms of processing that went on producing a result is readily available. The same set of links and persistent objects should also provide the back-bone for the on-the-fly reprocessing. To tag data and attributes in this very dynamical archive, context areas are introduced in the object attributes. Objects in this terminology are the persistent forms of the Object Oriented programming objects (used in the Python scripting language) that are the software counterparts of all OmegaCAM entities (of which a number are displayed in Fig. 1. Some of these contexts can be:

- Project, with possible values Calibration, Science, Survey, or Personal
- Owner, with possible values pipeline, developer, or user
- Strategy, with values Standard, Deep or Freq (monitoring),
- Mode, with values Stare, Jitter, Dither or SSO and
- Time, with time stamping.

These context areas can be used to partition off areas of the database for certain projects. It will allow individuals to maintain their own partition, but also for larger projects, like large Sky Surveys to maintain parts of the data with project wide defined levels of quality control. The context areas also facilitate public access and provide the mechanism to interface the database to public browsers such as envisioned by the Virtual Observatories.

The federated database makes the Object Oriented programming languages objects persistent. Therefore any creation of a persistent object in the pipeline automatically has a counterpart that will be stored in the database. Because all data processing (intermediate) products have been defined in the OmegaCAM data model, classes can be programmed in the Python scripting language. The Object Oriented inheritance is also available from the persistence implementation, usually in terms of object links. For each (persistent) class a number of methods are defined which directly interact with the federated database, thus insuring database integrity.

The object oriented scripting language Python is used throughout the project providing the glue between the different working environments, such as:

- access to the database, through SQL,
- various pipeline codings as built by the scientific programmers and
- scripts provided by the astronomer-user allowing them to run "own methods" provided it fits into the datamodel.

Actually, another way of putting this is "the system provides the possibility to operate user customized pipelines and still maintain persistency".

The actual implementation of the database connectivity from the Python scripting language allows for a 'file structure' implementation of the database environment as well, thus allowing the pipeline to operate on files in a directory structure, completely independent from a federated database. However, in this case many of the advantages of a global environment are lost.

The current planning is to have the system ready for data acquisition by the very end of 2003, to test and populate it in 2004 and to prepare it for further mass production for 2005 and beyond. The system will be deliverable to satellite nodes at other European locations.

References

1. Valentijn, E.A., Begeman, K.G.B., Boxhoorn, D., Deul, E.R., Rengelink, R., Kuijken, K.H., 2001: VST-SPE-OCM-23100-3050 OmegaCAM Data Flow System User Requirements, ESO Garching, `www.astro.rug.nl/~omegacam/documents`
2. Valentijn, E.A., Deul, E.R., Kuijken, K.H., 2001; Preston, Observing and Datamining with OmegaCAM, in The New Era of Wide Field Astronomy, ASP Conference Series, Vol 232, p 392, eds. R.G. Clowes, A.J. Adamson and G.E. Bromage `www.astro.rug.nl/~omegacam/documents`

The Canadian Virtual Observatory Project

David Schade, Pat Dowler, Daniel Durand, Luc Simard, Norm Hill, and Severin Gaudet

Canadian Astronomy Data Centre, Herzberg Institute of Astrophysics, National Research Council of Canada, Victoria, B.C. Canada

Abstract. The Canadian Virtual Observatory (CVO) project is now funded at a level that places it among the major Virtual Observatory (VO) projects in the world. The fact that the CVO is a project within the Canadian Astronomy Data Centre makes it unique in several ways. Our VO development efforts coexist with the efforts required to maintain an operational data centre. Although this implies the potential for competition for resources, it also means that we have the storage, processing, and database infrastructure in place to quickly develop and deploy services. We have the obligation to deploy new services to our users on a regular basis which rules out long periods of research. This is a valuable spur toward quick prototyping. Importantly, we have excellent access to data that we can process in order to upgrade it to a VO-compliant state and we can therefore make an important contribution to increasing VO content. It is not a bad place to be doing VO work.

1 Introduction

Within the Canadian Astronomy Data Centre the move toward a Virtual Observatory (VO) is seen as a natural evolution of the functionalities of a facility like ours. This view is probably shared by many existing data centers. We need to evolve toward the VO along a path that delivers a regular increment in the services that we provide to our users. At the same time we need to have clear goals and those goals include enhancing the ability of our systems to interact with those at other data facilities.

A working set of requirements for the Virtual Observatory includes:

- that it should allow seamless access to data across archive boundaries
- that it should allow queries that are based on science rather than technical considerations
- that it should provide access to well-characterised data
- that it should provide services that facilitate the integration of data sources and effective joint analysis

There are a great many things that need to happen before these requirements can be met in anything but a very limited sense. As a data centre whose mandate is to provide services of value to our users, CADC will undertake to make progress on as many of the tasks as possible that need to be done to move us closer to a functional VO. We are developing our storage and processing capability and enhancing our database capabilities. We have hired staff to think about the various problems of characterising and delivering datasets effectively. We are even processing selected archival data content into a form that is more suitable for the VO. We have designed and developed a working VO prototype that we believe has a number of interesting features and that we are certain has already taught us a great deal about what is needed in terms of delivering VO capabilities to the world.

2 Components of the Canadian Virtual Observatory Prototype

Our CVO prototype is a layer of software that is distinct from the CADC archives and is linked to the archives in well-defined ways. The data archives contain all of the technical details of the telescope and instrument and the details of the basic processing which removes instrumental signatures. For optical CCD data this includes bias subtraction and flat-fielding but these basic processing steps will vary for different types of data in different energy regimes.

Some, but not all, archival data are sufficiently well-characterised that they can be included in the Virtual Observatory. VO content can be pixelsets or catalogs (sets of derived parameters). Images that are published from the archives into the CVO must include in their headers all of the information that is necessary to execute further advanced processing within the VO environment. This is done so that it is not necessary to access the archives to gather information for processing and emphasises and reinforces the distinction between the archive and the VO. This standard also means that CVO pixelsets can be passed to other data centres and they will carry with them all of the information that other data centres require in order to execute specialised processing.

The CVO prototype itself is a set of three primary databases. The first is VoPix which includes high-level information that describes the pixelsets. The second, VoSrc is the source database and the final database is VoProc which links VoSrc and VoPix and describes the processing that was applied to VoPix in order to produce the content within VoSrc.

2.1 The VoPix Table

The VoPix table describes and allows queries to be executed on the pixel data (pixelsets) in the VO. An important goal of the VO is to allow queries that are based on scientific criteria rather than on technical or engineering details of the instrument or telescope. These details are more or less obscure to different users who have expertise in different energy regimes but it is rare to find a user who understands all of the details of every instrument. In pursuit of the goal of enabling science queries and enabling cross-archive queries we have developed a high level description of the data, in effect a data model that is an abstraction of the characteristics of a dataset.

The fundamental quantities that characterise an astronomical observation are energy, space, and time. Our spatial descriptors include the usual ones:

- Right Ascension and Declination
- Galactic latitude and longitude
- Ecliptic latitude and longitude.

In addition we include spatial span (detector size for a square CCD), spatial sample, (pixel size for a CCD), and spatial fill factor which provides a crude indicator of more complex geometry which may be present in some observations. These descriptors provide a very simple description of the spatial characteristics of an observation. The goal is that the descriptors constitute a set that is sufficient to describe the dataset accurately and that are simple enough to allow effective querying.

Our prototype characterises the time of an observation using start and end times, the time sample-size or bin-size, and the time filling-factor. These choices are driven by the recognition that observations may be simple exposures (time span equals end-time minus start-time, time sample equals the integration time, time fill-factor equals unity) or they may be composites or stacks of observations taken over a number of years where the integration time is much less than end-time minus start-time, and the time fill-factor is much less than unity. If the stack is composed of individual frames with widely varying integration times then there is no well-defined value of time sample and one would need to adopt a mean or median integration length as a characteristic time sample. In many real cases that the time-span is years and the characteristic time-sample is roughly one hour. In cases like this our crude characterisation of the time element of an observation provides valuable functionality. It describes the data and it allows for effective queries.

A flaw in our current scheme for characterise the energy sampling of an observation is that it is inordinately "optical-centric". This can be seen by the fact, for example, that we use pivot wavelength which is a characteristic used with optical and near-infrared filters. We are in the process of examining ways to make this more general. As it now stands the energy of an observation is characterised by min-energy, max-energy, energy-span, energy-sample, and fill-factor parameters for wavelength. This is general enough to provide a framework for broadband filters (fill factor 1 and span sample filter width) and for spectroscopic observations (sample spectral resolution). There will be cases where, for a specific scientific reason, broadband images in U and I might be combined so that the wavelength or energy span is very different from the filter width, the energy sample is the filter width, and the fill factor is 1/3.

Why are quantities like exposure time, filter name, and instrument name missing from this characterisation of pixel data? Users of WFPC2 may be intimately familiar with the F814W filter and they may understand the sensitivity that is implied by 4000 seconds of integration time with HST. Furthermore, they might fully understand the geometry of WFPC2. Science users whose primary training and experience have involved X-ray satellites may understand none of these things. Since one of our goals is to facilitate cross-archive and multi-wavelength access to data, we must use abstractions of the properties of an observation that are general enough to be understood by any qualified researcher. We have not achieved that goal with the current system but we are moving toward it and we are continually revising our system in order to approach that goal.

The ultimate goal of our method of characterising pixelsets and sources is to allow inter- operability so that other VO nodes can access the content at CADC. The problem of inter-operability is not ultimately a technical problem. The difficult part of the problem is developing a common system to allow VO nodes to understand the content that they are trying to access which resides at another node. It is a problem of understanding one another. Once that is achieved then access is a relatively easy technical problem.

There are two obvious approaches to solving the problem of understanding. The first approach is to develop a framework that characterises every internal detail of an archive or database and then to publish this so that VO nodes can use it as a translation table to understand the content of data archives and databases. This can be characterised as "giving the VO the ability to reach into" a data centre. This approach has the drawback of broadcasting complexity to every VO node. Complexity of the framework is accompanied by complexity in implementation and this breeds error.

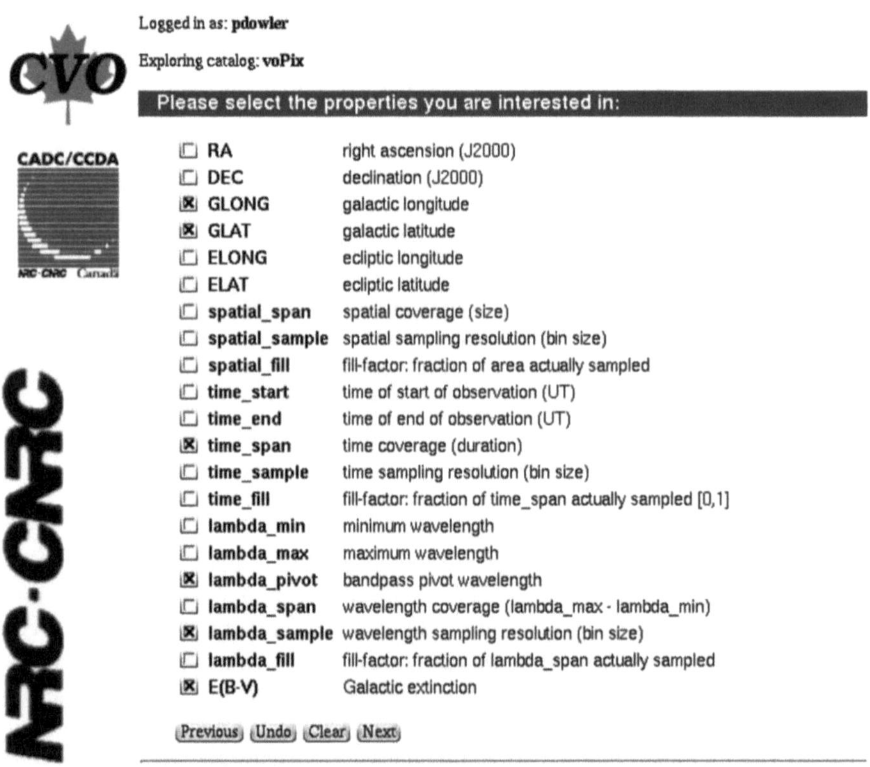

Fig. 1. The pixel table (VoPix) query interface

The second approach is for the data centres themselves to adopt a high-level description of their own content. This can be characterised as a data centre "offering up" their data to other nodes. This has the advantage of localising complexity. Furthermore, it is localised to a region where the expertise exists to handle it. Data centres are intimately knowledgeable about their content and how to process it effectively.

It is clear that a high-level abstraction of the content of data centers must exist. Such an abstraction of the properties of the datasets is one component of the scientific process. Therefore, since science can be done, and in fact is done, on archival data then it follows that the required abstraction exists. Furthermore, it is clear that such a high-level abstraction needs to exist and must be defined explicitly to make it possible for science- based user queries to work. We have chosen to try to develop this abstracted or generalised representation of our VO content because we need it to proceed with our prototype work. Our initial stab at the problem has yielded some success and shown us some problems and our second attempt will be much improved.

2.2 The VoSrc Table

The VoSrc table is a database of derived parameters for sources. A source is defined as a single observation or detection and the resolution or mapping of a source onto a physical object is a separate issue that is not addressed in the Source table. These parameters have been derived from the VoPix content using prescriptions defined through the VoProc table. The VoPix table provides all of the information about time and energy sampling.

The most general description of an observation of a source is the amount of flux (scaling) and the distribution of the flux (morphology) on the sky. Flux and morphology are the fundamental parameters and the energy and time sampling is already represented in the VoPix entry describing the data from which the source originates.

There are many different ways to measure the flux of an object. There are isophotal fluxes (with the isophotal level as a parameter of the measuring process), aperture fluxes, fluxes estimated through a modelling process, and a number of other approaches. In the case of morphology there exists an even richer variety of techniques to choose from for characterising the distribution of flux over the plane of the sky. It is true, fortunately, that there exist methods that are accepted by some fraction of the science community as being reasonable ways of measuring flux and morphology. It is also true, unfortunately, that these accepted methodologies tend to be different in one wavelength regime than another. Thus there are real problems in developing a general and widely-understood set of descriptors for source observations.

The present design for the source table includes position of a source, and several measures of flux. We include an extension index (effectively a star-galaxy discriminator) and a few parameters designed to measure morphology. Our present scheme is again inordinately optical- centric. We have initially relied on sExtractor (Bertin) to provide the source measurements for our prototype. We presently include a number of sExtractor-specific parameters.

Characterising the point-spread function (PSF) is another obstacle to creating a generalised scheme for characterising sources. In principle, the PSF is an instrumental effect that should be removed during the calibration process. But there exists no technique for achieving this that does not degrade the quality of the data in one way or another. In general the PSF varies from place-to-place on a detector and this is why the PSF is a source-specific quantity. Despite this problem it is possible and desirable to compute an "extension index" which, at least, defines (within some statistical uncertainty) whether a source is consistent with being a point source or not. But other statistical measures of the distribution of light are degraded by the effect of the PSF. A reasonable goal that we are pursuing is to define the PSF (or "beam width") at the position of a given source on the detector and providing access to this information to users.

2.3 The VoProc Table

The processing table (VoProc) links the content of the source table to the pixels that it was derived from and defines the processing steps that were used. It does not include all of the archive-level processing. A second function of VoProc is to act as a table of processing that could be done on the content of VoPix. Our design includes Discovery Agents that will match VoPix datasets with VoProc recipes and queue tasks for execution by our distributed processing system. When fully deployed this will result in new content for VoSrc which will then be inserted into that table.

3 Generating Data Content for the Virtual Observatory

We have a major project underway to process WFPC2 frames into a form that is close to being VO-ready. This project has been underway for several years as a collaboration between CADC and Space Telescope-European Coordinating Facility, and has recently grown to involve STScI doing quality assessment, interface design, and deployment which will occur on Friday November 8, 2002 simultaneously at all three sites.

This project has determined dither-pattern offsets for all of the WFPC2 associations and has produced stacked (or combined) images for about 20,000 of the associations with more to come. The stacked images are a great improvement over the currently-available individual images. Furthermore, these high-quality combined images form the basis for the CVO prototype. They are the best data products that we have for that purpose. When the CVO prototype is released it will not only demonstrate VO-like capability but will also give another valuable tool to our users.

Information Hub of the Russian Virtual Observatory

Oleg Malkov, Olga Dluzhnevskaya, Elena Kilpio, Alexander Kilpio, and Dana Kovaleva

Institute of Astronomy, Moscow 119017, Russia

Abstract. The ultimate goal of the Russian Virtual Observatory (RVO) initiative is to provide every astronomer with on-line access to the rich volumes of data and metadata that have been and will continue to be produced by astronomical survey projects. The information hub of the RVO has a main goal of integrating resources of astronomical data accumulated in Russian observatories and institutions, and providing transparent access for scientific and educational purposes to the distributed information and data services that comprise its content.

One of the general purpose data centres for astronomy is the Moscow Centre for Astronomical Data (CAD). CAD has been systematically collecting and distributing astronomical data for more than 20 years. The CAD staff will carry out the activities on construction of the information hub of the Russian Virtual Observatory.

1 Introduction

A virtual observatory (VO) is a collection of interoperating data archives and software tools which utilize the internet to form a scientific research environment in which astronomical research programs can be conducted. The VO consists of a number of data centres each with unique collections of astronomical data, software systems and processing capabilities. In the past months, three major international projects (NVO [1], AVO [2], ASTROGRID [3]) and a number of smaller ones have been funded to develop and realize the vision of using astronomical data repositories as virtual observatories.

Recently Scientific Council on Astronomy of the Russian Academy of Sciences strongly endorsed the RVO initiative with the Centre for Astronomical Data (Institute of Astronomy of the Russian Academy of Sciences) and Special Astrophysical Observatory (Russian Academy of Sciences) as co-ordinators. The RVO will be an integral component of the International Virtual Observatory (IVO), which will link the archives of all the world's major observatories into one distributed database, with powerful tools to optimise the extraction of science from the data.

The Russian contribution will be in the following areas.

- To provide Russian astronomical community with a convenient access to the world data grid.
- To unite Russian and former Soviet Union (fSU) data, to provide them to the rest of the world and to integrate them into the IVO.

- To take part in developing of software, techniques, standards, formats necessary for the establishment of the IVO.
- To use Russian instrumentation to provide observational data in remote mode when needed.
- To strengthen education and public applications of world astronomical data.

Centre for Astronomical Data contributes mostly in the first three points that will described below in three corresponding chapters.

2 Providing World Data for Russian Astronomers

The main activity of CAD, as an information hub of the RVO and as a national data centre, is to install and maintain mirrors of principal world databases. CAD holds now the mirrors of the largest astronomical electronic library ADS [4] and the most complete database of astronomical catalogues and data tables VizieR [5]. CAD serves also as a National Host of the IUE Newly Extracted Spectra database INES. One of advantages of this activity is that for many Russian institutions domestic Internet-traffic is cheaper than international one.

We continue our activity on mirroring of databases. Among others, we plan to mirror Vienna Atomic Line Database VALD [6] and Database of the Belgrade Astronomical Observatory BELDATA [7].

Some data resources that are not available on-line (mostly observational archives, catalogues and software) are kept on CD and distributed upon request.

Another CAD's activity is to read modern media currently supported by data archives and to distribute the data on lower-capacity media.

CAD also fulfills some other functions: browsing, visualization and cross-identification of catalogues ([8], [9], [10], [11], [12], [13], [14], [15], [16]) review and expert evaluation of data sets [17]; analysis of user requests.

3 Russian Data to be Integrated into the IVO

There are about 30 astronomical institutes and organizations in Russia. Many of them maintain extensive data archives. But the main value of Russian astronomical observational data is their large time scale of observation. Russia is the most extended in latitude country in the world - there are 11 timing zones in Russia, and it is situated almost on the opposite side of the globe to the most of the world astronomical observatories. This allows, for instance, to obtain the uninterrupted row of observation for variable objects.

CAD coordinates the activity of the Section 13 "Databases and Informational Environment" of Scientific Council on Astronomy of the Russian Academy of Sciences. The main goal of this section is to collect principal Russian astronomical data to make them available to the world astronomical community. The first stage of this program involves the organisations situated in the European part of Russia.

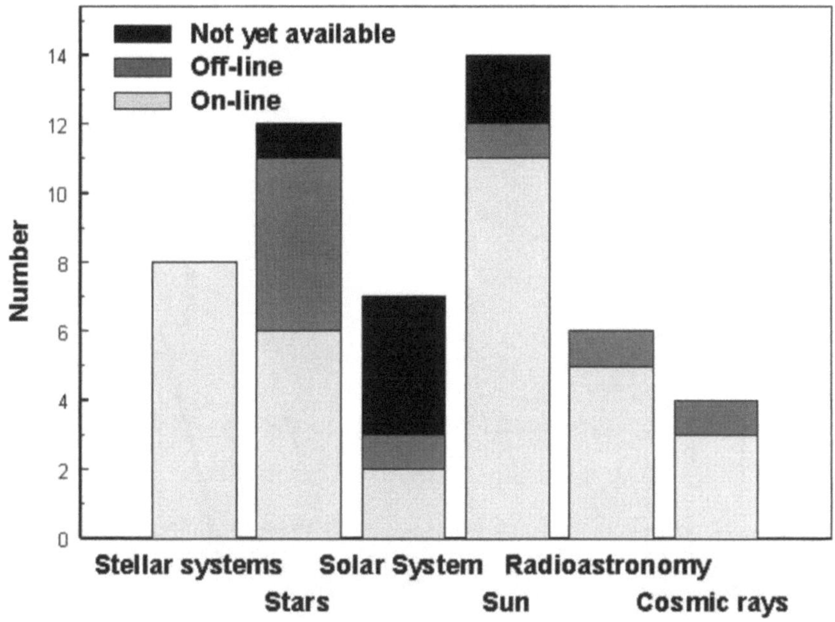

Fig. 1. Main Russian and fSU astronomical data resources

We collect information about all available (both Russian and some former Soviet Union) resources and classify them according types of observed objects or spectral range, namely: Stellar systems, Stars, Solar System, Sun, Radioastronomy, Cosmic Rays. Figure 1 presents distribution of the resources, kept in a dozen of Russian (see Fig. 2) and some fSU (Ukraine, Latvia, Kazakhstan, Tadjikistan) astronomical organisations, according to the types and to a degree of availability.

The list of Russian and fSU astronomical resources [18] is compiled for the first time and will be kept up to date. This list of resources (as well as other CAD services) is available on high performance CAD-servers at `www.inasan.rssi.ru/cad` connected to Internet through 2×2 Mb fiberoptics channels and back-up radio channel. We provide a convenient direct access to on-line resources and distribute some off-line resources.

CAD provides an access to electronic tables published in main Russian astronomical journals.

CAD (in collaboration with Russian astronomical organisations) produces machine-readable versions of catalogues, glass libraries, printed papers.

Another CAD work is a construction of catalogues and databases. The staff of CAD has significant experience in construction of astronomical catalogues. Some of them are:

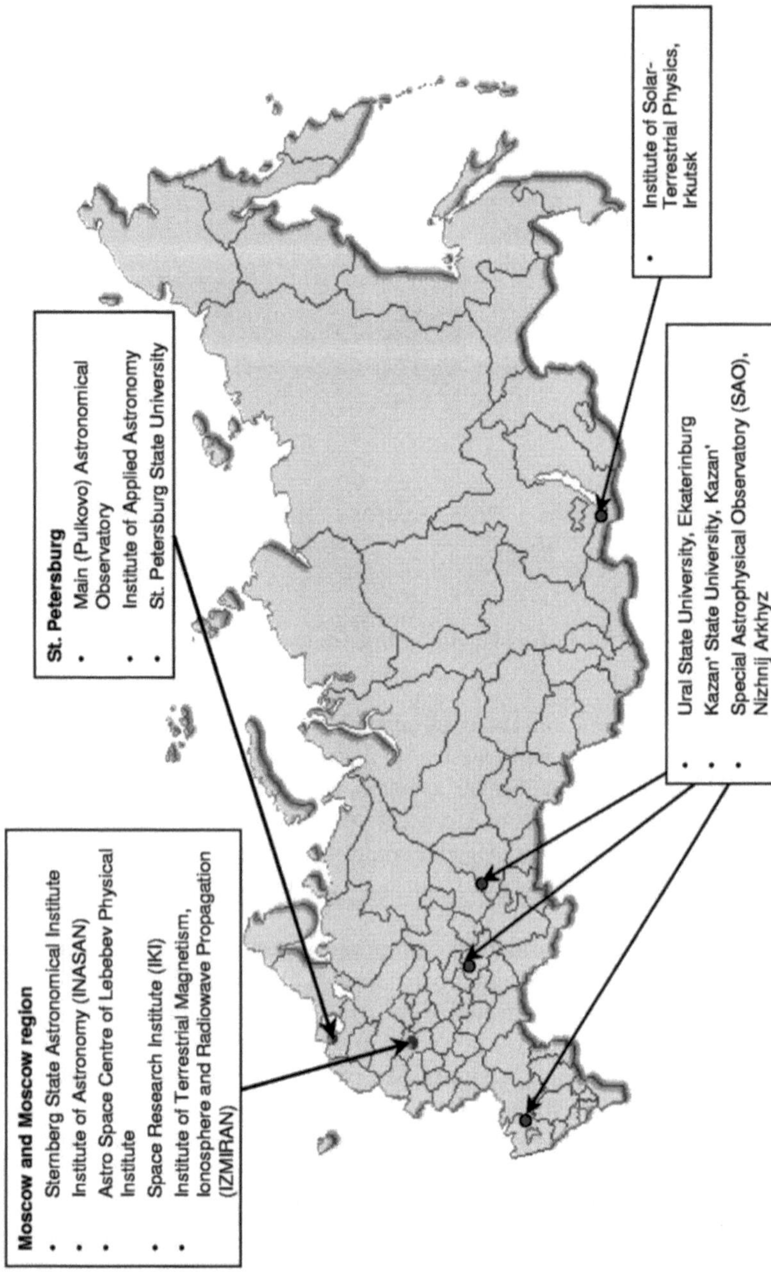

Fig. 2. Main Russian astronomical organisations, holding astronomical data resources

- Observational data in galactic star-forming regions [19],
- Stellar mass catalogue [20],
- Stellar polarization bibliography [21],
- Catalogue of astrophysical parameters of binary systems [22],
- Masses and ages of stars in open clusters [23],[24],
- Low-mass binaries database [25].

This experience allows us to provide scientific and technical support to authors of catalogues.

4 Standards and Formats Development

The RVO project starts a couple of years later than other principal VOs, and, therefore, should follow interoperability standards of various kinds already established. We must enable the open exchange of information and share our experiences among other VO projects.

To provide users with convenient links to Internet astronomical data resources a special collection of links to useful resources, called Internet Resources in Astronomy (IRinA), is compiled by CAD staff. This collection is based on an original multi-level classification scheme, supplied with search facilities and will includes reviews and expert analysis, comprehensive list of national resources and bilingual resource descriptions.

One of CAD immediate tasks in the frame of this project is standardization and unification of information on national resources (resource identifications, object names, units, etc.), their rating and completing of a (meta-)database of Russian astronomical resources.

Another our future goal is to construct interoperability tools, particularly, for national observational archives.

The project is supported by RFFR grant No 02-07-90391.

References

1. R. Hanisch: this volume
2. P.J. Quinn: this volume
3. A. Lawrence: this volume
4. G. Eichhorn, S.S. Murray, M.J. Kurtz, C. Stern, T. Karakashian: Astron. Astrophys. Suppl. Ser. **182**, 7109 (1993)
5. F. Ochsenbein, P. Bauer, J. Marcout: Astron. Astrophys. Suppl. Ser. **143**, 23 (2000)
6. N.E. Piskunov, F. Kupka, T.A. Ryabchikova, W.W. Weiss, C.S. Jeffery: Astron. Astrophys. Suppl. Ser. **112**, 525 (1995)
7. N. Milovanovic, L.C. Popovic, M.S. Dimitrijevic: Baltic Astronomy **9**, 595 (2000)
8. O.Yu. Malkov: Bull. Inf. CDS **31**, 187, (1986)
9. O.Yu. Malkov, O.M. Smirnov: 'Guide Star Catalog data retrieval software III'. In: *Proc. Astronomical Data Analysis Software and Systems III, Victoria, Oct 1993*, ed. by D. Crabtree, R. Hanisch, J. Barnes (ASP Conf. Ser., v. 61, 1994), pp. 183–186

10. O.Yu. Malkov, O.M. Smirnov: 'Classification of objects in the Guide Star Catalog'. In: *Proc. Astronomical Data Analysis Software and Systems IV, Baltimore, Sep 1994*, ed. by R.A. Shaw, H.E. Payne, J.J.E. Hayes (ASP Conf. Ser., v. 77, 1995), pp. 257–259
11. O.Yu. Malkov, O.M. Smirnov: 'Integrating the HST Guide Star Catalog into the NASA/IPAC extragalactic database: initial results'. In: *Proc. Astronomical Data Analysis Software and Systems VI, Charlottesville, Sep 1996*, ed. by G. Hunt, H.E. Payne (ASP Conf. Ser., v. 125, 1997), pp. 298–301
12. F. Pasian, P. Marcucci, M. Pucillo, C. Vuerli, O.Yu. Malkov, O.M. Smirnov, S. Monai, P. Conconi, E. Molinari: 'Integrating the ZGSC and the PPM at the Galileo telescope for the on-line control of instrumentation'. In: *Proc. Astronomical Data Analysis Software and Systems VII*, ed. by R. Albrecht, R.N. Hook, H.A. Bushouse (ASP Conf. Ser., v. 145, 1998), pp. 433–437
13. O.Yu. Malkov, O.M. Smirnov: 'Data retrieval software for the USNO-A catalog – another member of the -DARES family'. In: *Proc. Astronomical Data Analysis Software and Systems VIII*, ed. by D.M. Mehringer, R.L. Plante, D.A. Roberts (ASP Conf. Ser., v. 172, 1999), pp. 407–410
14. O.M. Smirnov, O.Yu. Malkov: 'XSKYMAP 2: a multi-catalog visualization facility'. In: *Proc. Astronomical Data Analysis Software and Systems VIII*, ed. by D.M. Mehringer, R.L. Plante, D.A. Roberts (ASP Conf. Ser., v. 172, 1999), pp. 442–444
15. O. Dluzhnevskaya, O. Malkov, F. Pasian: Baltic Astronomy **9**, 618 (2000)
16. O. Malkov, A. Tutukov, D. Kovaleva: 'The role of astronomical catalogues in modern theory and observation'. In: *Proc. Special Session of XXIV GA IAU, Astronomy for Developing Countries, Victoria University of Manchester, United Kingdom, Aug 2000*, ed. by Alan H. Batten (ASP, San Francisco, USA, 2000), pp. 291–302
17. O. Malkov: 'Large astronomical catalogues'. In: *Proc. Workshop on Automated Data Analysis in Astronomy, IUCAA, Pune, Oct 2000*, ed. by Ranjan Gupta, H.P. Singh, Coryn A. L. Bailer-Jones (Narosa Publishing House, New Delhi, India), pp. 1–14
18. O.B. Dluzhnevskaya, O.Yu. Malkov, A.A. Kilpio, E.Yu. Kilpio, D.A. Kovaleva, L.A. Sat: 'The Russian Virtual Observatory'. In: *Proc. Library and Information Services in Astronomy (LISA) IV conference, Prague, Jul 2002*, ed. by Brenda G. Corbin, Liz Bryson (U. S. Naval Observatory publications, 2002), in press
19. V.S. Avedisova: Astronomy Reports **46**, 193 (2002)
20. A.N. Belikov: Bull. Inf. CDS **47**, 9 (1995)
21. M.L. Belous: Bull. Inf. CDS **48**, 5 (1996)
22. O.Yu. Malkov: Bull. Inf. CDS **42**, 27 (1993)
23. A.E. Piskunov: Bull. Inf. CDS **19**, 67 (1980)
24. V.I. Myakutin, R. Sagar, U.C. Joshi: Bull. Inf. CDS **26**, 103 (1984)
25. D.A. Shpil'kina, O.Yu. Malkov: 'The design of the low mass binaries database'. In: *Proc. ESO Workshop, The Bottom of the Main Sequence – And Beyond, Garching, Aug 1994*, ed. by C. Tinney (Springer-Verlag, 1995), pp. 151–154

Early Virtual Science: Some Lessons for the AVO

Gerry Gilmore

Institute of Astronomy, Madingley Rd, Cambridge CB3 0HA, UK

Abstract. Experience with ASTROVIRTEL, scientific analysis of current large data sets, and detailed preparation for the truly huge future missions especially GAIA, provide important lessons for the Astrophysical Virtual Observatory. They demonstrate that the science cases are impressive, specifically allowing new thresholds to be crossed. The AVO is more than just faster cheaper better, it allows the new. The example of use of pre-explosion imaging of supernova to identify progenitors is used to illustrate some general challenges. Some non-trivial technical astronomical issues arise, especially astrometry, to complement the many technical implementation challenges. A critical scientific lesson is the need to quantify data quality. How are we to ensure the Virtual Observatory produces top science, and avoids being overwhelmed with mediocre data?

1 Introduction

The Astrophysical Virtual Observatory has an exceptionally strong science case. One example, identification of the pre-explosion state of core-collapse supernovae, is described below. This case illustrates some of the technical and strategic challenges which the Virtual Observatory projects have still to face.

There is a strong scientific case to identify and access appropriate archival data on the sites of supernova explosions. Since the candidate star is identified by its self-immolation, the only relevant data are archival! A project to obtain suitable new data to act as a future archive is underway with HST. In addition, much valuable data already exists. Some of this is in archives, and is well calibrated. Much is in private hands. The effort needed to use even excellent quality calibrated and published data is illustrated below, using as examples the searches by Smartt etal for the progenitors of SN1999em and SN2002ap.

One the most difficult of the astronomical challenges facing the integrated use of federated multi-wavelength multi-resolution archives involves source extraction, and astrometric cross-matching. Some examples are given below. The challenge is however obvious to all who have experience even with combination of HST WFPC2 and NICMOS images. Even in this case, with high-quality, high spatial resolution, stable, well-calibrated data sets, with only a one-half decade wavelength range, simple matching of optical and near-IR images of star clusters is not trivial. A nice example is available in figures one and two of Johnson etal (2001), who present PC and NICMOS images of two young LMC globular clusters.

A second challenge involves data reliability. The example below illustrates how even unusually high-quality data cannot be used beyond the range in which their systematic uncertainties become relevant to the science at hand. However, few (if any) data archives are calibrated well enough to provide this information, except in response to very specific questions and applications.

This raises the spectre of well-meaning providers of what are in fact data of limited calibration ensuring that virtual observatory data product users either produce defective science, or are overwhelmed with learning the limits of every individual data set accessed by the entire system. Might it be that data-archives need a quality-assessment check before they are 'eligible' for access? Or is it to be *caveat emptor?*

2 An Example Application: Identifying the Progenitors of Type II Supernovae

Supernovae are the evolutionary end points of all stars more massive than about $8M_{\odot}$. Predictions of the pre-explosion evolutionary status of these stars is a key test of stellar evolutionary theory. Supernova explosions additionally drive the chemical evolution of the Universe and play a major role in shaping the dynamics of the interstellar medium of gas rich galaxies. They are of crucial importance to fundamental studies of the evolution of galaxies and the origins of the chemical elements in the Universe.

The spectra of supernovae come in many different varieties, with the classifications based on the lines observed and the temporal evolution of these features. The presence of broad H I optical lines indicates a SN Type II classification, while those that do not show hydrogen are classed Type I. The SNe Ia are thought to arise through thermonuclear explosions in white dwarf binary systems, hence the progenitors are low-intermediate mass stars. All other supernovae including the Types Ib/Ic and all flavours of Type II are thought to be due to core-collapse during the deaths of massive stars. SNe II show prominent, broad H I lines in their optical spectra, indicating that the progenitor retained a substantial hydrogen envelope prior to explosion. SNe Ib/Ic do not show any significant signs of hydrogen in their spectra, although SNe Ib display pronounced He I absorption.

There is strong though indirect evidence that SNe II and SNe Ib/Ic are associated with the deaths of massive stars, as they are never seen in elliptical galaxies, are observed only rarely in S0 galaxies, and they often appear to be associated with sites of recent massive-star formation, such as H II regions and OB associations in spiral and irregular galaxies [23,5]. The Type II events are further seen to be split into subtypes (IIb, IIn, II-L and II-P). Leonard etal [9] discuss the widely held belief that core-collapse events can be ranked in order of their increasing hydrogen envelope mass at the time of explosion, which is - Ic, Ib, IIb, IIn, II-L, II-P.

This overwhelming, but still indirect, evidence implies that SNe II arise from the deaths of single, massive stars, with initial masses $M > 8-10M_{\odot}$, which have retained a substantial fraction of their hydrogen envelope. However there has been only one definite and unambiguous detection of a star that has subsequently exploded as a supernova of any type – that of Sk$-69°202$, the progenitor to

SN1987A in the LMC [25]. Prior to explosion this star was a blue supergiant of B3 Ia spectral type [24], which would correspond to $T_{\rm eff} = 18000\,$K (from the temperatures in [11]) and $\log L/L_{\odot} = 5.1$ (from the photometry in [24]), and an initial mass of $\sim$20M$_{\odot}$. The closest supernova to the Milky Way since then was SN1993J in M81 (3.63 Mpc), which was a Type IIb event. Ground based *UBVRI* photometry of the SN site before explosion was presented by Aldering etal [1]. The photometry of the progenitor candidate was best fit with a composite spectral energy distribution of a K0 Ia star and some excess UB-band flux suspected to be from unresolved OB association contamination (confirmed by recent HST observations). Neither the progenitor of SN1987A nor that of SN1993J is consistent with the canonical stellar evolution picture, where core carbon burning finishes and core-collapse occurs relatively soon afterwards ($\sim 10^3 - 10^4$ yrs) while the massive star is an M-supergiant.

Other attempts have been made to identify SNae progenitors on pre-explosion archive images, with little success in directly detecting progenitor stars. Fortunately, even an upper limit on progenitor brightness provides a mass-limit. An upper mass limit to the progenitor of SN1980K has been estimated to be $\sim$18M$_{\odot}$ [22], while only an upper limit to the absolute visual magnitude was determined for SN1994I [3]. Recently we [16] studied HST archive images of the site of the Type II-P SN1999gi which were taken before explosion. SN1999gi occurred in a young OB-association: however the the progenitor was below the detection limit of the available pre-explosion images. By determining the sensitivity of these exposures and comparing the estimated bolometric luminosity with stellar evolutionary theory, an upper limit to the mass of the progenitor was set at 9^{+3}_{-2}M$_{\odot}$.

3 SN1999em: Archival Experience with Excellent Published Data

SN1999em in NGC1637 was discovered on Oct. 29 1999 by the Lick Observatory Supernova Search [10] at an unfiltered CCD magnitude of $\sim 13.5^m$. It was soon confirmed to be a Type II and being a very bright event it has been studied extensively in the optical since then. It has been firmly established as a normal Type II-P event, having a plateau phase lasting approximately 90 days after discovery [9]. There have also been UV, X-ray, radio, and spectropolarimetry observations. [2] have presented model atmosphere fits to the early-time optical and HST-UV spectra, indicating that an enhanced He abundance is required to fit the data satisfactorily. They further use the very blue continuum of the early spectrum to determine a reddening. The expanding photosphere method (EPM) has been applied to SN1999em by Hamuy etal [7] to determine a distance to the host galaxy of 7.5 ± 0.5 Mpc, illustrating the possibility of using SNe II-P as luminous distance indicators. Chandra and radio observations of SN1999em have been used to probe the interaction of the SN ejecta with the circumstellar material, which are consistent with a mass-loss rate of $\sim 2 \times 10^{-6}$M$_{\odot}$ yr^{-1} and a slow wind [15]. Given the substantial interest in this bright supernova and the extensive multi-wavelength observations of the event it is of great interest to have direct information on the progenitor star. Further it would be desirable to have more detections of progenitor stars (as in SN1987A) in order to draw a meaningful physical picture of what causes the different varieties of core-collapse events.

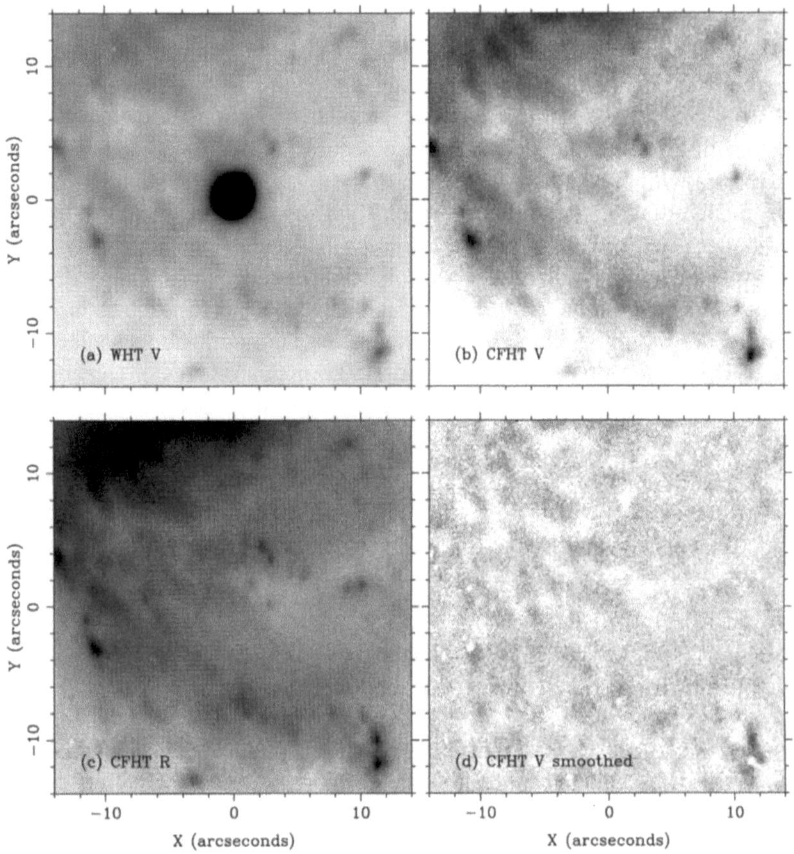

Fig. 1. (a): The position of SN1999em in NGC1637 in a transformed WHT post-explosion image. In this image the centroid of the SN is saturated but a further short exposure is used to measure it accurately, and is set at (0,0) in all frames. **(b)** and **(c):** The region of pre-explosion VR-band CFHT images. **(d):** An image with a smooth background removed and all PSFs from single stars subtracted. Sohn & Davidge (1998) catalogue a star with a coordinate of $(0.08'', -0.24'')$, and magnitude $V = 23.47, R = 23.33$ which is within the astrometric error of the transformation discussed in the text. However on close inspection there is no evidence for a point source at this position in any of the VRI bands. The two stars at (1.6, 1.9) and (2.9,0.2) are have $V = 23.97, 23.15$ and $V - R = 0.26, 0.10$ respectively. The detection limit is position dependent as the background varies considerably over small scales. This figure is from Smartt, Gilmore, Tout & Hodgkin 2002a

By chance there are optical images of this galaxy taken 7 years before SN1999em occurred in the archive of the Canada France Hawaii Telescope, maintained at the Canadian Astronomy Data Centre[1]. These high-resolution images were taken by Sohn & Davidge [20], who presented photometry of the luminous supergiant members of the galaxy. Amongst other results in their paper, a distance of 7.8±1 kpc is derived from the magnitudes of the galaxy's brightest stars. As SNe Type II are thought to have luminous supergiant progenitors, high quality pre-explosion images of nearby galaxies which resolve the brightest stars could allow direct detection of progenitors, or at least limits to be set on luminosity in the event of a non-detection.

In fact, Sohn & Davidge published photometry of a source coincident with the supernova position. Smartt etal (2002a) however presented an accurate astrometric determination of the position of SN1999em on the pre-explosion frames. They show that there is no detection of a point source at this position which could be interpreted as the progenitor. The detection limits of the exposures are determined, allowing bolometric luminosity limits and an upper mass limit to be determined for the progenitor star.

This discrepancy between different analyses of the same high-quality data using essentially the same photometric techniques illustrates one of the primary challenges for the Virtual Observatory: even high-quality data are often not suitably calibrated for use in a different application than that for which they were obtained, without considerable interactive analysis by an experienced astronomer.

3.1 The Challenges of Repeat Data Analysis

The galaxy NGC 1637 was observed on 5th January 1992 on the CFHT with the HRCam [12], with exposures of 900s, 750s and 600s in V, R_C, I_C. The material is publicly available through the CFHT archive at CADC[1]. The reduction, analysis and multi-colour photometry of the bright stellar objects in the field was presented by [20, hereafter SD98]. Their limiting magnitudes for detection, defined as the magnitude where DAOPHOT [21] predicts errors of ±0.5 or greater, are ~24.9, 24.8, 23.9 in V, R_C, I_C respectively. These data hence probe stars brighter than $M_v \simeq -4.9$, assuming the distance modulus from SD98, and their estimates for average line of sight extinction of $A_V = 0.34$. The image quality of the archive data are 0.7″ FWHM in all three bands. SD98 determined the colours for 435 objects in the frames which are simultaneously detected in all three filters. The CFHT HRCam used a 1024×1024 pixel Ford-Aerospace CCD mounted at prime focus, with 18μm pixels, corresponding to 0.13″ on the sky.

On 28th November 1999, Smartt etal obtained two $V-$band images of NGC1637 on the William Herschel Telescope on La Palma – 30 days after discovery of SN1999em. The AUX-port camera at Cassegrain was used, which has a 1024^2 Tektronix detector (ING CCD TEK2) at a plate scale of 0.11″ pix^{-1}. This was done through the ING SERVICE program and two exposures were taken

[1] `http//:cadcwww.dao.nrc.ca/cfht/`

(900s and 10s), during which the seeing was $0.7''$ FWHM. The considerable similarities between the cameras, telescope apertures and observing conditions in both cases mean the sensitivities of the pre and post-explosion data are very similar, and substantially ease joint analysis (see Fig. 1).

Astrometrically calibrating either of the two frames as they stand onto an absolute reference frame is not possible due to their limited FOV, and the fact that any isolated stars outside the main body of the galaxy which could be used as secondary astrometric standards are saturated in the deep CCD frames. However, given the similarity in the plate scales and the detection limits of the two data sets, Smartt etal were able to perform a simple geometric transformation of the WHT pixel array onto the CFHT array (similar to the method in Smartt et al. 2001a). First of all they identified ten bright, relatively isolated stars in both the WHT 900s V exposure and CFHT V frame, and measured the centroids of the stars on the WHT frame by fitting a model point-spread-function (PSF) to each using standard techniques in DAOPHOT within IRAF, using the pixel coordinates of the 10 stars from the tabulated photometry of SD98. A spatial transformation function was calculated, which fitted a linear shift, a magnification factor and a rotation angle. Polynomials of various orders were tried to fit the x and y mapping, but the results were no better than the simple scaling formerly described. The transformation function was applied to the WHT 900s frame, and both were trimmed to the common region of overlap (625×560 pixels, as shown in Fig. 1. This process left no residual systematic difference in the pixel astrometry between these two datasets. The mean offset in radial positions of the stars in the CFHT and WHT frames is $\delta r = 0.17'' \pm 0.13$.

This astrometric mapping process, already rather more elaborate than could be achieved by even careful use of WCS FITS headers, was further complicated by the wide dynamic range difference between the bright and the faint stars, which exceeds the detector dynamic range. The supernova itself of course was saturated, and so would be rejected from analysis by most pipeline processing systems. Thus a further short-exposure astrometric transfer image was also required. Using the stellar centroid method to check for offsets between the long and short frames proved problematic due to the low counts in stars in the short exposure frame; a significantly longer exposure would have led to saturation of the SN. However some stars in common could be matched, and indicated mean offsets of $(-0.01'',0.01'')$.

All these transformations required care and one-off interactive checking.

3.2 Re-Use of Published Photometry

The photometry list of SD98 reports the detection of star #66 (hereafter NGC1637-SD66) at $(0.08'', -0.24'')$ and the nearest other object is $2.5''$ away. Star NGC1637-SD66 is the only candidate for the progenitor in the existing photometry of SD98, at a distance of $\delta r = 0.25''$ from SN1999em. This does fall within the 1σ standard deviation of the differences in positions of the 106 matched stars, and hence is compatible with being coincident with the supernova position.

However on closer inspection this does not appear to be a reliable detection of a stellar-like object. In Fig. 1 the region around SN1999em is displayed from the CFHT V and R band images. There is no obvious resolved luminous object from a visual inspection and star NGC1637-SD66 is not apparently obvious (the results for the I-band data are similar). The position of the supernova appears to lie on a faint "ridge" (running diagonally left-right in the figure), and the detection limits of the image are highly position dependent given the variable background. In deriving their final photometric list, SD98 applied a background smoothing technique to recover faint stars against the varying galaxy background. Smartt etal repeated this method to determine if any sign of a single point source at the SN1999em position appears after background subtraction, following the steps described in [19]. The DAOPHOT package was used to fit model PSFs to the brighter stars in the images. These were subtracted from the data and a boxcar median filter of pixel dimension 25×25 (i.e. 5 times the seeing width) was applied to this subtracted image. This was assumed to be indicative of the varying background of the galaxy and was subtracted from the original frame. The PSF fitting routines within DAOPHOT were re-run on the resultant frame. The results from this for the V band are shown in Fig. 1(d), where the point sources subtract off quite cleanly apart from some objects which are not resolved but are broader than a PSF. Again there is no clearly identifiable point-source at the SN1999em position after the smoothing technique is applied, and no object is visible in the R and I frames either.

Smartt etal conclude that the progenitor of SN1999em is below the sensitivity limits of the pre-explosion VRI data, and that the star NGC1637-SD66 detected by Sohn & Davidge is a noise fluctuation which survived even their exceptionally careful photometric analysis.

4 SN2002ap: A More Complex Set of Archive Data

Supernova 2002ap was discovered by Yoji Hirose on 2002 January 29.4 UT in the spiral galaxy M74 [14]. It was discovered at $V = 14.54$, and at a distance of approximately 7.3 Mpc, may be the closest supernova since SN 1993J in M81 (at 3.6 Mpc). Several observers rapidly obtained spectra, reported that it appeared similar to the peculiar SN 1998bw but caught at an earlier epoch [13], exciting much activity. Later optical spectra of SN 2002ap indicate that it does appear to be a Type Ic, and its optical lightcurve appears to have peaked at approximately $M_V \simeq -17.5$, some 1.7^m fainter than SN 1998bw. Two popular theories for the origin of Type Ic supernovae are the core collapse of massive stars when they are in the WR phase, or the core collapse of a massive star in an interacting binary which has had its envelope stripped through mass transfer.

The host galaxy, M74, is a large and pretty spiral, which has been much imaged by many telescopes, including HST and Gemini-North. However, the supernova lies *JUST OFF* the field of view of all these high-quality studies. Archival analysis did identify one set of wide-field optical images of M74 taken before discovery of 2002ap. These images are from the Wide-Field-Camera (WFC) on the

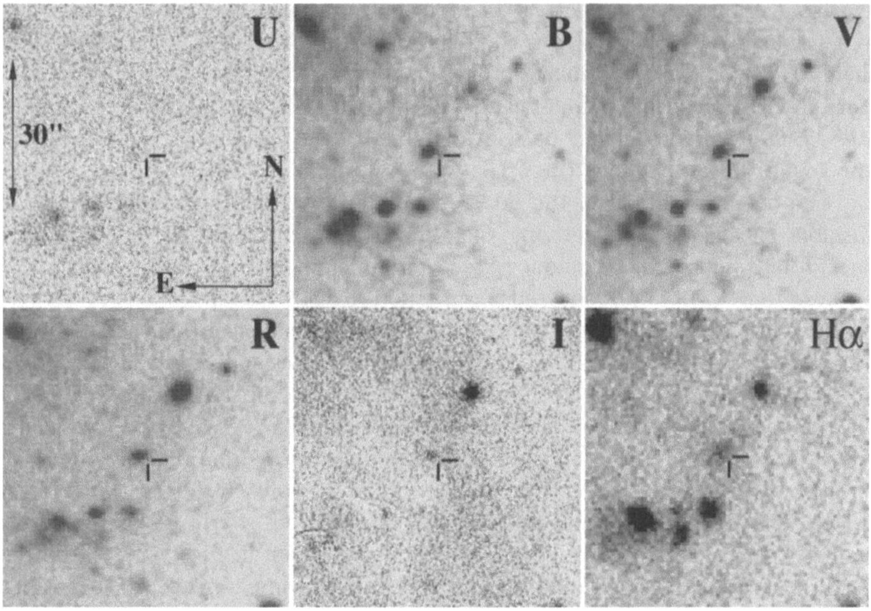

Fig. 2. SN2002ap in M74: prediscovery optical images, with BVRHα from the KPNO 0.9m and UI from the INT WFC. The location of SN 2002ap is at the centre of each frame, indicated by the orthogonal lines. The SN position is $2.31'' \pm 0.29''$ away from the nearby bright object detected in $BVRI$ (and marginally seen in U) i.e. it is clearly not coincident with this source. This figure is from Smartt etal 2002b

Isaac Newton Telescope (INT), La Palma, taken on 2001 July 24 through filters $UBVI$. The exposures were 120s in each of BVI and 180s in U. These were taken at the end of a night during the Wide Field Survey programme on Faint Sky Variability [6]. The WFC comprises 4 thinned EEV 4k×2k CCDs, with 13.5μm ($0.33''$) pixels. Repeat exposures of 120s in UVI were taken on 2002 February 2. The supernova core saturated in these frames, and shorter 2-10s exposures were taken with the telescope guiding continuously between the short and long exposures to determine an accurate position for SN 2002ap.

A second set of images was identified by searching the ADS for published studies of M74 which would indicate the existence of data, even though those data had not been archived. Such a study was identified, using data from the KPNO 0.9m with the Direct Imaging Camera taken on 1993 September 15 & 17. The authors were contacted, the tapes found, and *ad hoc* virtual access turned into real access. The subsequent analysis of these data followed the methodology outlined above, and is detailed by Smartt etal 2002b. Figure 2 shows the outcome, where yet again no progenitor was detected...

The galaxy M74 has been imaged by HST, Gemini, CFHT and WHT, however the supernova position does not fall on any of these images. All the publicly available archives have been searched for deeper, higher resolution images of

M74 but there are no superior images to those shown in Fig. 2 that include the pre-explosion site of SN 2002ap.

The important lesson for the present is that truly 'virtual' data, ie data whose existence can be deduced but which are not even physically in an archive, can prove of considerable science value.

5 Supernova Progenitors: The Future Virtual Observatory – Real Observatory Interface

We require data on more progenitors before we can be confident of the origins of the core-collapse SNe sub-types. Prompt and frequent multi-wavelength observations of SNe provide quite detailed information on the explosion and circumstellar material, and by inference on the mass-loss and envelope properties of the progenitor. However having high-quality archive images of SNe sites taken *prior to explosion* is the only robust way to set firm limits on the nature of the progenitor stars.

Table 1. Comparison of all information that is currently available from direct observations of the progenitors of core-collapse SN. The metallicity refers to estimates for the progenitor star, in the case of the spiral galaxies from measured abundance gradients and galactocentric radii of the SN. Mass refers to the *main-sequence* mass of the progenitor.

SN	Type	Mass	Z	Spec. Type
1987A	II peculiar	$20M_\odot$	$0.5Z_\odot$	B3Ia
1980K	II-L	$< 20M_\odot$	$0.5Z_\odot$	?
1993J	IIb	$17M_\odot$	$\sim 2Z_\odot$	K0Ia
1999em	II-P	$< 12M_\odot$	$1-2Z_\odot$	M-supergiant ?
1999gi	II-P	$< 9M_\odot$	$\sim 2Z_\odot$	M-supergiant ?
2002ap	I-c	$25M_\odot$	$0.5Z_\odot$	WR? binary?

Observations of nearby spiral and irregular galaxies within ~20 Mpc of the Milky Way allow the massive stellar content to be resolved. Multi-band images from the Hubble Space Telescope of all the face on spirals would be an excellent archive for future use when SNe are discovered. In the worst case this will allow limits to be set on the progenitor masses, as shown here and in Smartt et al. (2001), and should lead, in some cases, to definite identifications of progenitor stars. Already the HST archive contains approximately 120 Sb-Sd galaxies within ~20 Mpc which have observations of useful depth in at least 2 broad-band filters. There are a further 130 Sb-Sd spirals with exposures in 1 broad-band filter. Smartt etal have a Cycle 10 HST project to supplement the latter 130 galaxies with 2 further filters, and observe 120 more late-type spirals in three filters. This should give a total of ~370 Sb-Sd galaxies with HST observations. This number is steadily increasing each year, with data coming from projects with other scientific goals. This is supplemented with high-quality ground-based images from the well maintained archives of the ESO, ING, CFHT (and soon Gemini).

The various initiatives aimed at producing combined virtual observatories have, amongst many other applications, the unique historical aspect which is essential to SNe progenitor searches. One of the first of these (ASTROVIRTEL[2]) has already allowed us to search multi-telescope archives (HST + ESO telescopes) and use catalogue data as search criteria (e.g. LEDA). Along with some manual searching of the ING and CFHT archives, this suggests there are a further 100 spirals with ground-based observations of the quality presented in Figs. 1 and 2. Assuming a combined SNe II/Ib/Ic rate of $1.00 \pm 0.4\,(100\mathrm{yr})^{-1}(10^{10}L_{\odot}^{B})^{-1}$ [4], and that the galaxies in our archive have a mean luminosity $\sim 10^{10}L_{\odot}^{B}$, then one would expect $\sim 4.7 \pm 2$ core-collapse SNe per year in this sample. As the field-of-view of the WFPC2 on HST will only cover an average of 50% of the area of the optical disk of spirals between 10-20 Mpc, then an estimate of the number of SNe which will have pre-explosion archive material available is $\sim 2.4 \pm 2$ per year. Within a period of $3-5$ yrs we would hence expect the statistics presented in Table 1 to improve significantly. This is an example of unique science to be done with future Virtual Observatories.

6 Conclusions

Identification and analysis of high-resolution ground-based images of the pre-explosion sites of Type II supernovae provide unique information on the late stages of evolution of massive stars, the chemical evolution of the Universe, and the physics of feedback on galaxy evolution. Examples of the successful application of this method are presented above. From a Virtual Observatory perspective there are some important lessons to be learned:

- Even excellent quality and carefully derived data products (crowded field photometry in the specific example here) can prove unsuitable for purposes different than their original application. The scientific integrity of data retrieved by the Virtual Observatory 'system' must still be established by the responsible astronomer.
- A considerable amount of non-archived non-calibrated data exist in private repositories. While in some cases this can be of unique value, the work involved in retrieving, and especially in calibrating, old data suggests this is a worthwhile use of resources only in special cases.
- It is interesting to consider if the point above means that the AVO should restrict itself to accessing only well-described and major public data sets.
- A general challenge is to provide adequate astrometric cross-matching for different datasets. This raises the issue of different spatial resolutions, source dropouts, etc. The examples given here suggest that no complete general solution is feasible, except for imaging data from telescopes with extremely well-quantified optical systems.

[2] `http://www.stecf.org/astrovirtel`

References

1. Aldering G., Humphreys R.M., Richmond M., 1994, Astron.J., 107, 662
2. Baron E., et al., 2000, A+A, 545, 444
3. Barth A., Van Dyk S.D., Filippenko A.V., Leibundgut B., Richmond M., 1996, Astron.J., 111, 2047
4. Cappellaro E., Evans R., Turatto M., 1999, A+A, 351, 459
5. Filippenko A.V., 1997, ARAA, 35, 309
6. Groot P., et al., 2002, MNRAS, submitted
7. Hamuy M., et al., 2001, ApJ 558, 615
8. Johnson, RA, Beaulieu, SF, Gilmore, G, Hurley, J, Santiago, B, Tanvir, N, & Elson, RAW, 2001 MNRAS 324 367
9. Leonard D.C., Filippenko A.V., Ardila D.A., Brotherton M.S., 2001, ApJ, 553, 861
10. Li W.D., 1999, IAU Circ. No. 7294
11. McErlean N.D., Lennon D.J., Dufton P.L., 1999, 349, 553
12. McClure R.D., et al. 1989, PASP 101, 1156
13. Meikle P., Lucy L., Smartt S., Leibundgut B., Lundqvist P., IAU Circ., 7811
14. Nakano S., Kushida R., Kushida Y., Li W., IAU Circ., 7810
15. Pooley D., et al., 2002, ApJ 572 932
16. Smartt S.J., Gilmore G.F., Trentham N., Tout C.A., Frayn C.M., 2001, ApJ, 556, L29
17. Smartt S.J., Gilmore G.F., Tout, C.A., & Hodgkin, S.T. 2002a ApJ 565 L89
18. Smartt S.J., Vreeswijk, P., Ramirez-Ruiz, E., Gilmore G.F., Meikle, W.P., Ferguson, A., & Knapen, J. 2002b ApJL 572, L147
19. Sohn Young-Jong, Davidge T.J., 1996, Astron.J. 111, 2280
20. Sohn Young-Jong, Davidge T.J., 1998, Astron.J. 116, 130 (SD98)
21. Stetson P.B., 1987, PASP 99, 191
22. Thompson L.A., 1982, ApJ, 257, L63
23. Van Dyk S.D., Hamuy M., Fillipenko A.V., 1996, Astron.J., 111, 2017
24. Walborn N. et al., 1989, A+A, 219, 229
25. White G.L., Malin D.F., 1987, Nature, 327, 36

The Great Observatories Origins Deep Survey: A VO Test Case?

Robert A.E. Fosbury[1] and the GOODS teams[2]

[1] Space Telescope - European Coordinating Facility
[2] STScI/ESO/ST-ECF/JPL/SSC/Gemini/Boston U./U. Ariz./ U. Fla./UCLA/UCSC/IAP/Saclay/Yale/AUI: see
`www.eso.org/science/goods/`
`www.stsci.edu/science/goods/`

Abstract. The Great Observatories Origins Deep Survey is an effort to orchestrate limiting observations of a northern (HDF–N) and a southern (CDF–S) field, totalling some $300 arcmin^2$, with the most powerful telescopes over the widest possible wavelength range. Based on a SIRTF Legacy, an HST (ACS) Treasury, and ESO Large Programmes, many other data sets – including very deep X-ray observations with CHANDRA and NEWTON – will be coordinated to provide a public database with which to address fundamental questions concerning the formation and evolution of galaxies and AGN. The survey will provide a comprehensive picture of the history of galaxies from early epochs until the relatively recent past and will form the springboard for future investigations with the NGST and ALMA.

1 What Is GOODS?

The programme was conceived as an orchestration of limiting observations of the HDF-N and the CDF-S regions ($\sim 300 arcmin^2$ in total) with the most powerful telescopes over the widest wavelength range: 30 times larger in solid angle than the combined HDF-N and HDF-S. It is based on large programmes with SIRTF, HST, Chandra, Newton, VLT etc. The specific GOODS proposals all produce datasets and derived products for the public domain.

2 Primary Science Goals

The main scientific aims of the programme can be summarised as follows:

- Elucidating the mass assembly history of galaxies and tracking the formation of the Hubble sequence
- Performing a census of energetic output from star formation and supermassive black holes
- Making measurements or placing limits on the discrete source component of the extragalactic background light
- The discovery of high-z supernovæ to explore the cosmic expansion

3 Major Components

Many observations will contribute to the overall effort on the GOODS fields. Indeed, it is expected that these regions of the sky will become reference points for future observatories/missions such as ALMA and NGST. The major components of the effort at present are:

- SIRTF Legacy; PI: M Dickinson STScI – 647 hours IRAC + MIPS
- Hubble Treasury; PI: M Giavalisco STScI – 398 orbits ACS (plus existing WFPC2, NICMOS and STIS)
- ESO Large Programme, PI: C Cesarsky ESO – Imaging: 296 hours ISAAC, 40 hours WFI Spectroscopy: proposals for complete low-res spectroscopy of $\sim$6000 sources in CDF-S to R$\sim$25 ($\geq$30 nights MOS already allocated to other programs in CDF-S)
- Chandra 1 Msec on CDF-S and HDF-N; HDF-N to go to 2 Msec in CXO Cycle 3, – 5 Msec goal
- XMM-Newton 500 ksec CDF-S; 225 ksec HDF-N
- VLA, ATCA, Merlin, WSRT – Radio continuum imaging
- KPNO, CTIO, Subaru, Gemini, Keck, Palomar, MMT – Optical/NIR imaging and spectroscopy
- JCMT – SCUBA imaging of HDF-N
- GALEX – UV imaging

4 Data Products

All data and data products from GOODS programmes and its supporting observations will be made public and distributed as quickly as possible to enable community research.

A broad range of released products is expected from GOODS. These include V0.5, V1 and V2 data products resulting in photometric/astrometrically calibrated mosaics and catalogues; catalogues extending beyond the confusion limit (e.g., for MIPS); spectral catalogues and atlases; colour composite images; enhanced calibration files and models; software developed as needed for the programme; self-calibration; deconfusion; SED construction and analysis tools and public outreach materials.

Based on the original plans for the SIRTF data, the following series of releases are envisaged for all the major imaging datasets:

- Raw and basic pipeline processed frames: available from archive
- Version 0.5 – best-effort reduced images about 3 months after each campaign; mosaics if possible
- Version 1.0 – full release of image data products and initial catalogues, 3 months after last campaign
- Version 2.0 – final recalibrated data products and catalogues, 1 year after last campaign

Similar, but probably somewhat slower, release schedules are envisaged for spectroscopic data. In the case of the ESO imaging data, the reduction and release schemes are implemented by the EIS team.

5 GOODS and VO

What is the attraction of the GOODS project for the development of VO? GOODS is not huge in data volume but it is extremely rich in complexity. It consists of multi-wavelength imaging with different PSFs requiring resolution of confusion limits at long SIRTF wavelengths. Complex mosaicing strategies are needed, incorporating geometric distortion corrections. Photometric quality control must be rigorously applied. Catalogue production is needed for a wide variety of different purposes. The multi-object spectroscopy requires a major organisational effort for catalogue management, MOS mask creation, data reduction and archiving. There will be a need to manage the archiving of complex data products. In these senses, GOODS may make a useful VO prototype.

6 What GOODS Needs

In promising to provide all of these products for public rapid release, a heavy burden of responsibility is place upon the GOODS collaboration. What would help the team – especially in Europe where the programme receives no dedicated funding? Firstly, it needs people to pipeline process data, check quality, do photometric and astrometric calibration, build and check catalogues, develop dedicated s/w for special needs, design and implement archive data structures and descriptions, select sources and design MOS programmes. Secondly, it needs tailored archive utilities e.g., for derived data products, housekeeping tools for complex datasets, cutout servers, multi-instrument image-based comparison tools and a range of other tools whose functions are not, perhaps, yet recognised.

7 Conclusion

The GOODS programmes – or rather the ensemble of constituent programmes – are currently producing the first installments of a multi-faceted dataset which will remain a cornerstone of observational physical cosmology well into the next decade and the era of operation of major new observatories such as NGST and ALMA. The variety and complexity of the GOODS database make it a relatively harsh, but nonetheless rich, environment for realising the VO concepts. It must be appreciated, however, that the overall programme is conceived as a coherent whole with as high a degree of homogeneity as possible. In this sense it is easy fodder for VO. The real world of astronomical data will be far less homogeneous. So, if VO fails for GOODS, there is little hope of subsequent success.

We believe that the onus is on the GOODS teams to ensure that the datasets are well described and accurately calibrated. The test of the VO is whether or not it can react quickly and surely enough to contribute significantly to their scientific exploitation.

Visualizing and Analyzing Massive Astronomical Datasets with Partiview

Brian P. Abbott[1], Carter B. Emmart[1], Stuart Levy[2], and Charles T. Liu[1]

[1] American Museum of Natural History & Hayden Planetarium, New York NY 10024, USA
[2] National Center for Supercomputing Applications, University of Illinois Champaign-Urbana, Urbana IL 61801, USA

Abstract. Partiview is an advanced, real-time visualization tool for multi-dimensional datasets. Developed at NCSA, Partiview has cross-platform compatibility that allows its use in environments ranging from laptops to the 21-meter-diameter Hayden Planetarium dome, on scales varying from the solar neighborhood to large-scale structure. Current applications of Partiview include the simultaneous, fully interactive visualization of multiple datasets, including both observed and simulated data. The software is available at no cost from `www.haydenplanetarium.org` along with our Digital Universe data archive. A major priority of the ongoing development of Partiview is its integration as an analysis and visualization component of an International Virtual Observatory.

1 Introduction

In 2000, the American Museum of Natural History (AMNH) completed the Rose Center for Earth and Space which included a rebuilt Hayden Planetarium. As part of this renovation, AMNH/Hayden created one of the largest immersive theaters in the world serving not only as an education/outreach facility, but also as a scientific tool for visualization. With our 21-meter dome and through our web site, we have brought complex scientific visualizations and our three dimensional atlas, the Digital Universe, to millions of visitors.

The Digital Universe is a multi-dimensional atlas of the universe that one can use for data analysis or to simply explore the Milky Way or the extragalactic universe interactively. The software that allows the display of this atlas, Partiview, has been developed at the National Center for Supercomputing Applications (NCSA) and runs on a personal computer. Since April 2002, we have been distributing Partiview and our data to the public at no cost via our web site (`www.haydenplanetarium.org`) and have served over 10,000 downloads of our Milky Way Atlas. This paper will focus on these two interrelated projects, discussing their potential as an education/public outreach resource and scientific analysis tool for a Virtual Observatory.

2 The Partiview Visualization Software

Partiview is a software package that was designed to provide real-time, interactive data visualization for three dimensional particle data. Developed at the

National Center for Supercomputing Applications (NCSA), Partiview is actually the desktop version of a more versatile program called Virtual Director. These programs were designed to interactively record, edit, and play back a flight path through massive data sets inside a virtual environment like the CAVE.

In 2002, the Department of Astrophysics and Hayden Planetarium at AMNH was given a grant from the NCSA Alliance to distribute and help develop Partiview to aid in scientific visualization. We have contributed to the distribution, documentation, and data collection for Partiview. We are also making contributions to the software itself, which will eventually be open source.

Partiview differs from other data visualization programs in its ability to display and interleave multiple data sets at once. Seeing many data groups simultaneously, or even blinking one data set as you navigate in real-time among other data, provides a unique 3-D view that scientists often do not have access to. Partiview also differs from other visualization tools in the ease by which data are imported. There is no new programming syntax to learn, just a simple, concise command structure that tailors the display to your needs. Figure 1 shows Partiview with the 2dF galaxy and Sloan quasar surveys displayed.

Currently, Partiview runs on Linux, IRIX, Windows, and Mac OSX. The graphics output is OpenGL while the graphical user interface is written in FLTK. The efficiency of data rendering in Partiview depends on the graphics support as well as the complexity of the data. With our off-the-shelf laptop computer, we are able to display hundreds of thousands of particles and render them in real-time without any performance problems.

Time evolving data are also supported in Partiview. Namely, time evolving, three dimensional stellar dynamics codes have been accommodated for display and analysis in Partiview. Furthermore, Partiview can display other particle simulations such as galaxy collisions as well as polygonal surface models. Partiview, then, is not just a static particle viewer, but can accommodate dynamic simulations and theoretical computations, making this software an ideal tool for both theoretical and observational astronomers and scientists of all disciplines.

3 A Digital Universe

As part of the rebuilt Hayden Planetarium and newly-formed Department of Astrophysics, the American Museum of Natural History endeavored to create a Digital Universe that would not only take visitors on guided tours of the three dimensional universe but, more importantly, be a resource for the scientific community. Initially funded by NASA in 1998, our first goal was to build a Digital Milky Way Galaxy. We achieved this goal, building a statistical galaxy with over 800 million stars and other objects integrated with observed data sets.

Since then, we have expanded our interests to include the entire observable universe. We are collecting data on all scales and from many scientific disciplines. Additionally, many scientists have offered their data for use and distribution in the Digital Universe so that the scientific and education communities can share these data and view them within Partiview. We intend for the Digital Universe

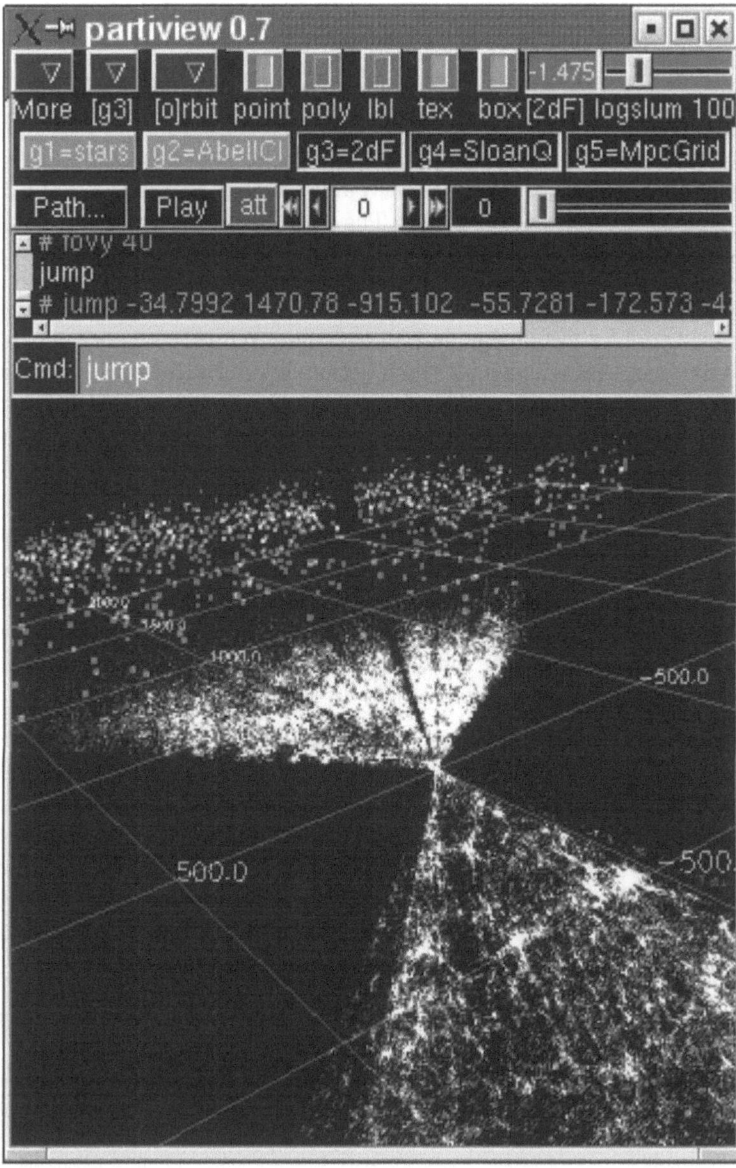

Fig. 1. A screen shot of Partiview. Shown (in glorious black and white) are the 2dF galaxy survey (white points) and the Sloan Quasar Survey (gray points). Our vantage point is such that the flat, fan-like portion of data extending out from the Earth (toward the top of the viewing area) is at a low angle of inclination while the opposite side (at bottom) is inclined at a higher angle, thereby revealing the structure of these data. The megaparsec grid is in the same plane as the data in the top of the viewing area. It should be noted that static images cannot capture the 3-D view, particularly when you are in motion

to be a Partiview data node for the Virtual Observatory, shared by and accessible to scientists, educators, and the general public.

4 Integration Into a Virtual Observatory

We see great potential for both Partiview and the Digital Universe to be integral components of a Virtual Observatory. The main benefits that are available right now include:

- **Data Visualization** Currently, Partiview is mature enough to be used as a robust tool for visualizing data in real-time. While there are many paths of development that will enhance its functionality in the future, as it stands now Partiview is more than capable of providing an interactive, multi-dimensional visualization of existing observed and theoretical particle data.
- **Data Distribution** We are currently distributing data for use in Partiview. These data are mainly targeted at education and outreach organizations. With our Milky Way Atlas and Extragalactic Atlas, we provide everyone, from student to scientist, a rare view of our universe. Most of these data sets are derived from publicly-available data that exist in archives, however, others are from direct contact with project scientists. While the means of data distribution are primitive right now, we intend to integrate data standards that will be updated as those of the Virtual Observatory evolve.
- **Education/Public Outreach Node** Although we are a research institution, the American Museum of Natural History has a strong commitment to education and public outreach. We provide complex scientific visualizations to millions of visitors per year – our outreach potential is among the best of any scientific institution. We are tapping this potential for our distribution of the Digital Universe and Partiview. As these become more strongly coupled with the Virtual Observatory, we will be ideally placed to bring complex data sets to a public audience in understandable ways.

While there are many improvements to be made, we feel Partiview is ready for integration into a Virtual Observatory right now. These improvements, which we discuss in the following section, will enhance the functionality of Partiview as well as improve access to an expanded Digital Universe data archive.

5 Future Enhancements

In the future, we intend to enhance both the Partiview software as well as the Digital Universe archive. These changes will take place in conjunction with development by other scientific groups who are also adding features to Partiview.

We will concentrate our enhancements to Partiview in several areas. One major area for improvement is the user interface. Currently, much of the functionality in Partiview is accessed via commands, while some of the more common tasks are found in a small graphical user interface (GUI). We plan to add functions and commands to the GUI which will make Partiview easier and more natural to use. We will also make improvements to the remote conferencing and collaborative modes of Partiview so that one Partiview could talk to many other instances of Partiview, broadcasting the view of multiple data sets over the Internet. In addition, we will provide more sophisticated data access capabilities into Partiview, allowing for on-line data access, conversion, and display. Finally, we will continue to update and improve the documentation and distribution methods for Partiview.

The Digital Universe, being less technically sophisticated, will require more development. We hope to first provide our data in sensible ways over the Internet. Currently, these are offered as ASCII files; however, once data standards are set, be it XML or some other data standard, we intend to integrate these standards so that our Digital Universe archive will be compatible with other standardized archives. We will also make data importation a smoother process, creating routines to convert data into the Partiview format. This will open up the possibility of on-the-fly, over the Internet visualization by loading and displaying data interactively.

We believe these enhancements will insure that Partiview and the Digital Universe, as part of a Virtual Observatory, will be a powerful visualization tool for scientists and a valuable outreach resource for transmitting science to a broad audience.

Acknowledgements

We would like to acknowledge the National Center for Supercomputing Applications Alliance for their support of the development and distribution of Partiview and NASA for their support of the Digital Universe.

e-Science and the Grid

Marcel Kunze

Forschungszentrum Karlsruhe, P.O Box 3640, D-76021 Karlsruhe, Germany

Abstract. Many areas of science involve access to distributed computing and data resources, specialized and expensive instruments for world-wide collaborations of scientists. There are many examples ranging from bioinformatics and astrophysics to collaborative engineering and earth observation. A new IT infrastructure called the Grid will enable scientists to maximize the science derived in such a fashion. The Web gave us easy access to html pages and information: the Grid will give us seamless access to a much wider range of distributed resources; it will enable us to form transient virtual organisations without compromising security or privacy.

1 Introduction

Scientific research and development has always involved a large number of people, working together with different expertise, extending the body of knowledge. Whilst in former times researchers worked on the basis of their own observation, there is now a move to collaborate in large teams who share results between widely dispersed participants. At present the predominant technologies to communicate are email and the Web, and researchers are increasingly relying on computer and communication services in their daily work. It can be foreseen that this trend continues and science will be increasingly done through distributed global collaborations enabled by the Internet, using very large data collections, terascale computing resources and high performance visualisation. Dr. John Taylor, the chair of the British science council, paraphrases this evolution by the term e-Science (in the sense of *enhanced* Science): "E-Science is about global collaboration in key areas of science, and the next generation of infrastructure that will enable it."

The computational Grid defines an infrastructure for e-Science that generally covers the distributed application domain of various disciplines, as for instance high energy physics and astrophysics, life sciences and medicine, environmental research, including instruments (accelerators, satellites, observatories etc.). Grid-Computing offers access to distributed data archives enabling sophisticated information analyses, federating various sources of information: As an example, in the event of the observation of a potential future supernova explosion scientists should be able to correlate the recordings from astronomy with the observations of the gravitational wave experiments as well as with the data of neutrino detectors from particle physics: The added value of such a comprehensive scientific methodology is obvious.

2 Grid Architecture

One of the most important factors for future success in basic research and industry will be fast access to data and IT resources distributed world-wide, independent of time and location. Information systems are no longer just defined by what they can process, but also where they can connect to. For example, the growing demands for computing power in simulation and engineering design projects is increasingly satisfied by "on-demand" sharing of CPU-cycles and data stores across distributed networks. The ultimate goal for these interconnected networks, or grids, is to make IT-power as commonplace and omnipresent as electricity [1].

On the basis of this original simplistic view, Grid Computing is today often more comprehensively defined as the safe, flexible, coordinated and common use of resources within virtual organizations that are formed dynamically from globally distributed institutions [4]. A resource in this context could be data and instruments, but as well in the broad sense persons and experts.

One of the most difficult problems with the architecture of Grid systems lies in the standardized definition of protocols and services, that guarantee the general interoperability of systems and user programs. The software tools that allow for seamless interoperation of resources worldwide are coined by the term middleware. The middleware represents the layer of Grid Computing, on which distributed applications can be developed in a platform independent way. Examples of such applications are large-scale simulation programs in high energy and astroparticle physics as well as access to distributed very large data bases in astronomy or medicine. The middleware for the implementation of a Grid must adhere to the following conditions [3]:

- Co-ordination and co-allocation of resources, without a central control instance
- Use of open standard protocols and interfaces
- Realization of nontrivial complex service qualities by connecting distributed resources

In practice these demands lead to the mechanism of Web services for resource aggregation: Web service functionalities offer e.g. directory services to locate resources, resources brokers for dynamic allocation, as well as schedulers for booking of resources. The security problem can be solved by the employment of authentication on the basis of ISO certificates. The latest development goes towards standardized Web services, that allow to set up a distributed application out of standardized components with standardized and open interfaces. The future industry standard Open Grid services Architecture (OGSA) [5] would furthermore allow to construct applications with components and services originating from the Unix world and from the Windows world, in any mix.

A local computer cluster with a centrally administered batch system however does not yet form a computational Grid (this would correspond to the situation that a customer would have to operate his electrical appliances at a certain power station). The current Grid middleware projects try to set up an environment that meets the criteria mentioned above.

3 Grid Infrastructure

There are many Grid projects funded by national agencies and the European Union to produce and demonstrate grid middleware. A subset of these projects is managed by or focused on the High Energy Physics community - for example DataGrid, CrossGrid, DataTAG, INFN Grid, GridPP, NorduGrid and others in Europe, PPDG, GryPhyN and iVDGL in the US, in addition to projects in Japan and elsewhere. All of these projects have their own reporting structures, goals and deliverables, set by the funding agencies. Long-term support for the middleware developed by these projects is unclear, but all build on top of the Globus ToolKit [2]. In addition there is considerable industrial interest in Grid middleware, and several well-established companies are investing heavily in this area (e.g. IBM, SUN, HP, MicroSoft).

As an example of a large-scale Grid infrastructure project the LHC Computing Grid project will be discussed: The current research program at CERN requires the construction of the so far world-wide largest particle accelerator, the Large Hadron Collider (LHC). With the LHC the origin of mass is to be examined, one of the most fundamental open questions of science. Doing so, CERN expects an immense amount of data, which must be stored and processed, and thousands of scientists in the whole world will co-operate with the analysis of the collected data. The idea is to distribute the data on regional centres in a transparent way: In the MONARC study [6] it was shown that a multi-tiered hierarchical system with different levels fulfils the requirements. In this model CERN as Tier-0 stores and reconstructs the raw data. The analysis as well as the production of simulated events takes place at regional Tier-1 centres, followed by national Tier-2 centres, local computer centres (Tier-3) and personal computers (Tier-4).

In order to meet the computing challenge the LHC Computing Grid project aims at providing a Grid production environment. The actual work is divided into four areas of concern:

- *Applications*
  Providing the basic tools and infrastructure for applications software development, and the organisation of projects to implement the common solutions for the experiments
- *Fabrics*
  The management of the large computing cluster at CERN, sharing of technical information with systems management staff at other large regional centres, regular re-evaluation of the basic technologies that are used to build Grids
- *Technology*
  Negotiating the provision and maintenance of the software tools (middleware) needed to interconnect the regional centres to form a global grid. The

base assumption is that initially these tools will be developed by projects that have been funded independently (e.g. DataGrid, Globus, GryPhyn etc.)

- *Deployment*
  Organising and operating the pilot global grid service. This will involve a great deal of coordination between the managers of the regional centres and the people responsible for running data production in the experiments. In addition this comprises software distribution, running the grid infrastructure and providing user consultancy and support.

Building on some solid experience with the ROOT framework and object databases, the POOL project aims at a hybrid event store, using a relational database for meta data, and storing the physics data itself with ROOT I/O in sequential files [7].

4 Knowledge Discovery

If the Grid is to support a wider range of applications, both scientific and industrial, then integration of database technologies into the Grid will become important. Metadata describes the information about the data, and it is important as it adds context to the scientific data, increasing confidence people have in the data they use, and making it possible to locate data without knowing exactly where it is stored. Provenance data on the other hand provides information on the source and subsequent history of processing. It includes information on data creation, source, owner, what processing has taken place, what analyses have been applied, what results have been produced, and the level of confidence in the quality of information. Currently, the use of metadata and provenance data in Grid applications is very simple - it is mainly for the mapping of logical names for datasets into the physical locations where they can be accessed. However, as the Grid expands into application areas like astronomy, more sophisticated systems and tools will be required. The result is likely to be a *Semantic Grid* that is analogous to the idea of the *Semantic Web.*

Concerning data, information and knowledge the *Semantic Grid* infrastructure could be conceptualized in three consecutive layers:

- *Data*
  This layer deals with allocation of computational resources, and deals with large volumes of data and fast networking. It builds onto the physical grid fabric infrastructure, which may as well comprise scientific equipment. Data in this context is understood as a sequence of uninterpreted bits and bytes.
- *Information*
  This layer deals with the way that information is represented, stored, accessed and maintained. Information is understood as data equipped with meaning.
- *Knowledge*
  This layer is concerned with the way that knowledge is acquired, used, retrieved and published to support particular e-Scientific subjects. Here knowl-

edge is understood as information applied to achieve a goal, solve a problem or take decisions.

The degree to which the various layers are important and utilised in a given Grid application will be domain dependent, but all will have some element of all three layers in them.

It is another aim of Grid-Computing and e-Science to promote the open publication of scientific data. If this is realised then it is expected that many of the advances to flow from the Grid will come from applications that combine information from multiple data sets. This will allow researchers to combine different information on a single attribute to gain a more complete picture, and to aggregate similar information about different entities. Furthermore, if a generic and standardized provenance service would be provided, operations could be offered to locate data with a particular provenance, or the provenance of all data returned by a query could be identified, thus avoiding duplicated effort in the construction of Grid applications and data analysis.

References

1. I. Foster, C. Kesselman: *The Grid: Blueprint for a New Computing Infrastructure* (Morgan Kaufmann Publ., 1999)
2. *The Globus Project* (`http://www.globus.org/`)
3. I. Foster: *What is the Grid?* (`http://www.gridtoday.com/02/0722/100136.html`)
4. I. Foster, C. Kesselman, S. Tuecke: *The Anatomy of the Grid: Enabling Scalable Virtual Organizations* (International J. Supercomputer Applications, 15(3), 2001)
5. I. Foster, C. Kesselman, J. Nick, S. Tuecke: *The Physiology of the Grid: An Open Grid Services Architecture for Distributed Systems Integration* (June 22, 2002. `http://www.globus.org/research/papers/ogsa.pdf`)
6. *MONARC Phase 2 Report* (CERN/LCB 2000-001, `http://monarc.web.cern.ch/MONARC/docs/phase2report/Phase2Report.pdf`)
7. *The ROOT Framework* (`http://root.cern.ch/`)

National Virtual Observatory Architecture

Reagan W. Moore

San Diego Supercomputer Center
moore@sdsc.edu

Abstract. The National Virtual Observatory is assembling web services, data analysis pipelines, and grid software to support astronomical research. The initial goal is to facilitate access to existing sky surveys and provide standard services for manipulating catalogs and image collections. The long-range goal is to support analyses of entire sky surveys and enable applications that examine multiple collections. The layers of software infrastructure required to provide these goals are examined.

1 National Virtual Observatory Testbed

The National Virtual Observatory will provide mechanisms to discover, access, and analyze images and information stored in all-sky surveys. There are multiple ways to characterize the desired capabilities, in terms of interactivity, scale of analysis, and type of discovery capabilities. Interactivity requirements vary from interactive web services, to sustained processing pipelines, to large-scale analyses that require the massive computing power of the National Science Foundation Teragrid. The scale of analysis varies from the interactive picking of individual images from a sky survey or a small number (say 10-100) records from a survey catalog, to applications that will require processing of an entire image collection. Finally, discovery capabilities range from requests that can be satisfied by a single sky survey, to requests that require the integration of results from multiple sky surveys across multiple data encoding formats.

The challenge in designing an architecture for the NVO is to recognize the importance of each of the different usage models, while minimizing the number of software layers needed to implement the desired capabilities. The architecture should also maintain existing capabilities already provided by many of the sky surveys and collection catalogs. The NVO architecture approach is based upon identifying common web-based services that can be easily implemented as new capabilities on existing systems, identifying the data grid requirements needed to federate existing collections to support massive computations, and identifying the information and knowledge management systems that will unify discovery across the catalogs.

The system architecture is designed to build upon each of these three sets of requirements, starting through the addition of simple web services for picking images and records, and the development of a common set of semantics for describing physical quantities across the collections. Larger scale computational jobs will then be supported by migration onto the supercomputer resources provided by the NSF Partnership for Advanced Computational Infrastructure. The

PACI partnership is implementing a Teragrid that couples high performance clusters at each of four sites via 10-Gbit/sec network links. The major all-sky surveys are being replicated onto resources attached to the Teragrid to enable large scale analyses. The web services environment and the large-scale analysis testbed will then be augmented with production pipeline systems that interface to the grid batch computing services.

The NVO architecture is based upon the ability to off-load large computational jobs onto an NVO testbed, which is being built on top of the Teragrid. In turn, the Teragrid is being assembled by integration of the Globus Grid Toolkit that is being developed through an NSF National Middleware Initiative, with data management software that is used to implement digital libraries, data grids, and persistent archives. The data management infrastructure is based upon the San Diego Supercomputer Center Storage Resource Broker, which is used to build data sharing systems, data publication systems, and data preservation systems. The data sharing systems make it possible to manage personal collections, while the data publication systems are used to provide discipline wide access to data collections. The Teragrid will also include knowledge management systems that organize relationships between the semantic terms used to describe physical quantities in the all-sky surveys. The Uniform Content Descriptors are used as tokens that can be associated with attributes in collections. Attributes that have the same associated token are assumed to be equivalent. For collections that do not have shared tokens, ontologies are used to define additional semantic terms that can be used to build relationships between catalogs.

2 Architecture Requirements

Three exemplary applications are being used to drive requirements for the NVO system architecture, based upon a Brown Dwarf Search, a Galaxy Morphology analysis, and a Gamma Ray Burst analysis. The system components that are needed to support each application are listed in Fig. 1. An attempt has been made to identify common requirements across the applications. The technologies that are being used to support the applications include web-based access through the VOTable catalog access system, web-based access through the SRB collection management system for access to replicas of the 2-Micron All-Sky Survey and Digital Palomar Observatory Sky Survey, and web-based image cut-out services. Requirements are also coming from applications that require the reprocessing of the 2-Micron All Sky Survey. Together they exemplify the requirements for the picking and large-scale analysis environments, but do not address the batch pipeline processing environment. For pipeline processing, the Montage software from IRSA is being evaluated for its ability to interact with batch systems in grid computing environments.

Brown Dwarf Search	Galaxy Morphology	Gamma Ray Burst
Information repository abstraction for characterizing interactions with catalogs in databases	Distributed joins across multiple catalogs	Mechanism to specify relationships within a complex VOTable as a concept space
Standard query language for interacting with entire catalogs.	Catalog Cone Search and positional cross-matching of extended sources	Cone search across multiple catalogs
Standard information interchange format based on VOTable.	Mechanism for formatting data into a standard catalog format - VOTable	Standard information exchange format based on VOTable
Web based portal to control input parameter selection.	Portal for defining input and processing steps.	Portal for registration of interest in desired events.
Standard access mechanism based upon SOAP protocol and WSDL services.		Support for interactive display of results, including parsing of VOTable format.
Service registry for publishing service descriptions, and supporting a service directory	Service registry for catalogs	Service registry for both catalog and image data
Method registry for methods that can be applied against data models.		Support for distribution of notifications to registered users.
Standard display mechanism for results.	Support for publication of new tables	Support for publishing results
Procedural support for composing results from multiple catalogs.	Support for visualization services including contouring.	Support for tracking astronomical events
Interactive response time for browsing tables and columns in databases.	Support for analyses of images, including galaxy morphology analysis	
Integrate existing systems and tools	Grid computing interface to support high-throughput analyses	Support for mapping from current format to standard format
Persistent interactive access to image archives	Image retrieval from image archives, including partial images or cutouts	Image retrieval from image archives, including partial images or cutouts
Modular design to enable rapid prototyping.		Support for exception handling.

Fig. 1. Typical Application Requirements

3 System Components

The Global Grid Forum is developing descriptions of system architectures that integrate web-based access with grid computing. Fig. 2, taken from the characterizations published by William Johnston in the Grid Protocol Architecture working group, defines the set of software layers that can be used to support applications running on a grid. The components include:

- Portals - the user interfaces that manage interactions with the multiple NVO services. Portals encapsulate domain knowledge by associating explicit mediators or transformation programs with access to the chosen catalogs and archives. Portals also provide interfaces to control image processing pipelines.
- Analysis and support tools - the NVO supplied applications for processing image and catalog records. These tools provide capabilities such as conversion to a standard data format, visualization mechanisms for 2D and 3D displays, metadata extraction mechanisms for populating attributes in catalogs, and algorithms for processing data.
- Application support platforms - these typically provide an environment for interactive execution support. Examples include Java execution, Corba ob-

ject management, DCOM integration, WSDL/SOAP services, and Open Grid Service Architecture services.

- Grid support environment - the set of collective operations used to build a common data management and job execution environment across administration domains. Grid services manage distributed state information.

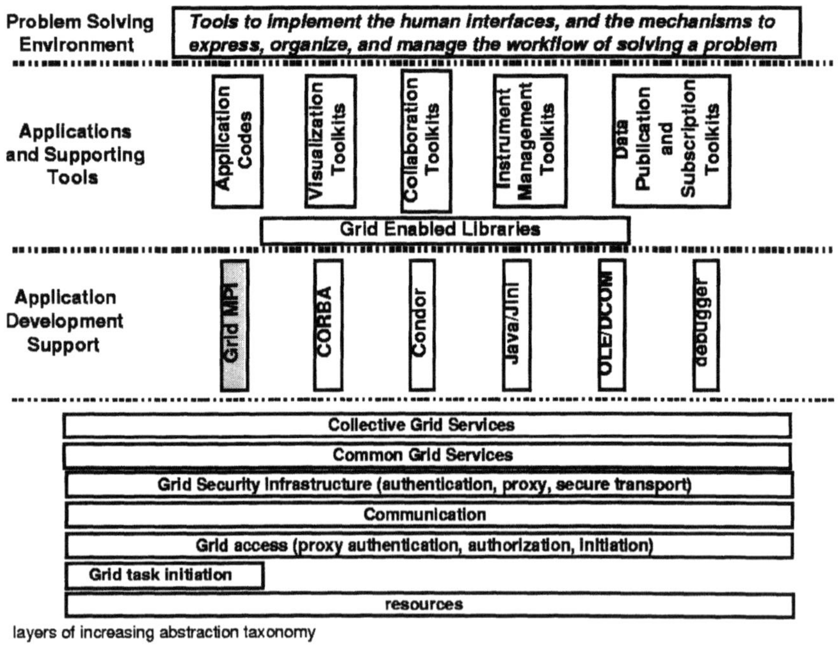

Fig. 2. Application Support Environments

The capabilities provided by the grid support environment can also be expanded, into the collective grid services that manage distributed state information, the common grid services that provide fundamental capabilities, and the grid fabric that represents the resources that are available within the sites that are linked by the grid environment. These capabilities are shown in Fig. 3, taken from the characterizations published by William Johnston in the Grid Protocol Architecture (GPA) working group.

For the NVO testbed, both data management and collection management services are needed. In Fig. 3, these services are encapsulated in the boxes at the collective grid services layer under the title "Data Cataloging".

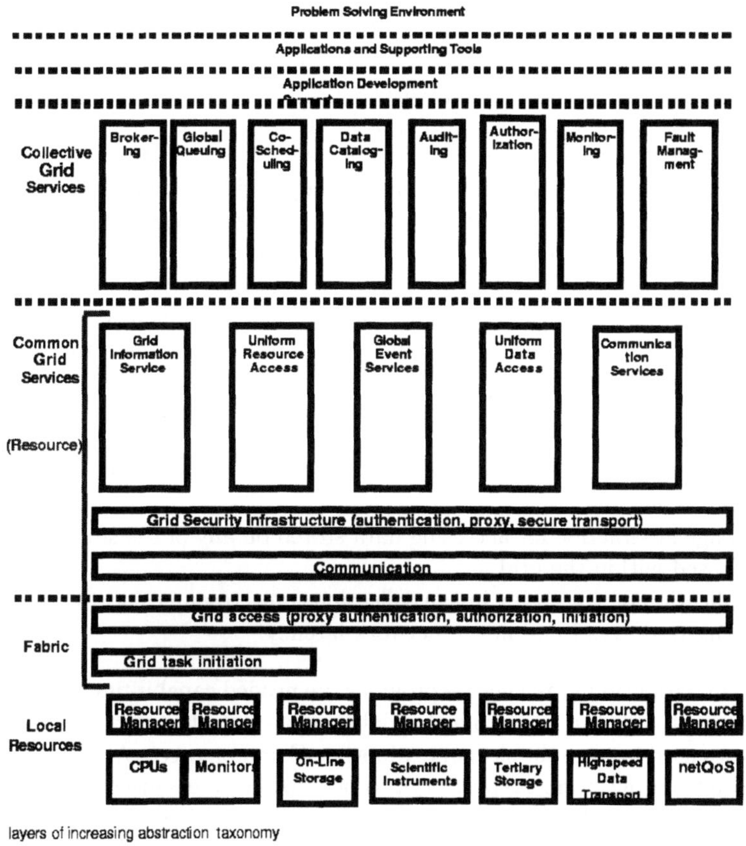

Fig. 3. Infrastructure Components

4 NVO Testbed

The NVO has implemented multiple services that integrate web-based access with grid-based computation. It is quite interesting to note that some of the web interactivity requirements are actually enabled by grid computing infrastructure. An example is support for image cut-outs on the DPOSS sky survey. Each image within the collection is about 1-GB in size. A web service that supports generation of image cut-outs would retrieve the appropriate image, process the image to reduce the size to the cut-out desired by the researcher, and then transmit the reduced image over the web. This process requires the transmission of the 1-GB image file to the processing server.

To create an interactive service, the ability to generate the image cut-out needed to be applied directly at the storage repository where the image collection was stored. Data grid technology provides latency management functions, such as the ability to aggregate I/O commands into remote proxies that can be executed directly on the storage repository under data grid control. For the cut-out service, the associated data subsetting commands were encapsulated into a remote proxy that was then invoked when a image cut-out request was made. This eliminated the need to move the entire file, greatly speeding up the interactive service.

A second example shows the integration of image processing pipelines with grid technology. Image processing pipelines are typically based on a hardware system that has a maximum processing capacity. Grids provide a way to redirect processing requests onto alternate platforms. For jobs that are being queued through a batch system, this integration is straightforward. The batch jobs are queued onto the remote system where processing capability exists to handle the request. This approach puts additional data handling demands upon the grid. The movement of the batch job should not require any changes to data access mechanisms within the batch job. This implies the use of grid infrastructure to provide:

- Logical name space supporting global, persistent identifiers for sky survey image collections
- Storage abstraction for accessing data stored on any of the storage repositories used within the grid
- Replication of sky survey image collections onto resources near the grid computing resources to minimize network latency

Again data grid infrastructure provides the needed capabilities. Logical name spaces are provided to map from the global persistent identifiers to the physical file names used to store the sky survey images. Common access APIs are provided that map from the storage repository protocol to the access mechanism preferred by the application. Replication is used to provide multiple sources for the sky survey images. The source for the images is chosen to provide the lowest latency access.

Grid technologies also manage authentication and authorization across the multiple administration domains under which the original sky surveys are managed. Authentication is managed by use of an inter-realm authentication environment based upon PKI certificates. The Generic Security Service API is used to map from the PKI authentication environment to the local Unix, Kerberos, or DCE authentication environments.

The major area of development for grid technologies is the integration of knowledge management environments to support discovery across multiple image collections that use disparate attributes. The types of knowledge include not only logical relationships that describe how semantic terms are related, but also procedural relationships that describe the processing steps that are applied to create derived data products such as image cut-outs, and structural relationships that describe data encoding formats. The ability to transform from one data encoding format to a standard data encoding format can be viewed as a set of structural relationships that can be imposed on an image or astronomy data set by a remote proxy. The characterization of data encoding format transformations as knowledge is a major area of research.

In summary, the NVO testbed is composed of:

- Portals that explicitly integrate web services to data resources and analysis procedures. Examples are cone search, VOTable catalog query, simple image access.
- Process management systems that control data processing pipelines. The process management systems map from a description of the steps used to create a derived data product to the grid computational and data resources
- Data access layer for application of methods on data models. The methods can be executed within the portals, or within the process management systems, or within the data grid as remote proxies. The data access layer encapsulates the knowledge needed to map from the data encoding format to the desired representation, and basically provides the support for transformative migrations of data.
- Data grid for managing distributed collections. The persistent replication of data between multiple resources can be managed through a logical name space.
- Persistent data caches for interactive access to sky survey image collections. Because of the rapid decrease in disk system prices, it is now feasible to keep 10-TB collections on line on disk farms. This makes it feasible to provide interactive access to data picking environments for the retrieval of arbitrary image cut-outs from a survey.
- High performance disk caches for data analyses. The ability to process collections relies on the ability to keep data close to the computational resources. One goal of the NVO testbed is to support data intensive analyses at the rate of 3 GB/sec, or ten Terabytes per hour from high performance disk cache.
- Computational grid. The distribution of image processing steps onto remote resources can be automated through grid technology.
- Persistent archives. The original sky surveys are replicated across multiple platforms to guarantee that the collections will survive through all possible types of data loss. Data grids are used to provide infrastructure independence, meaning that the replicas can be managed across arbitrary choices for storage architecture or media.

5 Ontologies

The NVO testbed can be viewed as a set of software infrastructure layers that enable the processing of images from all-sky surveys. A more aggressive viewpoint is to consider the NVO testbed as a set of hierarchical ontologies. Each level of the ontology hierarchy specifies a different set of naming conventions that are applied to the images as they are manipulated. The NVO testbed provides a way to map between the different naming conventions. The processing steps that are applied are shown in Fig. 4.

It is worth noting that there is a tight coupling between the data model and the transformations that can be applied to physical quantities. This represents

Transformative Operation	Ontology — Name space
Manage transformations between physical quantities	Concept spaces — the relationships used to describe transformations on physical quantities
Global naming convention	Data grid — the logical name space that is used to create global persistent identifiers, and map to physical file names
Organize discovered physical quantities	Catalogs — the attributes used to label the digital entities to support discovery, the semantic labels (UCDs) used to describe physical quantities.
Turn bits into physical quantities	Data model — the encoding format (structure) of the digital entity, the semantic terms used to label the parts of the structure, and the processes (methods) that can transform the structure

Fig. 4. Ontology Hierarchy in the NVO Testbed

one of the major challenges to the NVO testbed, as it implies one can choose to manage the knowledge required to do the transformations within the methods that are applied to the data, or within a concept space that is managed independently of the data. Within the grid community, both approaches are being pursued. The experiences of other disciplines in the use of grid technology to federate collections can help the NVO testbed resolve where knowledge should be managed.

Acknowledgements

The characterization of the NVO architecture is being developed by Doug Tody (NRAO), Roy Williams (Caltech), John Good (IRSA), Bruce Berriman (IRSA), Ray Plante (NCSA), Eva Deelman (IAI/USC), and Carl Kesselman (ISI/USC). Implementations of the NVO architecture are being done by Roy Williams, Alex Szalay (JHU), John Good, George Kremenek (SDSC), and Ray Plante. This project is supported by the NSF cooperative agreement on the National Virtual Observatory AST0122449.

Authentication and Authorization Architecture for AstroGrid and the Virtual Observatory

G. Rixon, T. Linde, E. Auden, N. Walton

[1] Institute of Astronomy, University of Cambridge, Madingley Road, Cambridge CB3 0HA, U.K.
[2] Dept of Physics and Astronomy, University of Leicester, Leicester LE1 7RH, U.K.
[3] Mullard Space Science Laboratory, Holmbury St. Mary, Dorking, Surrey RH5 6NT, U.K.

Abstract. Although the bulk of the data useful to the Virtual Observatory are in the public domain and have uniform, read-only access from archives, there are several important cases where appropriate access varies from user to user. We outline these cases and list the desirable feature of an access-control system to cover them. AstroGrid is building such a system and we describe its architecture: X.500 names and X.509 certificates for user identity; a Public Key Infrastructure for authentication of identity; and the Community Privilege Service for sharing authorization information.

1 Why is Access Control Needed?

The vast majority of the data to be used on the Virtual Observatory have a very simple access policy: they are in the public domain, meaning that anybody on the planet can read them but only archive staff can change the physical copies in the archives. This is the access model supported by the WWW and used in existing on-line interfaces to astronomical archives. Why then should the Virtual Observatory need any special arrangements for access control?

One should note that the 'special arrangements' are for granting different kinds of access to different users. There is always access control: notably, archive data may not be altered by outside users. The question could be rephrased 'why cannot all access to the VO be anonymous?'

The simplest model of access control works only as long as the VO is made stateless. If data exist only inside an archive or outside the VO (i.e. on the user's desktop) (and are in the public domain), then there is never a case where the VO has to allow different levels of access to different users. However, AstroGrid's research suggests strongly that the VO is not stateless. It needs to retain results of queries and computations internally to be input into further operations: i.e. the VO needs to provide 'workflow'. Therefore, the VO has states where it maintains data extracts and half-complete processes on behalf of individual users. These resources are not in the public domain and access to them must be controlled on a per-user basis.

There are also archive data, not yet in the public domain, that the VO may need to use. Observations within their proprietary period should be available to the observers through the VO. Data available to researchers of certain nationalities (e.g. from ESO member-states) should be treated as such. The VO should respect and implement the existing access policies of its constituent archives.

It may be desirable to publish data on the VO. I.e., there may appear long-lived data-sets that are readable by the entire community and to which the authors retain write access. By extension, there may be data with access shared among a small group of researchers.

Finally, some major resources of the VO, such as data-mining engines, may be over-subscribed as telescopes are now. Access to these facilities would then need to to be rationed.

2 Desirable Features of an Access-Control System

Access control should be as transparent to users as possible. It should be be possible to log in to the VO as a whole and then use one's identity in any part of the VO without needing to enter further passwords ('single sign-on'). The system should not be unduly slowed down by access-control checks.

Where no controlled resources are being used, and where no results are being cached inside the VO, it should be possible to work anonymously, as on the WWW. It should be possible to move smoothly from anonymous to identified usage if anonymous access proves insufficient during a session.

The system must be secure against accidental and malicious misuse.

Identities of users must be globally unique. There must be no risk that confusion of names gives users too little or too much access.

Resource providers – operators of archive sites and system managers – must retain control of the resources they contribute to the VO. However, providers should not be loaded down with the management of individual accounts for all the VO's thousands of users.

Users must retain control of private data that they use in the VO. They must also be allowed to share that data with collaborators in a way that they control.

Any access-control arrangements must not restrict the VO's ability to run jobs with minimal attention from the user. It must be possible for one part of the VO to call another in the user's name without the user being required to authorize all such calls explicitly.

3 Proposed Architecture

3.1 Identities

We express identities in the VO as distinguished names according to the X.500 standard for directory services [2], e.g.

/C=UK/O=es-grid/OU=ast.cam.ac.uk/CN=Guy Rixon.

These names embed nationality, organization, organizational unit and common name. They form a controlled namespace and guarantee globally-unique names.

The distinguished names are encoded in certificates of identity to the X.509 standard [3]. Each user has a certificate and sends a copy to a service as the first step in establishing identity.

The certificates are issued by a Certification Authority (CA). The CA checks the identities of applicants for certificates and digitally signs the certificates to eliminate the chance of forgery. AstroGrid uses the UK's central e-Science CA; hence, our distinguished names include /C=UK/O=es-grid.

We expect that other participants in the VO will use other CAs. We hope that only a limited number of CAs will be involved, ideally one per participating country. This would make it easier to strong checks on the security of the CAs which ultimately determines the security of the system.

3.2 Authentication

We cannot tolerate that a user is prompted for a different password for each VO site, nor do we wish to share password files between sites (it is awkward and potentially insecure; service providers may not accept such a scheme). Instead we use a Public Key Infrastructure (PKI) [1] for authentication, where public-key encryption of a challenge phrase replaces exchange of passwords as the secret that proves an identity.

We needs a PKI that allows easy delegation of rights by a service to subsidiary services. The Grid Security Infrastructure [4] does just this, and the certificates issued by our CA are set up to work with GSI.

In a normal PKI, a service receiving a certificate from a client as part of a request cannot use that certificate to authenticate to a subsidiary service because it, the initial service, does not have the private key matching the certificate. In a GSI system, the initial service is allowed to use the given certificate to sign a new proxy certificate for which it does have the private key. The initial service can then send the proxy certificate to other services and can authenticate to those service in the user's name. This 'delegation by impersonation' is central to GSI and allows easy delegation of task inside the VO. In fact, even the certificate passed by the user to the initial service is a proxy, made and signed by the user using his or her primary certificate. This means that the user's primary certificate is only used briefly and occasionally for signing the first proxy; at other times it is password-protected for greater security. Thus, the only password a user needs in the VO is the one protecting the primary certificate.

Services in AstroGrid are web-services. They are invoked with messages in XML according to the SOAP standard. When a service is invoked that requires the user's identity, the client software talks first to a special web-service port that establishes a security context by passing in the user's certificate. Subsequent messages in that context are authenticated by digitally signing them; checking a digital signature is quicker than the full GSI operation that sets up the context. These web-service techniques are part of the Open Grid Services Architecture [7]. They exploit the IT-industry standards WS-Security [5] and XML-Signature [6].

3.3 Authorization

We expect AstroGrid to have a few hundred users and the VO to have as many as 10,000. We do not expect service providers at data-centres to manage separate accounts for all these users, but we do want to support different privileges for each user.

We estimate that there may be of order 10^8 separate resources in the VO if each data-set is counted separately. Each resource can have a number of possible permissions: e.g. file-like permissions such as read, write, create and delete; or service permissions such as maximum run-time or greatest number of rows output from a database. We do not want to manage all this detail centrally, nor do we want to prevent the service providers from managing their own resource.

Furthermore, most access rights are granted to users by their rôle in the astronomical community, not by their individual identity. National affiliations are an important set of rôles. Observers named on observing proposals are another. Informal collaborations of users are yet another. We have to support these rôles without requiring the service providers to manage the knowledge of who has which rôle.

We therefore provide a central authorization service in which the distinguished names of users and the users' rôles can be recorded. These rôles we call 'grid rôles' because they have the same meaning in all parts of the VO. We hope that it the membership of grid rôles may be managed on behalf of the users and service providers by officers in the community: e.g. the committees that allocate telescope time can record the membership of rôles representing observing proposals; national funding agencies can manage membership of rôles representing national affiliation; administrators at university departments can manage rôles representing members of their staff. Services needing to check access rights can ask the authorization service whether a named user has a named grid-rôle.

All grid rôles map to some local rôles at individual archive sites (e.g. Unix UIDs) and the local rôles map to specific privileges on actual resources. Our system allows service providers flexibility as to where these mappings are made. Some service providers may choose to keep the mappings local and private, and to read only membership of grid rôles. I.e., the providers can delegate the management of users to the community, but manage locally the details of privileges. Other providers may want to record specific privileges in the central authorization database, from where their services can read back the information. As an intermediate position, providers can register a privilege that is just the name of a local rôle and handle locally the mapping from local rôle to privileges.

Our authorization service is called Community Privilege Service (CoPS). It is derived from the Community Authorization service (CAS) [8], but differs in the way that authority data are exported. CAS requires the user to extract the authorization, in a special form of proxy-certificate, before invoking the restricted services. CoPS is designed to be called directly by other services. This has the advantage that the user does not need to know *a priori* which authorization data are needed by any particular service; the service can determine that itself after it is invoked.

CoPS is a web service with authentication. This allows it to accept on-line updates to the authorization database without compromising security. Service providers can add or remove privileges and authorities by connecting to CoPS via the Grid. In appropriate cases, end-users can also update CoPS over the Grid in order to share their privileges with collaborators.

3.4 Restricted and Free Services in Pairs

We seek to to implement the existing access policies of archives; we hope by this to make data more accessible rather than less accessible. Hence, we want no new restrictions on existing data and no security software where there are no restrictions.

To achieve this, we propose building services for the VO in pairs. In each pair, one half serves anonymous users and the other requires users to identify themselves. The identified-usage part enables the aspects of the service – e.g. write access, or read access to proprietary data – that cannot safely or properly be enabled in the anonymous half. We expect that the calling software can either choose the right service from the context of the job, or that it can switch smoothly to identified operation if it gets insufficient anonymous service.

4 Scaling up to the Global Virtual Observatory

The architecture described above is for the AstroGrid project, which is of lesser scope than the Global VO. However, we believe that AstroGrid's needs are comparable to those of the VO in everything but scale. We intend that our access-control architecture be capable of scaling up to serve the VO in its mature form.

References

1. A. Arsenault, S. Turner: *Internet X.509 Public Key Infrastructure: Roadmap* (IETF 2002).
2. *CCITT Recommendation X.500: The Directory*, (CCITT 1988).
3. *CCITT Recommendation X.509: The Directory – Authentication Framework*, (CCITT 1988).
4. I. Foster, C. Kesselman, G. Tsudik, S. Tuecke: 'A Security Architecture for Computational Grids'. In *Proc. 5th ACM Conference on Computer and Communications Security* (1988) pp. 83–92.
5. B. Atkinson, G. Della-Libra et al.: *Web Services Security (WS-Security)* (IBM and Microsoft, 2002): `http://www-106.ibm.com/developerworks/library/ws-secure/`
6. D. Eastlake III, J. Reagle, D. Solo: *XML-Signature Syntax and Processing* (IETF 2002).
7. I. Foster, C. Kesselman, J. Nick, S. Tuecke: *Computer*, **35(6)**, 35, 2002.
8. L. Pearlman, V. Welch, I. Foster, S. Tuecke, C. Kesselman: 'A Community Authorization Service for Group Collaboration'. In *Proc. 3rd International Workshop on Policies for Distributed Systems and Networks* (2002).

VO Activities at the Harvard-Smithsonian CfA

Giuseppina Fabbiano

Harvard-Smithsonian Center for Astrophysics
60 Garden St., Cambridge MA 02138, USA

Abstract. This paper summarizes the VO activities at the Harvard-Smithsonian Center for Astrophysics (CfA). Cfa has expertise covering virtually all branches of observational astronomy and of astrophysical research, as well as data and information management (the Chandra X-ray Center - CXC, and the Astrophysical Data System - ADS). At CfA, we have a leading role in the US National Virtual Observatory (NVO) team for the development of the VO Data Model(s), in collaboration with the European Astrophysical Virtual Observatory (AVO). To this end, we are validating our design with a local prototype, which will federate the CfA data archives, starting with the X-ray (Chandra) and optical (Telescope Data Center - TDC) archives.

1 The Harvard-Smithsonian Center for Astrophysics

The Harvard-Smithsonian Center for Astrophysics (CfA) is a joint institute composed of the Harvard College Observatory (HCO) and the Smithsonian Astrophysical Observatory (SAO). The CfA is unique worldwide in having, at one site, major, fully panchromatic (radio-waves to TeV energies), data holdings, together with the expertise to exploit them. The CfA has involvement in radio (e.g., VLBI and the Square Kilometer Array project; the wide field CO and HI surveys), millimeter (SMA), IR (including IRAC on SIRTF), optical (MMT, Magellan; with the newly developed Megacam 16,384x18,432 pixel CCD array and the Hectospec providing 300 spectra/exposure), UV, X-ray (Einstein, Chandra), and higher energies astronomy. There is a Solar & Stellar group (SOHO, Trace), and a Planetary group. CfA theory groups are also involved in numerical simulations. Theory data include: supersonic 3-dimensional MHD simulations of molecular cloud; N-body hydrodynamic cosmological simulations; accretion disk models; emission line emissivities (APED); stellar atmospheres.

The co-location of these data and expertise provides a unique opportunity for a local testbed of VO prototypes, and simplifies the development, speeding the iteration process cycle.

2 Data Center and Archival Activities at CfA

2.1 CfA Involvement in NASA Data Center Activities

The High Energy Astrophysics Division (HEAD) of the CfA has had experience with data management issues and has been involved in NASA data center activities since the mid-1970s. These activities include: pipeline processing of satellite data for image reconstruction, calibration and systematic extraction of source properties; data distribution and archival; development of analysis software packages; and user support. More recently, the CXC (`http://cxc.harvard.edu`) at SAO has developed and operates the Chandra Data Archive (CDA), and is at the forefront in developing astronomical Data Model software with extensible kernel, and axis/mission-independent analysis tools (e.g. Sherpa and the CIAO package). HEAD is also a partner in the HEASARC (the High Energy wavelength specific NASA archive center), and manages the ADS (Astrophysics Data System) project. ADS has developed, maintains, and operates the very successful astrophysical bibliography query service (`http://adswww.harvard.edu`).

The CXC and ADS are both part of the NASA Astrophysics Data Centers Executive Council (ADEC), and are involved in data centers interoperability efforts within this body. An example is the implementation of data-literature links between ADS, the CXC, and the Space Telescope Science Institute (STScI) archives, which allow users to find out which papers have been written on given archival data, and conversely give access to data from ADS bibliographical searches.

2.2 Data Archives at CfA

The CfA data repositories are in various states of archival readiness, paralleling that of the astrophysical community at large. They range from the 'VO-ready' CDA, to the soon-to-be-opened optical-IR archive operated by the Telescope Data Center (TDC), to smaller surveys created and maintained by small groups or individual scientists.

The Chandra Data Archive (CDA) - The CDA contains all Chandra calibration and satellite data, from raw telemetry to all levels of processed data products from the CXC Standard Data Processing pipelines. Because of NASA policy, most Chandra observations are proprietary for 1 year, and then become publicly available. The CDA manages both proprietary and public data. The current data product volume is of order of 1.5 TB. The CDA can be accessed via an application running at the user site (ChaSER) that can be downloaded from the CXC web page, or directly via the WWW (WebChaser). The public primary data products can also be downloaded from an anonymous ftp site. To facilitate European access to the public archive, the CDA has established a mirror site at the University of Leicester, UK. Future mirrors in India and in Italy are under discussion.

The Optical-IR Telescope Data Center (TDC) - The TDC maintains archives of spectroscopic data from SAO telescopes (`http://tdc-www.harvard.edu`). Uniform pipeline processing is performed before inclusion of data in the TDC. While some of the data are publicly accessible, most are still proprietary, pending the adoption of a CfA-wide data release policy, which is imminent. Data from the new MMT will also be managed by the TDC.

TDC data include:

- Z-Machine: all 27,171 spectra (August 1978 - September 1993) are publicly available on line and searchable by position, name,
- FAST: 5,400 of 88,166 spectra (from January 1994 to now) are public. The rest is ready, and awaits policy decision.
- 7,026 Z-Machine and 4,488 FAST spectra are accessible through the Updated Zwicky Catalog search page.
- MMT Blue Channel Spectrograph: 8,784 spectra archived (October 1980 - October 1994), but not yet released.
- Echelle data (all archived but not released): 120,559 spectra from Oak Ridge (from July 1982); 69,008 spectra from Whipple (from December 1988); 24,507 spectra from MMT (April 1981 - January 1998).

Starting this summer (2002), the CfA has begun a pilot project to evaluate the best way of digitizing the Harvard Plate Collection. This unique collection consists of 400,000 glass photographic plates, from both Northern and Southern hemispheres, taken over a 104 years time span (1885-1989). The uncompressed data volume of the digitized plates is estimated to be of the order of 80 TB.

Other Data Repositories - While the CDA and the TDC are established archives, other data repositories exist at CfA, which are being considered for inclusion in our VO testbed. These include for example the solar TRACE data, the CO survey of the Milky Way, theoretical simulations, the multi-wavelength COMPLETE survey of Galactic Molecular Clouds (a SIRTF legacy program that will include also ground-based IR, mm, and radio data). Moreover, the CfA is developing and will operate the Sub-Millimeter-Array (SMA) data center and archive.

3 CfA and Virtual Observatory Activities

3.1 The CXC Data Approach and VO Readiness

The data management and software design approach of the CXC grew upon 20 years of data center experience by the CfA/HEAD and by the X-ray astrophysical community. This approach leads naturally to VO-ready archives and to VO-friendly software. CfA/HEAD pioneered the uniform pipeline processing of entire mission datasets, starting with the UHURU data (the first X-ray astronomy satellite). CfA/HEAD developed the pipeline processing of the HEAO-1/A3 experiment, the Einstein Observatory data, and of the ROSAT HRI. The same approach was applied to the Hubble Space Telescope data, by scientists that moved to STScI from CfA/HEAD. This experience has been applied to the development of the complex Chandra Standard Data Processing (SDP) system (Fig. 1).

An innovation in the Chandra SDP, which makes easier both data retrieval and reprocessing, is the introduction of well defined data product levels, where

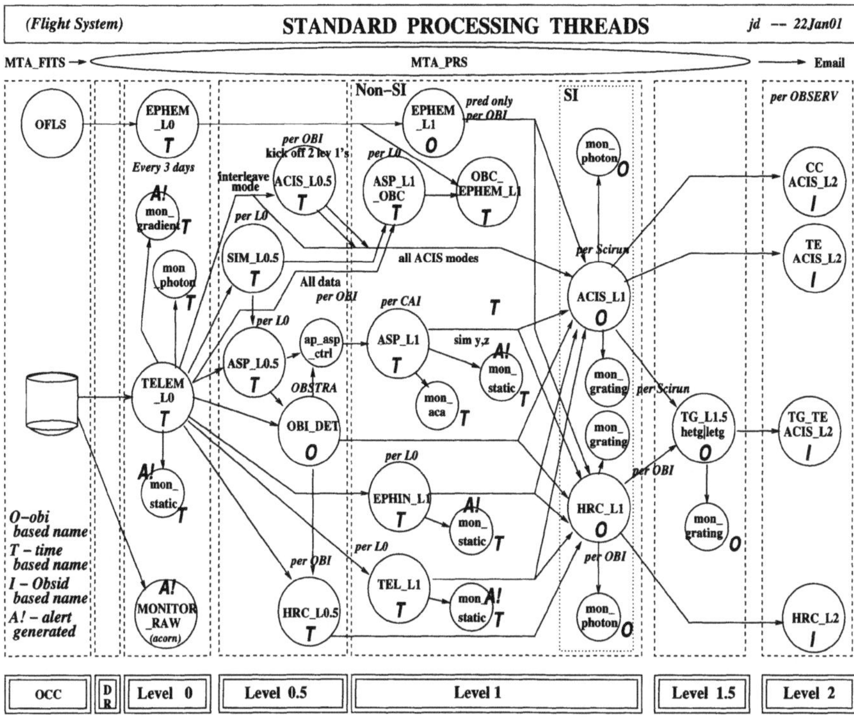

Fig. 1. The CXC Chandra Standard Data Processing pipelines. Each bubble in this diagram represents either a pipeline or a group of pipelines. Telemetry is sent to the SDP from the Chandra Operation Control Center (OCC)

each level depends only on data from the preceding level. Figure 2 shows the levels of processing that generate these products. In the SDP software, the high energy HEASARC OGIP standards were used, both for the FITS (Wells et al. 1981) keyword conventions and for the calibration database (CALDB). This approach has led to a clear definition of archival products and to clear connections between data products and calibration. Both are essential for a VO-ready archive.

The management of calibration data for Chandra is particularly complex, due to the multiple detector and mirror element combinations, which affects both imaging and grating observations. The complexity is handled in the software that accesses the CALDB, which abstracts the instrumental/telescope response and isolates it from individual FITS files (the ARDLIB). The ARDLIB abstraction addresses the signal-to-physical-quantity transformation for Chandra data (see next Section) in a mission-independent fashion (Davis, 1999). The ARDLIB interface is a CXC head-start in generalizing from the files in a VO database to their physical interpretations. This generalization would allow customized processing of the VO for specific or unanticipated needs (e.g., re-interpretation of colors as fluxes for a different assumed model).

The CXC has developed a user analysis tool-kit, CIAO, which can be downloaded from the Web (`http://cxc.harvard.edu/ciao`). Although aimed principally at the Chandra data analysis, this tool-kit was developed as multi-mission analysis software, and can be considered a prototype of the type of analysis software that will have to be developed to support VO science. Users can use both CIAO tools and other tools (e.g. the HEASARC FTOOLS) in a single analysis session. CIAO tools use the Chandra Data Model Interface, which allows them to operate on HEASARC FITS files (the interface of the FTOOLS), but also supports a more wide set of file formats. CIAO is freely available on the Web (both executable and source files), and is supported by on-line manuals, tutorials and documentation.

3.2 CfA Focus: the VO Data Model

What is a data model? The data model is an abstract representation of the data, that provides the basis for separating the specifics of file formats from the data analysis tools. A data model achieves this separation by structuring the metadata. A software tool that is data model aware will learn the structure of any data model compliant file that it opens, as well as the coordinate systems and units on the data. The CXC built a data model because, even though FITS is the archival format, there were various analysis file formats in X-rays. Once this was done though, we could also access optical Hubble (Forster et al. 2001) and infrared ISO data with the same tools. Because of the data model, the CXC tools had become inherently multi-wavelength, and so easily extensible to other archives and new missions.

Even if every individual wavelength archive had a fully functioning data model, the VO requires a single unifying structure that these data models fit into; otherwise we will have a Babel of mutually incomprehensible archives. This structure is the VO Data Model (VODM). There are three levels to the VODM: (a) the abstract object description, which provides a way that archive centers can describe the structure of their data at the level of content and meaning (rather than format); (b) a definition of the canonical VO format for these objects, which may be realized with server-side translators written (in any language) by individual data centers; and (c) the client-side DM library and toolkit, which operate in terms of the abstract object description on files in at least the canonical format, and provides hooks for the translators from individual specialized formats.

The VODM structure is embodied in metadata keywords in physical disk or memory formats. Most of these keywords already exist in the archives; the data model merely gives them a higher level meaning, enhancing their power without necessarily having to alter the archived files. The Chandra experience shows that adding a small number of additional keywords can greatly enhance scientific analysis capabilities.

Designing the VODM involves defining object classes for astronomical data, which needs a deep understanding of these data and of the individual ways of the astronomy/astrophysics community. Jonathan McDowell's talk in this meeting discusses in more detail the VODM.

4 The CfA VO Prototype

To validate our VODM design in a realistic environment, we are in the process of setting up a CfA VO prototype. This CfA-VO, which we hope will evolve into a full-fledged VO node and portal, is also intended as a real internal CfA facility that will facilitate access to and comparison of the different CfA data sets for the CfA scientists and Harvard students. At the same time, we intend to give public community access to these data, via the CfA-VO portal. The CfA already provides widely used data and services to the community (the CDA and the ADS), so we already have an existing user base that will be ready to make use of our extended services. Feedback from these users will provide the best measure of the success of our prototyping efforts, and will guide further development.

Figure 2 shows a schematic of the CfA-VO architecture. In its general lines it resembles the architecture of the US-VO approach, and this is not accidental. After a user has submitted his query (in user language), the query interface will translate this query into QML (Query Model Language), and dispatch grid searches. Exploders resolve the query into subqueries and dispatch them to the archives (e.g. X-ray data queries, or the X-ray component of a general data query will go to the CDA). At each targeted archive a site specific extractor (SSE) receives and translates the query into archive-specific LQM (Local Query Model) to generate the request. Archival data are then retrieved and translated from the LDF (Local Data model Format) into a DMP (Data Model Protocol) compliant format. Generic data format filters (GDF) then process the DMP data from the various subqueries, and data fusers concatenate the data from the various archives for return to the data access layer (DAL) and the user.

Following the successful approach used for the development of the large and complex Chandra data system, we will test our design with partial prototypes that will exercise key threads though the CfA-VO testbed. These will be demonstrated to the community at scientific meetings. The first such demo is scheduled for the upcoming 2002 ADASS meeting in Baltimore. A successive approximation approach will be applied to get to the final working system.

5 Education & Outreach

The Education Department and the Chandra Outreach Group (with the award winning public Web interface `http://chandra.harvard.edu`) provide in situ expertise on the outreach and educational aspects of this project. CfA has strong links with the nearby headquarters of AAVSO, the world's most well known amateur astronomy organization. The AAVSO are part of the CfA VO team and will

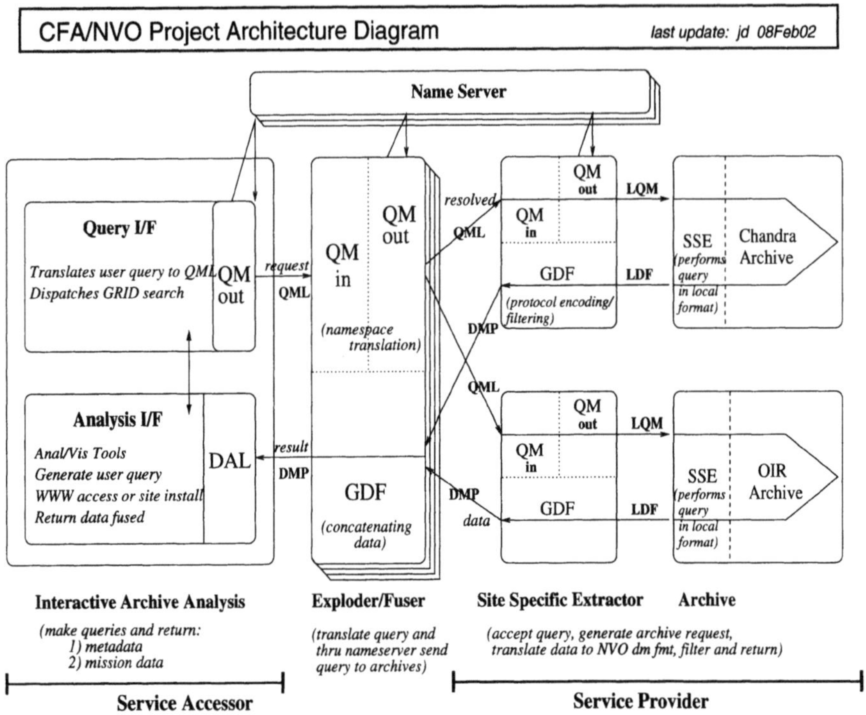

Fig. 2. CfA-VO testbed and prototype architecture diagram. While only the Chandra and the Optical-IR TDC archives are shown in this figure, the testbed will be extended to all available CfA data holdings

provide the first opportunity for the amateur community to directly contribute to a professional data archive. CfA also has a long tradition of involvement in science popularization, and one of the team members is a contributing editor to Sky and Telescope magazine. We are presently evaluating a E&O VO demonstration.

6 Conclusions

We are still at the beginning of this project, but gathering momentum. We are following several paths: (a) We are working internally at the CfA to ensure that a growing fraction of the rich Harvard and Smithsonian astronomical and astrophysical data holdings are archived and accessible on line; (b) We are part of the large NVO team and collaborate with the AVO team in several areas, the VODM, metadata definition, testbed, and VO science; (c) We continue our inter-operability work within the NASA ADEC forum; (d) We are building a CfA VO prototype to link the CfA archives and test our VODM design; (e) We follow the CXC-proven design validation and development approach based on increasing complex prototypes and demonstration threads, that we will make available to the community.

Note: Parts of this paper were also presented at the August 2002 SPIE meeting in Hawaii, and can be found in the proceedings of that meeting.

The work presented is the result of a team effort, including: Jonathan McDowell and Mark Cressitello-Dittmar (VODM); Mike Noble and Ian Evans (VO infrastructure and testbed); Janet DePonte Evans (software and project management); Arnold Rots, Doug Mink and Mike Kurtz (archives and metadata); Alyssa Goodman, Robin McGary, and Alice Argon (science and data). The larger CfA collaboration also includes Bob Kirshner, Martin Elvis, Janet Mattei (Director of the AAVSO), Philip Sadler, Tanya Szlateva (of Boston University), Roy Gould, Kathy Lestition and Roger Brissenden. We thank Rick Harnden and Dave Huenemoerder for comments on this paper.

This work was funded by the NSF (ITR/IM+AP: Developing the National Virtual Observatory Data Model, Award No. 0105601; and Building the Framework of the National Virtual Observatory, Cooperative Agreement No. AST-0122449). We also acknowledge partial support by the NASA contract NAS 8-39073 (CXC).

References

1. D.C. Wells, E.W. Greisen & R.H. Herten, 1981, Astron. & Astrophys. Suppl., 44, 363-370. `http://fits.gsfc.nasa.gov`
2. J. E. Davis, A framework for the development of multi-mission software, in ADASS IX, eds. N. Mauset, C. Veillet, and D. Crabtree, 1999
3. K.Forster, P.J.Green, T.L.Aldcroft, M.Vestergaard, C.B.Folz, P.C.Hewett, Emission Line Properties of the Large Bright Quasar Survey, 2001, ApJ, Suppl., 134, 35.

The Prototype TNG Long-Term Archive and its Interactions with the Italian GRID Project

Fabio Pasian[1], Leopoldo Benacchio[2], and Riccardo Smareglia[1]

[1] INAF - O.A. Trieste, Via G.B.Tiepolo 11, I-34131 Trieste
[2] INAF - O.A. Padova, Vicolo Osservatorio 5, I-35122 Padova

Abstract. A pilot project is currently being carried out, aiming at demonstrating the feasibility of technical solutions for the construction of the Long-Term Archive (LTA) for the "Telescopio Nazionale Galileo" (TNG). In this project, coordination at the national and international levels (OPTICON, ASTRO-WISE, AVO) is being pursued with the aim of participating in the definition and construction of the international multi-wavelength Virtual Astronomical Observatory.

Recently, the Italian Ministry of Research has approved and funded a project to set the framework for an Italian Grid for research. The LTA prototype is planned to be a data-providing node of such a grid.

1 Introduction

The "Telescopio Nazionale Galileo" (TNG) is a 3.5 m Nasmyth telescope, based on the optical structure of ESO's NTT, located on the Roque de los Muchachos in La Palma, Canary Islands. TNG has seen first light in 1998; after a long integration - testing - commissioning phase, all of the first-generation instruments were operational in 2000. They are:

- an optical imager (OIG);
- a low-resolution imager/spectrograph (DOLORES);
- a high resolution spectrograph (SARG);
- a near-IR camera/spectrograph (NICS).

The telescope is also equipped with adaptive optics (ADOPT@TNG), and is complemented with a seeing monitor (DIMM), and a Meteo tower. The observing mode adopted up to now is based on assisted observations; starting with the 7th observing period (spring 2003), TNG will be offered in service/queueing mode.

The telescope and instruments control environment is tightly connected to the data handling and archiving, which is unified for all instruments following the TNG standards, and has been designed as a part of the "Archiving at the Telescope" (AaT) system (Pasian, 1996). Data follow a predefined path, where the information is handled and stored homogeneously, and the integrity and completeness of data are preserved. The AaT system has been active on the mountain since the TNG first light (Smareglia et al., 2000).

2 The TNG LTA Pilot Project

A pilot project is currently being finalized, aimed at demonstrating the feasibility of technical solutions for the construction of the Long-Term Archive (LTA) for the TNG. The natural implication of this activity is the implementation of a prototype LTA which is necessarily interfaced to the TNG AaT system mentioned above.

The end of the pilot phase is foreseen to be the end of June 2002, thus the core of the system has been already implemented and is undergoing testing and verification.

The project was originally Financed by the National Consortium for Astronomy and Astrophysics (CNAA). Since CNAA is ceasing its activities at the end of June, the results will be submitted to the newly-founded National Institute for Astrophysics (INAF), which is replacing and extending CNAA's responsibilities.

From a different perspective, an INAF-nominated board on "computing, archives and networks" is currently analyzing, *inter alia*, all issues related to archiving of data acquired from ground-based telescopes, and will produce a report in early July. The TNG LTA pilot project is meant to provide important information as an input to the board.

2.1 Scope of the Pilot Project

The main task assigned to the pilot project has been defined when the announcement of opportunities was originally issued. The project is to provide suggestions, plans and prototypes to build an archiving system having a number of purposes:

- allowing the *long-term storage of scientific and technical data* from the TNG;
- guaranteeing *access* by the TNG Observatory staff and by the scientific community to original and derived data;
- testing practically the feasibility of deriving *value-added information* from the stored data;
- providing tools supporting life cycle of *observing proposals*;
- providing *interoperability* with a number of existing international data centers and archives;
- guaranteeing *harmonization with other Italian projects* dealing with archiving of data of astrophysical interest.

2.2 Status

Up to now, quite a number of basic issues in the implementation of the prototype TNG LTA has been tackled and solved (Smareglia et al., 2002).

The *data model* of the TNG and consequently the data items needed to populate the archive, on the basis of the requirements defined by a dedicated WG set up in 1994, have been defined. The *data interface* between the archives at the telescope (AaT) and the LTA has been defined and implemented. After having designed the *database structure* in Oracle, the database management system chosen for the prototype LTA development, *ingestion tools* for TNG data (both technical and scientific) have been implemented, and the LTA is now being populated regularly.

As for the retrieval of data, a *user interface* (UIF) is being finalized on the basis of requirements defined within a group of participants in the project. A module for the *servicing of user requests* (including retrieval of telemetry and subsets of FITS files) has been designed and is being completed. *Visualization tools* are being finalized as a part of the UIF. They cover both image display and X-Y graphs, offering standard options (including zooming, panning, etc.).

A test was made to analyze the efficiency of the access to the prototype LTA *database in a parallel environment*, and in particular to verify the speedup achievable with an SMP system with shared memory using the parallel query feature available in Oracle.

Benchmarks of *file retrieval efficiency* for on-line, quasi-on-line and off-line data files are being carried out. *Network tests* will be made with a test user group when the prototype is completed.

A *data mining* activity is being carried out on the Technical DB. This is basically aimed at two separate goals:

- finding relations/correlations among housekeeping parameters, through the unsupervised use of neural networks;
- statistical analysis of the H/K parameters giving guiding error on the 3 axes, to analyze the mechanical behaviour of the telescope, and the possible presence of oscillations under specific conditions.

A full report on the TNG archives system is in preparation (Pasian et al., 2002)

2.3 Interoperability Issues

The interoperability of the prototype TNG LTA with other archives is an important issue. The problem is being tackled by direct participation of pilot project members in the Science WG of the AVO (a project funded by the EU to provide a prototype multiwavelength virtual astronomical observatory for the scientific community) and in the "Interoperability WG" of the OPTICON project financed by the EU (which is discussing standards to be eventually used in the Virtual Observatory context.

As a basic level of interoperability with other archives, a name resolver using the SIMBAD facility has been installed, and the TNG electronic Catalogue of Observations has been made available at CDS, and testing of its compatibility with CDS tools are being made.

The prototype LTA already uses standards common in the astrophysical community (e.g. FITS, ASU). The plan is for the operational LTA to use the additional standards (e.g. VOTable) being defined within the "Interoperability WG", thus aiming at being compatible with AVO developments. For the time being, the pilot project members are making tests with VOTable.

Interoperability is definitely an important issue, and a more accurate definition of its needs and practical solutions need to be found in the operational TNG LTA.

2.4 Lessons (Being) Learned

Logically enough, it was found that data quality is a real issue: some inconsistencies were discovered in data (especially in scientific exposures); some basic observatory information is still missing from the archive (e.g. instrument documentation, proposals, log of maintenance, observation schedule, etc.).

Data consistency and quality could be easily improved by slight modifications to observing procedures. Such changes have been checked to be possible, and suggestions will be provided to the TNG observatory management staff.

Neural-networks-based analysis of technical and engineering data has proven to be effective in identifying some possible problems in the overall Telescope + Instruments system (Longo, 2002). Such analysis will be used to improve knowledge of the system status, thus data quality.

Finally, it became evident that an archive must be developed in synch with the observatory: the late development of a TNG LTA (even if at a prototype level) led to a late discovery of inconsistencies in the data.

3 Grid-Related Activities in Italy

The Italian astronomical community does not currently plan build its own "AstroGrid", specifically dedicated to astronomy, but has rather chosen to coordinate efforts within two nation-wide initiatives to foster a Grid infrastructure: the project for a *national e-Science Grid*, and at the European level the *IG-BIGEST* project.

3.1 The Italian Research Grid Project

A proposal has been submitted in October 2001 to the Italian Ministry of Education, University and Research (MIUR) under the *"Strategic Programme on Enabling IC Technologies for Knowledge Society"*, with the purpose of setting the framework for an Italian Grid for research. The title of the project was chosen to be "Enabling Platforms for High-Performance Computational Grids Oriented to Scalable Virtual Organizations.

The objectives of the project are multiple, depending on the competence of the various bidders. There is R&D activity in the *Grid technological development project* carried out by CNR and INFN; there is *research on optical networks* led by CNIT. The item of interest for our community is led by INFN and refers to the deployment of an Italian eScience Grid infrastructure for Astrophysics, Biology, Computational Chemistry, Geophysics and Earth Observation. The various communities provide applications to act as demonstrators. The integration in the international environment is carried out by INFN, following requirements from the various scientific fields of interest.

The astrophysics community has proposed four demonstrators as its contribution to the Italian eScience Grid project:

- distributed computing
- accessing and processing VST data
- accessing archives and databases (*inter alia* the prototype TNG LTA)
- accessing observational facilities as "data nodes".

At the time of writing, we have been informed that the project has been financed for $\sim$ 8 MEuro. The division of funds within the collaboration has not yet been decided; it is however expected that about 1/15 of the total ($\sim$ 0.5 MEuro) will be the share of funding available to the Italian astrophysical community.

3.2 IG-BIGEST

There is also a new initiative, leveraging from MIUR proposal, and aiming at the common development of an Italian Grid for Business, Industry, Government, EScience and Technology (IG-BIGEST). This activity constitutes part of the Italian participation to the Expression of Interest for an Integrated Project called "Enabling Grids and E-Science in Europe" (EGEE) in the context of the EU 6th Framework Programme.

All major Italian Scientific National Research Institutions, Universities and Computing Centres reached general consensus to take part to the large European Initiative EGEE having CERN as leading partner and INFN as the organization representing the Italian Grid eScience community, and to have only one large integrated project aiming at the development and deployment of the Grid Infrastructure in Europe, as Geant does for networking.

The high bandwidth network needed for the project is provided in Italy by the national research network GARR-G, which is connected at the European level to Geant. High performance computing infrastructures are provided by national research Institutions, Universities and super-computing centers. Italian participants in the project are $\sim$ ten National Institutes, a couple of computing Centers and about twenty Universities. The formal participation of the various institutions and of the joint Italian community to the EU project has been finalized.

4 Conclusions (and Roadmap?)

The TNG LTA Pilot Project is finalizing its development. The result is a LTA prototype demonstrating the feasibility of having archives which are compliant with the international VO standards. Given the fact that the project has been given coordination duties (of national archives and with international activities), there is a high probability that all future Italian archives (the operational TNG LTA, VST, the archive of the Italian time at the LBT, etc.) will be VO-compliant.

In the meantime, the Italian e-Science Grid is starting its prototype phase, and an operational version of it is likely to become a reality. The participation of the astronomical community to the grid-related projects, and the above-mentioned feasibility and interoperability of national archives is being reported to the INAF board on computing, archives and networks, which will provide recommendations to the national funding authorities. The goal is to put the Italian community in a position to join the Phase-B of the EU AVO project.

Acknowledgements

The TNG LTA pilot project has been fully financed by CNAA; A. Zacchei, A. Caproni, G. Longo, U. Becciani, C. Gheller, N. Lama, M. Lodi and J.C. Guerra have contributed to its development. Italian Grid eScience activities are coordinated by Mirco Mazzuccato of INFN–Padova who is warmly acknowledged for support and advise.

References

1. F. Pasian: 'Archiving TNG data'. In: *Astronomical Data Analysis Software and Systems 5*, ed. by G.H.Jacoby, J.Barnes (Publications of the Astronomical Society of the Pacific no.101, 1996) pp. 479-488
2. R. Smareglia, F. Pasian, C. Vuerli, A. Zacchei: 'Operating the TNG data handling and archiving'. In: *Astronomical Data Analysis Software and Systems 9*, ed. by D.Crabtree, N.Manset, C.Veillet (Publications of the Astronomical Society of the Pacific no.216, 2000) pp. 149-152
3. R. Smareglia, U. Becciani, A. Caproni, C. Gheller, J. C. Guerra, N. Lama, G. Longo, F. Pasian, A. Zacchei: 'The pilot project for the TNG Long-Term Archive'. In: *XLVI Congresso Nazionale S.A.It.*, Conference at Padova, Italy, April 10–12, 2002, (Memorie S.A.It.), in press
4. G. Longo: 'Artificial Intelligence Tools for Visualisation and Data Mining in Large Astronomical Databases'. In: *Towards an International Virtual Observatory*, International Conference at Garching, Germany, June 10–14, 2002, ed. by P. Quinn et al. (Springer), this volume
5. F. Pasian et al.: 'Archive systems for the TNG', in preparation

Federation and Fusion of Astronomical Information: Standards and Tools for the Virtual Observatories

Daniel Egret and Françoise Genova

CDS, Observatoire astronomique de Strasbourg,
11 rue de l'Université, F-67000 Strasbourg, France

Abstract. We present a review of the current organization of astronomical archives and data centers in terms of data federation. We point the challenges faced, in the VO context, for developing the necessary standards and tools to make data federation and data fusion possible at a general level, covering all potential VO resources.

1 Introduction

Typical scenarios of the Virtual Observatory (VO) imply to cross-match surveys, mission logs, observational catalogues, or personal files, to collect all pieces of information about an object, or a set of objects, to build samples of astronomical objects, and/or to discover rare objects in a multiwavelength space.

In order to make all these actions possible, the VO will have to effectively propose tools for collecting and correlating data from several diverse and distributed systems: hence the need for data federation and data fusion. Selecting data sets by criteria other than position might imply to develop specific query strategies for the management of distributed computer resources.

Let us give here two definitions. We call *data federation*, compiling a comprehensive set of data relevant to the same objects or phenomena, extracted from archives and databases, possibly heterogeneous and distributed on different sites. *Data fusion* implies to go one step deeper in the *semantic* description of the data so that relevant pieces of information be immediately compared, merged and/or correlated.

In the following, we present an overview of current solutions, based on interoperability tools for data federation, and metadata dictionaries and standards for data fusion.

2 From Resource Discovery to Data Fusion and Visualization

Solving a complex query, involving access to diverse astronomical resources, may typically require the following steps:

1. Resource discovery: what are the resources that can provide relevant information?

2. Resource location: address and query syntax of the resources?
3. Query processing
4. Presentation of the federated answer
5. Data fusion
6. Data visualization.

In the following, we discuss these subsequent steps, and we review the present status and future prospects.

2.1 Resource Discovery

Discovering the resources that can provide relevant information is typically the first step in the processing of a complex scientific query. We can categorize the potential resources as follows:

- On-line archives of observatories, surveys, or space missions: typical examples are Hubble Space Telescope (STScI, CADC, ST-ECF), EIS and VLT (ESO), Sloan Digital Sky Survey, Chandra (NASA), 2MASS (IPAC), ISO (ESA). NASA centers, organized by wavelength, are the natural hosts of NASA space missions and related archives.
- Object databases: typical examples of reference database organized per astronomical objects are SIMBAD and NED.
- Federated databases: a typical example is VizieR (Ochsenbein et al. 2000), which can be characterized as a collection of all published astronomical catalogues and data tables. Brought together into a common description system, these tables share the same query mechanism.
- Data centers are generally physical organizations that may offer dedicated services for helping select relevant resources: see, e.g., HEASARC, CDS. One could imagine also virtual organizations federating several distributed services.
- Abstract services (ADS) and electronic journals.
- Yellow-page services, general query engines (such as Google), and resource lists (such as AstroWeb) may, finally, provide pointers to other kinds of resources and more focused information sites.

In the VO context we expect all these resources to be publicly available on-line. For a discussion on current evolution of data curation, and why science projects are becoming data publishers, see Gray et al. (2002).

Generic resource discovery services will be an essential part of the VO, as we cannot expect all users to be familiar enough with the growing complexity of the international network of astronomical data and information services. Resource discovery will have to be managed through VO portals, taking benefit of metadata dictionaries and careful indexing of existing resources.

2.2 Resource Location

The second step consists in locating the most recent (and possibly the closest, when mirror services are involved) version of the selected archive, survey, or table — identified through the previous step, or from a-priori knowledge.

This is a real concern for resource and service managers and providers, because of the large number of partners involved in data publication and data curation. Assuming the information is available from a web site, or a web service, the point here is not only to find the correct resource locator (URL), but also the description of the service, its query syntax, query parameters and results format.

In the recent years, a tool has been designed by CDS, and distributed for this purpose in the astronomical community: the *Générateur de Liens Universels* (GLU; Fernique et al. 1998). The GLU distributed dictionary is a knowledge base which includes metadata information describing astronomical information services.

It will be of the VO responsibility to maintain an inventory of potential resources, and to provide guidelines for new data curators. New solutions for metadata management, and service registry are currently being investigated by the VO consortia (see e.g., Hanisch 2002a, and Cone Search, below). For an example of first guidelines for data curators see Hanisch (2002b).

2.3 Query Processing and Interoperability

Query Processing After the first two steps, the user (assisted, in the future, by the VO system) shall be able to divide the original complex query into 'atomic' queries to be submitted to several archives or databases.

Submitting queries (or sub-queries) to several distributed heterogeneous systems, and building an integrated response out of the 'atomic' answers, imply to go one step further in the description of these systems (metadata management), and to agree on common protocols and exchange formats.

User identification, authentication, management of parallel query processing, are typical questions that are presently managed on a case by case basis, and shall benefit, in a very near future, of the emerging Data Grid (see e.g., Moore 2002; Williams 2002a; Rixon 2002).

Since the early projects such as AstroBrowse and ISAIA which aimed at providing a generic cone search facility (Hanisch 2000), new protocols and standards have been made available.

The NVO project has recently organized a simple *Cone Search*[1] where services can register (Hanisch 2002a).

New solutions based on Web services — using Simple Object Access Protocol (SOAP) and Web Service Description Language (WSDL) — are now being discussed (Szalay et al. 2002), and prototype modules have recently been made available in the astronomy domain (e.g., SkyQuery 2002).

[1] `http://www.us-vo.org/metadata/conesearch/`

Interoperability A central aspect of the VO is the interoperability: we do not want to simply accumulate results coming from heterogeneous databases and information services. We want them to exchange information, as part of the query processing, so that a really global response is constructed.

Interoperability makes full use of the Internet: a very early example of interoperability is the SIMBAD or NED name resolver which has been included in most of the on-line archive systems since several years. The user is requested to type an astronomical object name, and the archive system transparently queries SIMBAD or NED for retrieving the corresponding position on the sky to be used in the further archive query.

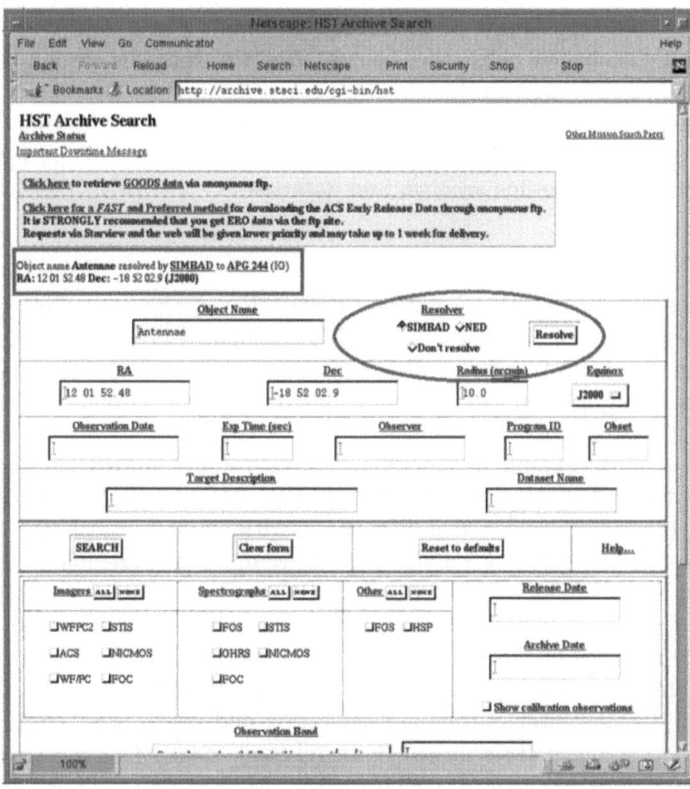

Fig. 1. Example of a name resolver

We now need to go much deeper in interoperability functions, in order, e.g., to be able to select a sample of objects in a database, according to the result of a query performed in a survey archive.

Interoperability of astronomical archives, surveys and databases is one of the present challenges faced by the VO projects: an international working group has been set up to build agreements on protocols and standard formats that will make it all possible. A first result has been the rapid adoption of the VOTable standard format (see below 2.5).

2.4 Data Federation

Data federation and presentation is the next step. It consists in presenting to the user, in an organized way, the results of the multi-site searches: number of records, pointers to datasets, data, data description, and documentation.

Presently generic databases or information systems, such as SIMBAD and NED for the astronomical objects, or VizieR for the catalogues present information summaries and dataset descriptions that can be used to identify the relevant pieces of information. While such databases cover most (if not all) of the published literature, they cannot be considered as complete at a time when projects are becoming their own publishers (survey archives, space or ground-based observatories, unpublished data).

Significant efforts are now expected in order to generalize the application of common standards, for integrated response in normalized form (format, units, etc.) to most of the on-line archives and databases.

2.5 Data Fusion

Data fusion implies to get an accurate description of the data. This is one of the first objectives of the emerging VOTable standard (Williams et al. 2002b, Ochsenbein et al. 2002) which includes a semantic description of the information items, using an ontology currently based on Unified Content Descriptors (Derriere 2002, this conference).

Let us give an example here: assume you receive an information item specifying that the radial velocity of a galaxy is $RV = 123.4$. This can be described on a syntactic point of view as a REAL F5.1, with column header RV. Now, we need to go further and get a *semantic* description, making possible to explicitly specify the unit (km/s), and the meaning (RV stands for Radial velocity, which is a data type listed under name so-and-so in the UCD). With such a description we are now able, for instance, to compare it to another information item for the same astronomical object, this time expressed as a redshift.

A good demonstrator of the data federation and fusion is presently the Aladin interactive digitized sky atlas (Bonnarel et al. 2000). This service relies on a database of digitized images (DSS, MAMA, 2MASS, etc.), and provides functionalities for overlaying, on the image field, information from external databases or archives (Fig. 2). In the framework of the Astrophysical Virtual Observatory (AVO) project, new features have been developed allowing more 'data fusion': colour composition of images from diverse origins, astrometric registration and resampling (Fernique et al. 2002, this conference).

2.6 Data Visualization

In addition to data fusion, the Virtual Observatory will need a set of tools (which can play the rôle of virtual instruments) for visualizing information from

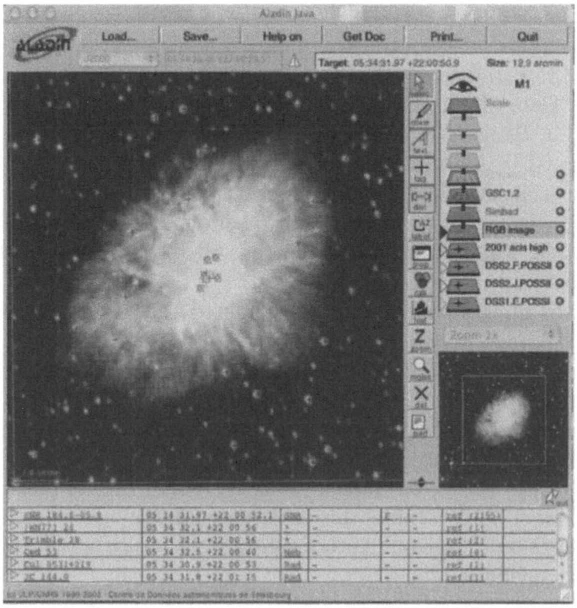

Fig. 2. Sample screen from Aladin showing an image composition of the Crab nebula, with overlaid GSC and Simbad entries

characteristic features of the retrieved data, relationships between information items, and more specifically wavelength and time dependencies of the data.

First examples of such data visualization tools are, e.g., the Spectral Energy Distribution proposed from the NED database, or the computation of image contours (CDS/ALADIN, IRSA/OASIS). Additional tools, available or in progress, concern on-the-fly extraction of point sources in an image, sky maps, histograms, or colour-magnitude diagrams.

Providing the user with the facility for navigating within multi-dimensional plots, modifying the points of view, selecting samples or outliers will be a powerful tool for data mining and data discovery.

This will imply also to create links between actual data and models or simulated data in order to help confront theories with observations.

3 Conclusion

At the end of the current VO deployment, we expect VO portals to provide resource discovery tools, full documentation and library, metadata dictionaries, normalized query engines.

As an intermediate step, science demonstrations planned within the main VO projects, for the coming year (2003) will use data and Information fusion as one of the key concept illustrating the efficiency of emerging VO solutions for solving complex queries.

References

1. F. Bonnarel,P. Fernique, O. Bienaymé, et al.: Aladin, A&AS 143, 33 (2000)
2. S. Derriere: Data Mining Facility, in this conference (2002)
3. D. Egret, R.J. Hanisch, F. Murtagh: A&AS 143, 137 (2000)
4. P. Fernique: GLU, ADASS VII Proceedings, ASP Conf. Ser. 145, p. 466 (1998)
5. P. Fernique et al.: Aladin 1.3, this conference (2002)
6. J. Gray, A. S. Szalay, A. R. Thakar, C. Stoughton, J. vandenBerg: Online Scientific Data Curation, Publication, and Archiving, Microsoft Technical Report 2002-74, in press (2002)
7. R.J. Hanisch: Integrated Access to Distributed Data and Information Services, *Computer Physics Communications*, 127, 177 (2000)
8. R.J. Hanisch: National Virtual Observatory, in this conference (2002a)
9. R.J. Hanisch: Resource and Service Metadata for the Virtual Observatory, `http://www.us-vo.org/` (2002b)
10. R. Moore: Data Grids for Collection Federation, in this conference (2002)
11. F. Ochsenbein, P. Bauer, J. Marcout: VizieR, A&AS 143, 23 (2000)
12. F. Ochsenbein et al.: VOTable, this conference (2002)
13. G. Rixon: Authentication and Authorization architecture in the AstroGrid and the VO, in this conference (2002)
14. SkyQuery: `http://www.skyquery.net/`
15. A.S. Szalay, T. Budavári, T. Malik, J. Gray, A. Thakar: Web Services for the Virtual Observatory, Microsoft Technical Report 2002-85 (2002)
16. A. Szalay, et. al,: The SDSS SkyServer, SIGMOD 2002, pp. 570-581 (2002).
17. R. Williams: Grids and the Virtual Observatory, in press (2002a)
18. R. Williams, this conference (2002b)

Data Models for the VO

Jonathan McDowell

Smithsonian Astrophysical Observatory, 60 Garden St, Cambridge, MA 02138, USA

1 What Is a Data Model?

The key to making the Virtual Observatory work is the definition and adoption of interoperability standards. One form of such standardization is to agree on exactly what we mean by the data objects we all deal with - images, spectra, coordinate systems, etc. This standardization and abstraction process is called data modelling. We have been using data modelling techniques for several years to develop the CIAO analysis system for the Chandra X-ray Observatory [1].

A data model [2] is a recipe to describe 'how is my data different from (the same as) your data?'. By different, I mean in terms of abstract information content rather than specific byte format. Suppose you have an simple 2D image of part of the sky, you can store it as a FITS file or a GIF image and you've got the same information – until you add a coordinate system to the FITS file or a color table to the GIF. Now suppose the image was made by mosaicing four chips and you have a FITS image with one extension per chip. There's new information – your display program may show the same picture but you have retained the information of which part of the sky is observed with which chip. By elaborating a data model that describes astronomical images we ask: what questions can I ask about an astronomical image? In this case, for instance, how close is this star to the edge of a chip?

The data model describes the information content, and the metadata protocols that Ray Plante discusses in the following paper describe the way that content should be formulated and tagged – the boundary between the two is a bit blurred. The data model may also describe the access functions ('methods') for the data.

For many VO uses, catalog federation is all you need to do and that can make do with a fairly simple data model – although the issue of sky coverage is tricky, and Arnold Rots addresses that in his paper. For VO applications that work with image and spectral data directly, data fusion work, a good data model is much more critical.

In the context of the VO, the VO consortia will use the data model to design the metadata, making sure the most general image can be represented. A data provider will use the image data model to map their data to the standard VO representation, and tell us which questions their images can and can't answer.

Data models also allow you to compare disparate types of object. All astronomical data has some commonalities (the need for keywords, coordinates). A data model for images or for spectra can be considered as a special case of an astronomy object; This is just another way of saying that we shouldn't implement coordinate systems for spectra and images in two totally different ways.

2 Not So Easy: A Real World Example

Let's get specific and look at the example of four images I got out of four different archives (Fig. 1, Table 1) – a ground based telescope, HST, Chandra and ISO. Not one of these is a simple FITS image, so when you get them back from the VO if they're in their present form it's a lot of work to combine them – **no one software tool will operate correctly on any two of these images!**

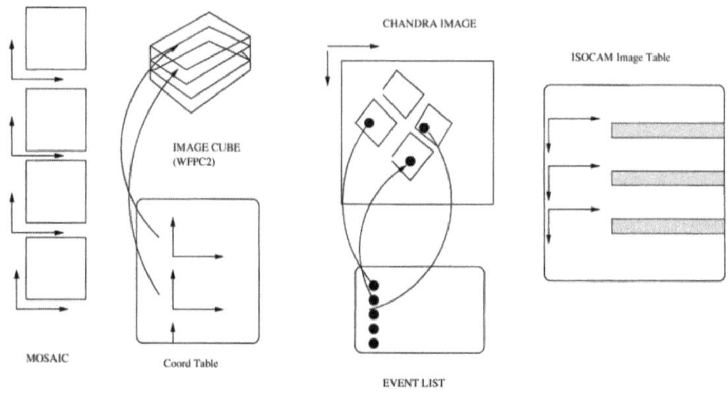

Fig. 1. Different FITS implementations of a mosaic image. The arrowed axes represent coordinate system metadata

Table 1. Structure of mosaic images in different archives

	FLWO Mosaic	HST WFPC2	Chandra ACIS	ISOCAM
HDU Structure	4 images	1 3-D image	Event pixel table	Table of images
Coordinates	WCS keys	Special table	WCS keys	WCS columns in table
Bandpass keywords	FILTER	PHOTPLAM	DSVALn	WAVELENG
Duration	EXPTIME	EXPTIME	LIVETIME	DATE-OBS, DATE-END

It's not just a matter of different keywords – it's easy to map those. It's a matter of different approaches to encoding equivalent information content. For instance, X-ray data doesn't have just a start and stop time, it has a table of multiple start and stop times, making it harder to answer the question 'did this star flare during my observation'. WFPC data has standard WCS coordinates on the 3-dimensional image cube, but they're misleading about all but the top plane and the real WCS values are stored in another table, while the logically equivalent ground based and ISO mosaic images take two further different approaches to encoding the same information.

It's not enough to map each of these examples directly to a VOTable [3] in their current structure. That won't capture the fact that they contain the same kind of information – we have to define how to give a uniform structure to these data. So for the VO we need a standard for mosaic images, a standard for coordinate systems, perhaps a standard for timing information.

Some things are more important than others – it'll be much more common to need the wavelength than the observing location. We can model the main things first and develop more standards as time goes on: the VO will become aware of the answer to more and more questions.

The data model will let the archive provider figure out what they have in the common language of the VO: 'mosaic image with one coordinate system per image'. The metadata standards will tell them how to represent such a thing in, e.g. a VOTable. VOTable has sets of nested tables, like most formats it has keywords, and it has a simple object for defining celestial coordinate frames. There are things FITS has that it doesn't yet, like coordinate transforms and image axes, and there's no structure to the header. In FITS, you have a set of images or tables each with a set of keywords. There's more implicit structure created by defining objects with groups of keywords. We will need a more sophisticated and explicit structure to describe the data we will analyze with the VO.

3 Modelling Data and Metadata in the VO

The NVO group is discussing a VO data model in which a dataset (for instance as represented by a table in a VOTable) will contain a set of columns and/or images with a set of metadata descriptors (Fig. 2; the model is presented in more detail in a discussion document available in the NVO document repository, currently at `http://bill.cacr.caltech.edu/cfdocs/usvo-pubs/files/vodm003.ps`). Where in FITS we have simple keywords as the building block for the header, in the VO each of the metadata descriptors is a whole object – this may be as simple as a keyword or it may be quite complicated, and as an object its definition may be extended as time goes on. Also, descriptors may be attached not just to the dataset as a whole, but to individual columns or images or even to other descriptors. I call out a special case of coordinate descriptors here attached to the image, since they are so important. In the figure, 'hypercube data' refers to the N-dimensional image data or the table column data, as appropriate.

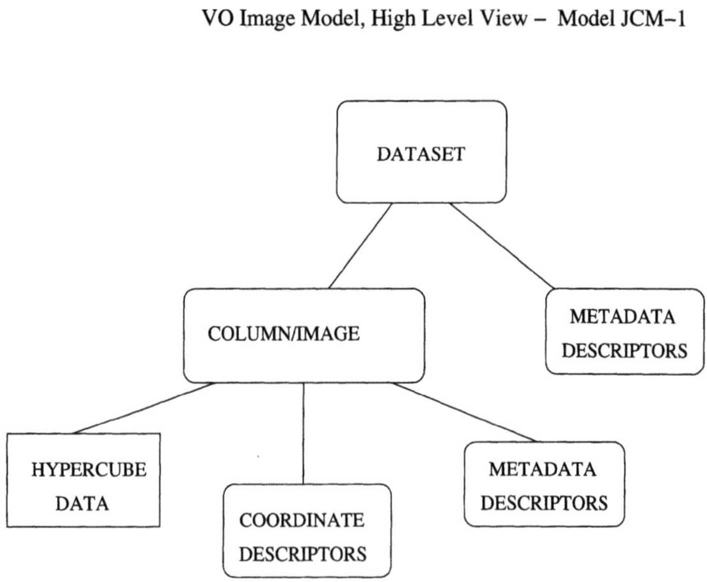

Fig. 2. Proposed overall image/table model

In Fig. 3 I've illustrated a set of metadata objects we have identified as worth modelling. Let me focus on three related objects to conclude this presentation.

- Data Quality: this comes up in all data analysis systems. We should define a common approach to describing bad pixel masks, quality flags, observing interval interruptions, and other kinds of lacunarity. Quality isn't just on/off as the definition of 'bad' may depend on the science; exposure depth folds in here too.
- Data Subspace is a concept we introduced in the Chandra data analysis system CIAO, to unify the answer to the question 'what range of time, energy, sky was this dataset taken from?'
- Data Fidelity is a new idea that I'm proposing here, it's slightly different: what level of correction has been applied to the data, where on the slider bar from raw, instrument-space data to unreliable, heavily modelled calibrated data do you lie? In the VO we'll eventually need to be able to specify this at some level. (In the discussion, Andy Lawrence pointed out that 'fidelity' is used with a different meaning in other astronomical contexts, and so a better name is solicited).

The way forward is to talk about these issues and compare the different approaches used by different archives and data analysis systems.

This work is the result of extensive discussions with the CfA VO team, the NVO collaboration members, and the CDS/Strasbourg VO team.

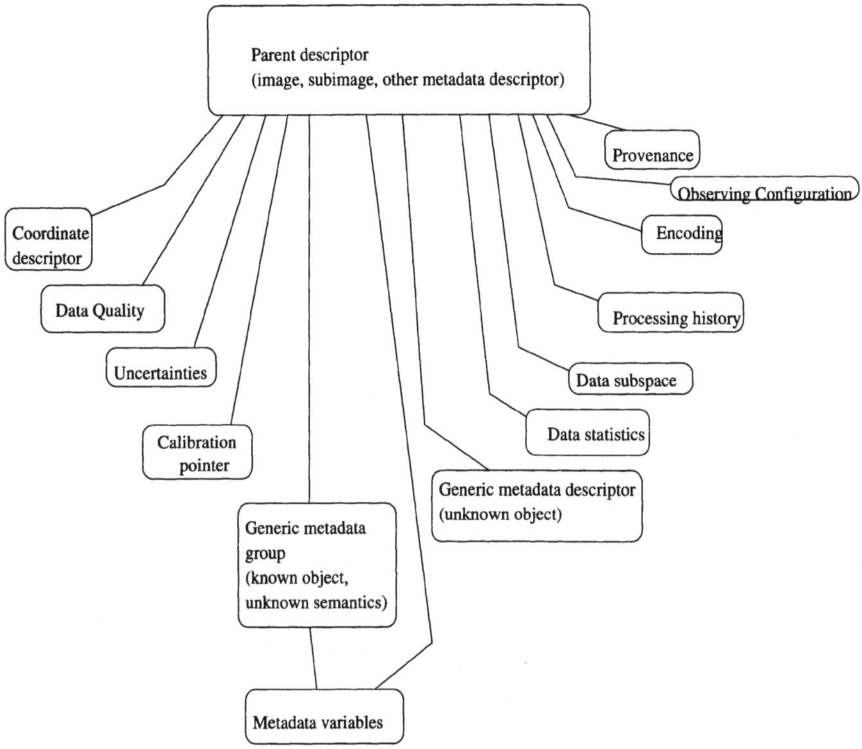

Fig. 3. Objects we will need to model

References

1. J. McDowell: Proc. SPIE 4477, 234 (2001)
2. A. Farris, ADASS 2, 145 (1993)
3. R. Williams et al., `http://cdsweb.u-strasbg.fr/doc/VOTable` (2002)

Scalable Metadata Definition Frameworks

Raymond Plante

National Center for Supercomputing Application, University of Illinois Urbana–Champaign, Urbana, IL 61801

Abstract. The VO will be an evolving environment that will enable increasingly complex interactions between resources. Naturally, expanding capabilities will drive the need for increasingly detailed information to be exchanged in the form of metadata. To encourage a smooth evolution of metadata standards that won't eventually collapse under its own weight, our long-term vision should include the establishment of a structured metadata definition framework. In this paper, I propose a foundation for the development of such a framework that is extendable yet scalable. The goals of such a framework are to unify the use of metadata across the different contexts of a VO application, clarify how the metadata should be bound to specific coding schemes and software interfaces, provide a manageable framework for extending metadata schemas, and specify a structured representation of metadata definitions upon which schema-independent software can be built. I outline several principles upon with a framework could be developed. I expect that such a framework will be critically valuable for establishing a community-based process the evaluation and adoption of metadata standards by the global VO community.

1 Motivation

The motivation for establishing a standard framework for metadata definition is part lessons from past schema definition efforts and part promise of emerging XML-based technologies. A natural consequence of defining metadata for a broad community can be a large and complex standard. Examples from the Z39.50 community include the BIB-1 ($\sim$100 terms) and GEO-1 (over 300 terms). A large standard has real implications for the cost of supporting it, both for those that define the standard and those that must implement it.

On the bright side, XML-based standards such as XML Schema, WSDL, and RDF provide an opportunity to automate the management and implementation of a metadata standard. They enable us to envision the publishing of dictionaries of metadata within which their meanings, range of values, and interrelationships are encoded in a structured way such that generic software can read the dictionary and understand how to use it. The existence of such generic software provides the hope for making a structured framework *scalable*; that is, as we define more metadata, covering more detailed concepts, the cost of supporting the metadata standards grows little or not at all.

This presentation is an overview of a white paper (Plante [1]) developed as part of the US-NSF-funded NVO project. In addition to motivating the need for a general metadata definition framework and enumerating its goals, it explores an approach one might take to develop such a framework (including a some rough XML mark-up) in more detail than can be covered here.

2 The Role of Metadata in a VO Application

Metadata come into play in many ways within a VO application. These include:

1. **Resource description.** This is expected to come in two forms. First is the use of metadata to provide general descriptions about what a resource is: its type (archive or service), its provenance, and its coverage of some concept space (e.g. sky, frequency, sensitivity). This information would typically be used for resource discovery. The second form might describe the interfaces it supports and perhaps how it supports it.
2. **Metadata exchange with a service.** In general, a service takes an input and returns an output. The most obvious example for the VO is a search service (such as the cone search). The same concept may be used in both the query and the response but not necessarily in the same format or syntax.
3. **Data access through a data model.** Metadata is used to communicate what the data represent, how to access it, and how to make logical sense of it.
4. **Semantic analysis.** This refers to a class of contexts in which a process makes use of the relationships between metadata to do some transformation. For example, one might want to translate a search query from metadata expressed by the user – say, as a simplified example, a sky cone search – into metadata the a data resource supports – say, a sky rectangular search.
5. **Data dictionary.** Traditionally, a data dictionary is useful primarily to real people: for developers, it provides a reference for building support for metadata into software; for users, it might serve as part of a help document for using a metadata-driven application. A structured format for the dictionary allows it to be used to configure software automatically (see §4.1).

3 Goals of the Framework

The goals are as follows:

1. *To unify the use of metadata across the different contexts.*
   In particular, we want to ensure that a concept used in one context is the same as in another context, even if they are rendered in different ways. For example, the meaning of the term *waveband* as well as the definition of named wavebands – *radio, infrared, optical, etc.* – should be the same whether we are describing a data resource in a registry or a set of query parameters.
2. *To clarify how metadata can be bound to various encoding schemes and software interfaces.*
   A structured definition of metadata will elucidate the most obvious way to render the metadata in a wire protocol or store it in memory. Where it is

not obvious, it should be possible to describe the binding explicitly. (This is the strategy used by WSDL.) This will make it easier to have have the different renderings look as similar as possible.

3. *To provide a manageable framework for extending schemas or creating new ones.*
   "Manageable" can mean several things:
   - It's not necessary to define all the metadata for the entire application domain from the top down in order to make use of the metadata in an automated way.
   - It is possible to use multiple (complementary) metadata schemas in a single application without difficulty. This allows, then, one to partition the concept space in logical and practical ways; applications can then pick and choose which parts to use.
   - It's not necessary to get the definitions correct the first time if the schema can be easily changed in the future.
   - It's possible to create new, specialized metadata integrated with "standard" VO metadata for specialized applications involving only a few resources. It is thereby possible to test out candidate schemas for VO "standard" status.
4. *To enable the development of schema-independent software for supporting metadata.*
   For many applications that use metadata, it's not necessary that the meanings of the metadata be hard-coded into the software. For example, we can conceive a package that can be configured to convert a query using a standard VO syntax into a form natively supported by a data resource. If this package has access to the metadata definition, this software can be made schema-independent. This is a powerful way to reduce the cost to curators of supporting new metadata as part of a standard search service.

4 An Approach to a Metadata Framework

The key feature of the metadata definition framework proposed here is the ability to encode the definition of metadata as an XML document. Traditionally, one might consider such a document as a *schema definition* or a *metadata dictionary*; however, we need to encode more than just the prose definitions and the syntax associated with the metadata. We would also like to define the supported operators that can be used to do such things as compare metadata values with test values. Finally, we may wish to include semantic relationships between metadata and descriptions of how the metadata can be rendered in different formats or contexts. Thus we might think of the XML document as a *metadata manual.* In this document, use of the term *metadata dictionary* implies this greater scope of information.

4.1 Principles

This document does not present a complete framework design but rather a foundation or initial approach to the design of such a framework. Thus, it is worth spelling out a few guiding principles upon which this approach is based.

Leveraging Existing Standards The framework should leverage off of existing standards where they are appropriate. The key ones are XML Schema, RDF, and WSDL. "Leveraging" usually means making it possible to use software that complies with these standards naturally within the framework. XML Schema and RDF address to important parts of a metadata dictionary: structural definitions and relationships. A framework that can integrate both kinds of declarations would be ideal.

The key to effective leveraging may lie with two emerging XML-based technologies: RDF Schema (RDFS) and DAML+OIL. How similar these frameworks are to the one being motivated here is currently under investigation. Clearly, it makes sense not to reinvent a wheel where it already exists; thus, *if RDFS/DAML+OIL can be used to meet the goals outlined above, then we should adopt them to encode our metadata dictionaries.* The appropriateness of these frameworks may not just be a matter of whether they can meet the goals outlined above but rather whether we can sufficiently emphasize the features important to our framework. For example, if the resulting XML document cannot be easily recognized as a metadata dictionary (either by software or a developer), we may find the technology too cumbersome to use effectively.

Sensible integration into WSDL will also be needed; in particular, it will be useful to combine the use of metadata described in our dictionary dictionary into WSDL descriptions of services. For example, for a cone search service, it would be useful to indicate the range of values of the searchable metadata that will return none non-empty results – that is, the coverage of the underlying database.

Schema Partitioning It should be possible to partition the concept space into separate schemas. For example, we can define the space-time metadata within a single schema. Another schema may handle resource provenance; while another, specifically address optical filters.

Mixing Multiple Schemas It should be possible to use multiple schemas in the same application when they have all been defined within the framework. This means there needs to be a way to uniquely identify schemas and the metadata they define. This is normally done through the use of namespaces.

Definition Styles Organizations (e.g. the NVO) can adopt specific requirements on how metadata can be defined. For example, they may wish to exclude the use of attributes as part of the definition of a metadata element.

Use Across Different Contexts A particular metadata dictionary document can be reused, with perhaps small changes, for a different purpose within a different context. This idea was explored in related study under the ISAIA project (Hanisch [2]) called FlexQuery (Plante, Guillaume, & Mehringer [3]). It demonstrated how a service provider could add markup to a data dictionary document to describe how to convert queries using standard metadata into the service's native metadata. Similarly, a client could use the data dictionary document to automatically configure a client GUI to the service (see Fig. 1; cf. [1]).

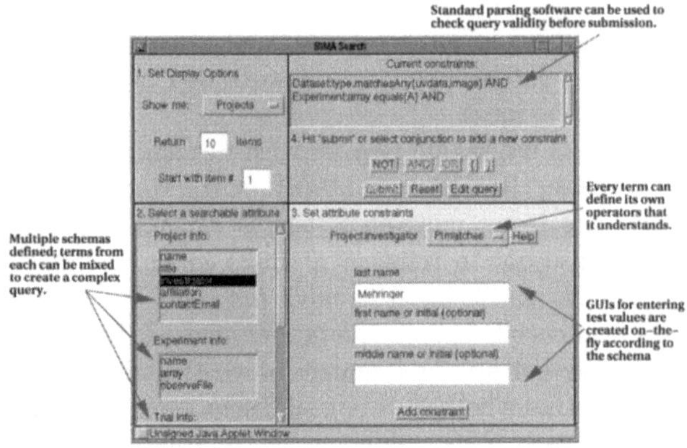

Fig. 1. The FlexQuery demonstration search client (taken from Plante et al. 2002 [5])

4.2 A Structure for an XML-Based Metadata Dictionary

The white paper [1] goes into greater detail about how an XML-based metadata dictionary might be structured to meet the goals via the principles described above. In overview, it suggests the following sections:

1. *Identification.* This section establishes a namespace for the terms the document defines. It also identifies the organization responsible for the definitions as well as any revision information.
2. *Syntactic Types.* Like XML Schema, it is important to associate metadata with value types. Like an WSDL document, this component allows one define new types or import existing types defined in other schemas.
3. *Dictionary.* This section itemizes specific terms being defined. It associates with each term a type, a human-readable definition, allowed values, and the definitions of any controlled values. It also defines any special operators (e.g. "less than", "contains") associated with each term.
4. *Semantic Relationships.* This describes how a particular metadatum is related to other metadata. The RDF resource–property–value model (also referred to as "subject–predicate–object") is an appropriate model for doing this.
5. *Bindings.* Similar to the Bindings section in a WSDL, this component describes how the schema can be bound to different encoding schemes and software APIs.

5 A Community Process for Adopting Standard Schema

The structured format provides a convenient vehicle for proposing, testing, and adopting new metadata schemas in a community-based way. The advantage of the structured format is realized with existence of schema-independent software that can test out the use of the metadata in existing applications.

Consider the following scenario. Imagine that three institutions wish to establish an application that interoperates across their respective resources which requires new metadata not covered by a current VO standard. They can create a new metadata dictionary that defines the new metadata and share it amongst themselves, to be used in addition to existing standard schemas. Using standard VO metadata software, they can integrate the new metadata into their application with little or no additional programming needed (depending on the nature of the application). If their application is successful, they can decide to propose the new schema be accepted as a VO standard. Or, if they don't expect the new metadata to be applicable outside their narrow focus, then they can choose not to propose it as a standard.

The VO may impose certain requirements or restrictions on how the metadata are defined. The first obvious requirement would be that the schema be submitted in the metadata dictionary XML markup format; however, other requirements might be imposed as well, such as metadata types should not include XML attributes. A testing and evaluation period can be set; the existence of software that understands the dictionary would accelerate this process. If the proposal is accepted, then the provenance information is changed to those of the VO community, and the document is registered in the VO dictionary registry.

Proposals for correcting or extending existing schemas would proceed in a similar way. Here, the use of revision numbers will be critical to managing changes of this type. Again, the ability to use existing metadata software will be critical to identifying problems of backward compatibility.

References

1. Plante, R. 2002, "A Scalable Metadata Framework for the Virtual Observatory", an NVO white paper, `http://bill.cacr.caltech.edu/cfdocs/usvo-pubs/files/fw-draft2.pdf`
2. Hanisch, R. J. 2000, in ASP Conf. Ser., Vol. 216, Astronomical Data Analysis Software and Systems IX, eds. N. Manset, C. Veillet, D. Crabtree (San Francisco: ASP), 201
3. Plante, R., Guillaume, D., & Mehringer, D. `http://monet.astro.uiuc.edu/~rplante/topics/FlexQuery/`
4. Hanisch, R. J., "National Virtual Observatory", this volume.
5. Plante, R. L., Guillaume, D., Mehringer, D., & Crutcher, R. 2002, in Astronomical Data Analysis Software and Systems XI, in press.

Space-Time Metadata for the Virtual Observatory

Arnold Rots

Smithsonian Astrophysical Observatory,
60 Garden Street, MS 67, Cambridge, MA 02138, USA
Université Louis Pasteur, Observatoire Astronomique de Strasbourg,
11, Rue de l'Université, F-67000 Strasbourg, France

Abstract. We discuss the hierarchy of objects that are to contain the metadata pertaining to space and time for the Virtual Observatory. The design needs to be sufficiently general that it can adequately represent the many coordinate systems and conventions that are in use in the community. On the other hand the most basic requirement is that the space-time metadata for queries, for resource descriptions, and for data be complete and self-consistent. If these two requirements are met, the VO should be able to respond to a query in any coordinate system and deliver data from depositories with a variety of native coordinate systems.

1 Introduction

There are almost no astronomical data that do not have coordinate information attached to them. Particularly the space-time coordinate metadata form the foundation for interpreting and combining datasets. Ironically, the number of coordinate systems that are in use, both for space and time, is large. One of the major reasons is that not only the position of the object studied enters into the metadata, but also the position of the observatory, while both spatial positions vary with time. As a result, the coordinates of an observation are often expressed in a system that is *convenient* for the observer, not necessarily one that allows easy comparisons with other observations. In defense of this custom it should be noted that such an approach involves the least number of transformations and hence protects against loss of information – provided that all pertinent metadata are preserved.

Consequently, if we are striving for seamless comparison and combination of data in the VO, we need to ensure that the space-time coordinate metadata are taken care of properly. Taking into account that is is taking FITS more than 20 years to get this right (that is, the World Coordinate System – WCS – specification), we should be prepared for a complicated job that will require great care in order to be done right.

It should be pointed out that space and time coordinates cannot be separated. That is to say, there are numerous examples where undoubtedly it does not matter, but we would neglect the subtle and not-so-subtle aspects of the space-time coordinate space at our own peril since cases where the space-time link does matter are far more numerous than is generally understood.

A related, but slightly independent issue is the definition of arbitrary regions. This is badly needed for a full implementation and will be an integral part of the space-time coordinate metadata, but it is something that can be encapsulated independent from the rest of the specification.

In the following we shall deal with the requirements, the basic building blocks and toplevel structures, the regions, and we shall assess the current status. The components are thought of as XML elements, partly because that is a likely implementation, partly because it provides a handy definition tool. As far as contents is concerned, we will build on the framework developed for the ISAIA project (see [1]).

2 Requirements

The requirements for space-time coordinate metadata can be spelled out in a surprisingly simple form:

1. All aspects of the space-time specification need to be encapsulated into a single structure.
2. That structure needs to be rigorous. To satisfy this requirement it needs to be:
   - Complete
   - Self-consistent
3. The structure needs to support a full coordinate description in four contexts:
   - Resource description
   - Search query
   - Catalog data
   - Observational data, or simulated observational data

Requirement 1 is just a formal rule that does not present much of a problem. The real challenge is in requirement 2 which drives the actual design specification, as derived from requirement 3. Requirement 2 is at the heart of the Space-Time Coordinate metadata for the VO; it cannot be emphasized enough that satisfying this requirement is absolutely essential and will determine whether we will ever be able to combine disparate datasets from different archives. A closer look at what is required for implementing the four contexts in requirement 3 offers the necessary insight in what the metadata need to provide:

- Resource description:
  coverage, resolution, error
- Search query:
  search area, resolution, error
- Catalog data:
  position, coverage, resolution, error
- Observational data:
  position, coverage (FOV), resolution, error
  Note that this context differs from the others in that it requires two positions: that of the data and that of the observatory.

3 Structures

We will define the very basic structures that are needed to build the top-level structures and then define the contents of these basic structures in more detail.

3.1 Basic Structures

The collection of basic building blocks is very simple:

- CoordSystem
  This contains a set of attributes or elements that specify a space-time coordinate system in full detail.
- Coords
  This structure defines a specific position in the coordinate system. It consists of an appropriate number of coordinate elements, each of which provides information on the coordinate value, resolution, and error.
- CoordArea
  This structure defines a specific area in the coordinate system.

There are two types of compound structures that are made up of basic structures:

- CoordLocation
  This is a combination of a CoordSystem and a Coords structure that, obviously, gives a full (complete and self-consistent) description of a position in space-time.
- CoordCoverage
  This is a combination of a CoordSystem and a CoordArea structure that specifies an area in space.

These structures are not really defined, as such, but we shall use them in the definition of the top-level structures in the next subsection, for the purpose of simplifying the discussion.

3.2 Top-Level Structures

In this subsection we shall demonstrate how the contexts defined in requirement three can be built from the basic structures defined in the previous subsection. The collection of basic building blocks is very simple:

- Resource Description
  Simply consists of a **CoordCoverage** structure that indicates what part of the space-time universe is covered by the data in the resource (archive, data center, catalog, etc.). Actually, it is really defined in the opposite sense: the resource guarantees that it does not hold any data outside the area specified by the CoordCoverage element. This is in order to accommodate archives of pointed observations. In addition, this structure may need to include a **Coords** element to provide information on resolution and errors.

- Search Query
  This also consists simply of a **CoordCoverage** element that defines the area to be searched by the query. In addition, this structure may need to include a **Coords** element to provide information on desired resolution and errors.
- Catalog Data
  This consists of a **CoordLocation** structure for the actual catalog data that are transmitted, but should also contain a **CoordCoverage** structure to indicate the region that is actually covered by the provided data. This is important for analyses that involve cross-correlations and source densities. In many cases it will be the intersection of the catalog's Resource Description and the Search Query.
- Observational Data
  This structure needs to contain two compound structures: an **ObservatoryLocation** (a **CoordLocation** element) and an **ObservationLocation** (a **CoordLocation** and a **CoordArea** element). The CoordArea element in the latter provides information on the observation's field-of-view (FOV).

3.3 Detailed Design

To give the reader a taste for what information needs to be contained in the various structures, or elements, we provide a list, without much further detail, of the attributes or elements (and their default values, when appropriate) that are to be included in the basic building blocks. One should be aware that this design is still in a state of flux and the information below should be taken as illustrative, rather than authoritative.

- CoordSys
  - CoordFrame (ICRS)
  - CoordRefPosition
  - TimeFrame (TT)
  - TimeRefPosition
  - CoordFlavor (SPHERICAL)
  - PlanetaryEphem (JPL-DE405)
- Coords
  - Time
  - CoordElement (up to 6)
    * Value
    * Error
    * Resolution
    * Size
    * Unit
  - CoordFile (may replace the CoordElements)
- CoordArea
  - TimeRange
  - Spatial area; one of the following:
    * Region
    * CoordInterval
    * CircleOrSphere (contains a Coords element)

3.4 Regions

We need a solid convention for specifying regions in spatial coordinate systems. Currently, there are several conventions implemented in the CXC (Chandra X-ray Center), Aips++ (NRAO), IRAF (NOAO), and HTM (SDSS). These conventions form a disparate collection, due to the very different requirements in the four systems. It is, in our opinion, quite feasible to define a syntax that encompasses the features of all conventions currently in use. The real challenge is in implementing the operations that the region elements require.

At this point we can define the requirements for a *region* convention and these are listed below:

- Applicability: one needs to be able to define a *region* in all WCS (World Coordinate System) types, spherical as well as Cartesian; however, it is probably wise to restrict them to WCS types (as opposed to expressing a *region* in pixel coordinates).
- Shapes: in principle, polygons (concave and convex; with great circle and parallel sides), ellipses, and sectors are sufficient; rectangles, circles, and points can be derived from these, though there may be an advantage in being able to specify these directly.
- Operations between shapes or *regions* (i.e., binary operations): union (OR), intersection (AND), XOR, difference, and disjoint (boolean) are required.
- Operations on a *region* (i.e., unary operations): calculate area, interior and exterior circles and rectangles.
- Conversions: translate a *region* from one coordinate system to another.

The issue that makes the *region* implementation so difficult is the handling of polygon sides that run along parallels in spherical coordinate systems.

4 Status

The definition of the Space-Time Coordinate metadata has progressed to a point where one can experiment with it, though substantial changes are still to be expected. In particular, there are some issues surrounding the question whether coordinate axes should be specified as vectors (arrays) or as individual scalars. This is not quite as simple as it looks since we need to accommodate the auxiliary information on errors, resolution, etc., as well. Another question concerns the issue of whether we can devise a general mechanism to introduce coordinate axis information into VOTables. And we need a mechanism to refer to columns in FITS binary tables in a manner that is similar to the referencing of columns in a VOTable. We are confident that these issues can be resolved in the near future.

The issue of the *region* specification is far more prickly. Suffice it to say that more discussion is needed.

References

1. A.H. Rots: 'The Space-Time Profile for ISAIA'. In: *Virtual Observatories of the Future, Conference at Caltech, 13-16 June 2000*, ed. by R.J. Brunner, S.G. Djorgovski, A.S. Szalay (APS Conf. Ser. 225 2001), p. 213

VOTable: Tabular Data for the Virtual Observatory

François Ochsenbein[1], Roy Williams[2], Clive Davenhall[3], Daniel Durand[4], Pierre Fernique[1], Robert Hanisch[5], David Giaretta[6], Tom McGlynn[7], Alex Szalay[8], and Andreas Wicenec[9]

[1] CDS, Observatoire Astronomique, Strasbourg, France
[2] California Institute of Technology, USA
[3] University of Edinburgh, UK
[4] Canadian Astronomy Data Centre, Canada
[5] Space Telescope Science Institute, USA
[6] Rutherford Appleton Laboratory, UK
[7] NASA Goddard Space Flight Center, USA
[8] Johns Hopkins University, USA
[9] European Southern Observatory, Germany

Abstract. The VOTable format is now an agreed upon XML standard for representing tabular data in the VO context. Tabular data are an important part of the data exchanged between all participants in the Virtual Observatory: such representations include data extracted from catalogues and surveys, calibration parameters and data, results of modelling applications, etc. An accurate description of what these data actually represent – the metadata – is an essential requirement in the VO context where the data are coming from many sources and are to be processed by many applications, in fully automatic procedures.

VOTable is based on the previous experiences of FITS and Astrores, and is the result of extensive discussions; the developments and discussions may be consulted on the Internet.

1 History

Catalogued data, and more generally data in tabular form, play a fundamental role in the presentation of the astronomical results. This role has become more prominent in the electronic era, where data are being requested from a wide variety of data servers and archives for further analysis or processing, and the on-going Virtual Observatory project has an obvious need of getting and *interpreting* the data originating from many different horizons.

The interpretability of the data requires that *metadata* – data describing the data – are transmitted in association with the data; these metadata must moreover be accurately defined in order to enable robotic data processing.

In astronomy, the FITS standard (see `http://fits.gsfc.nasa.gov/`) includes two tabular formats, one in ascii format (Harten et al. [4]), and one in binary format (Cotton et al. [2]); both suffer from two limitations: FITS files are not adapted for streaming processes (the number of rows has to be known

```
<?xml version="1.0"?>
<!DOCTYPE VOTABLE SYSTEM "http://us-vo.org/xml/{\dots}...dtd">
 <RESOURCE name="myResource">
   <OBSERVER>William Herschel</OBSERVER>
   <SOURCE id="mySource">
      <STAR-NAME>Procyon</STAR-NAME>
      <POSITION equinox="J2000" epoch="J2000">
       <RA unit="deg">114.827</RA>
       <Dec unit="deg">+05.227</Dec>
      </POSITION>
     <COUNTS>
       <COUNT>4</COUNT>
       <COUNT>5</COUNT>
       <COUNT>3</COUNT>
     </COUNTS>
   </SOURCE>
   ...
 </RESOURCE>
```

Fig. 1. Example of a "classical" XML document

prior to the data delivery), and the metadata are not accurate enough for easily merging data from different origins.

The VOTable project started in October 2001, with discussions between developers of two projects: *Astrores* (Ochsenbein et al. [5]) and *XSIL* (Blackburn et al. [1]). The first VOTable draft (version 0.1) was prepared in December 2001. Extensive exchanges on the project occured during an "Interoperability meeting" supported by the VO Projects and the OPTICON European Network, and held in Strasbourg in January 2002; the first official release, Version 1.0 is dated 15 April 2002. VOTable documents can be found at `http://cdsweb.u-strasbg.fr/doc/VOTable/`; VOTable archives & discussion groups are available from: `http://archives.us-vo.org/VOTable/`

2 XML and VOTable

XML is a language designed for embedding user-defined markers *(metadata)* in textual documents *(data)*. XML has become widespread since its creation some 4 years ago, and several tools are now available that facilitate the exchange of documents: parsers for an interpretation of XML documents by applications, query tools for retrieving user-specified elements in XML databases *(XQuery)*, and converters for a transformation into other XML dialects *(XSLT)* or into HTML for visualisation *(XSL)*; XML documents are also easily encapsulated in messages *(SOAP)*.

An example of what we call a "classical" XML document is illustrated in Fig. 1: each parameter is enclosed within "tags" like `<RA>` or `<COUNT>`, and therefore can easily be retrieved in the XML hierarchical structure itself described in another document – the *XML-Schema* or the *DTD* (document type definition).

```
<?xml version="1.0"?>
<!DOCTYPE VOTABLE SYSTEM "http://us-vo.org/xml/VOTable.dtd">
<VOTABLE version="1.0">
  <DEFINITIONS>
  <COOSYS ID="myJ2000" equinox="2000." epoch="2000." system="eq_FK5"/>
  </DEFINITIONS>
  <RESOURCE>
    <PARAM name="Observer" datatype="char" arraysize="*"
   value="William Herschel">
      <DESCRIPTION>Give here the observer's name</DESCRIPTION>
    </PARAM>
    <TABLE name="Stars">
      <DESCRIPTION>Some bright stars</DESCRIPTION>
      <FIELD name="Star-Name" ucd="ID_MAIN" datatype="char"
     arraysize="10"/>
      <FIELD name="RA" ucd="POS_EQ_RA" ref="myJ2000" unit="deg"
             datatype="float" precision="F3" width="7"/>
      <FIELD name="Dec" ucd="POS_EQ_DEC" ref="myJ2000" unit="deg"
             datatype="float" precision="F3" width="7"/>
      <FIELD name="Counts" ucd="NUMBER" datatype="int"
     arraysize="2x3x*"/>
      <DATA><TABLEDATA>
        <TR>
          <TD>Procyon</TD><TD>114.827</TD><TD> 5.227</TD>
          <TD>4 5 3 4 3 2 1 2 3 3 5 6</TD>
        </TR>
        <TR>
          <TD>Vega</TD><TD>279.234</TD><TD>38.782</TD>
  <TD>8 7 8 6 8 6</TD>
        </TR>
      </TABLEDATA></DATA>
    </TABLE>
  </RESOURCE>
</VOTABLE>
```

Fig. 2. A VOTable example

This "classical" XML way of representing the data obviously means huge overheads in terms data volume, especially in the context of large tables where, by definition, all rows have an identical structure – a set of columns or *attributes* with constant characteristics, and presented in an immutable order. Another limitation inherent to this "classical" XML representation is the necessity of frequently introducing new *elements* (tags) or adding new qualifications to the existing tags to follow the constant evolution of the data contents – that would mean in practice frequent changes in the schema definitions.

The VOTable schema follows the classical tabular presentation which can be found in the printed catalogues, where each column is described as accurately as possible only once, and is followed by the actual data grouped in rows made

datatype	Meaning	FITS	Bytes
"boolean"	Logical	"L"	1
"bit"	Bit	"X"	*
"unsignedByte"	Byte (0 to 255)	"B"	1
"short"	Short Integer	"I"	2
"int"	Integer	"J"	4
"long"	Long integer	"K"	8
"char"	ASCII Character	"A"	1
"unicodeChar"	Unicode Character		2
"float"	Floating point	"E"	4
"double"	Double	"D"	8
"floatComplex"	Float Complex	"C"	8
"doubleComplex"	Double Complex	"M"	16

Table 1. The primitive `datatype` in VOTable

of columns, each column conforming strictly to its description. A full VOTable document, illustrated in Fig. 2, is made of:

- a *metadata part* describing the data, consisting of a set of <FIELD> and <PARAM> specifications.
- a *data part* made of rows, each row being made of the ordered set of fields described in the metadata part; this data part is also called the *serialization* of the data, and VOTable allows several possibilities for serializing the data (see Section 4).

3 <FIELD> and <PARAM>

In the example of Fig. 2, the four columns of the table are described by four <FIELD> elements; each field is described by the following properties:

- a **name**, which represents the column label or header;
- a **ucd** (Unified Content Descriptor) which is basically a standardized categorization of the column contents; UCDs are useful for automatic data comparisons or data mining – see Derrière [3];
- a **unit**, which makes use of a standardized vocabulary with well-defined syntax rules in order to enable automatic unit interpretations and conversions;
- a **datatype**, which represents the computer type required to store the column contents; the allowed data types are detailed in Table 1;
- a **width**, which specifies the number of characters required for a textual representation of the quantity stored in the column;
- a **precision**, which gives an estimate of the number of significant digits – important for a textual representation of the quantity stored;
- an **arraysize**, to be used for columns that do not contain just an atomic quantity; conventions exist to express multidimensional arrays, and to define a variable size for one of the dimensions.

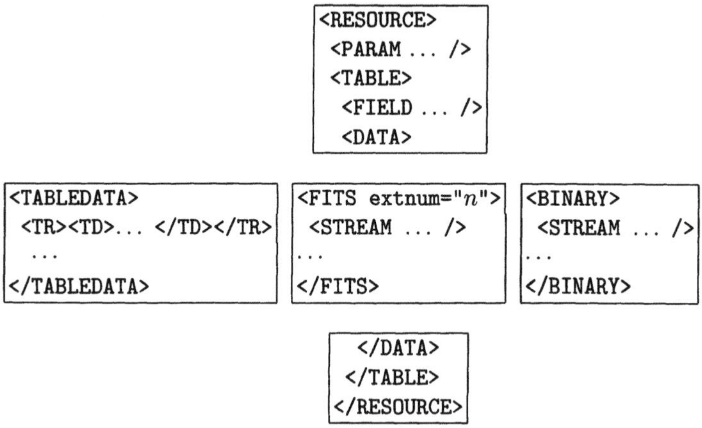

Fig. 3. The VOTable stream

A <PARAM> element, similar to <FIELD>, was defined to store values which can be viewed as *constant columns*, such as the observer's name in the example of Fig. 2; other typical examples could be the observation date in a table containing the results of one night's observation, or the mass of the prototype star in a grid of evolution models.

It is worth noting that VOTable was designed to be compatible with existing FITS binary tables – all the TTYPEs existing in FITS, and detailed in column 3 of Table 1, have a VOTable equivalent.

4 Data Serialization

One of the major aspects of VOTable is to allow several alternatives for the data serialization; this is illustrated in Fig. 3. Following the metadata description (top box of Fig. 3), the data can be presented in one of the 3 modes:

1. as an XML table introduced by the <TABLEDATA> tag, each row starting by <TR> and each column by <TD>, as in the example of Fig. 2;
2. as a reference to a FITS binary table extension file, introduced by the <FITS> tag; the <STREAM> element defines the exact location of a remote file, via a `href` attribute – meaning that existing FITS binary tables can be reached by adding a VOTable wrapper;
3. as a *binary* stream of bytes introduced by the <BINARY> element; the actual data stream may be remote when the <STREAM> element includes an `href` attribute, or may be embedded in the document as a coded stream of bytes when the <STREAM> element contains an `encoding` attribute.

5 Servers and Tools

Several databases are already delivering VOTable documents; as a first example, over 50 services offering a "cone search" around a position are available from `http://skyserver.pha.jhu.edu/VOConeProfile/`. General databases like SIMBAD are also providing VOTable output.

Tools interpreting VOTable documents are also being developed, and VOTable parsers are now available in Perl, Java, and C. VOTable validators, and XSLT translators of VOTable into HTML are being developed.

6 VOTable Appendices

The VOTable specification contains as appendices two extensions which existed in *Astrores*, but need further examination:

- The `<LINK>` convention describes how to get the related data, like explanations, or related images, spectra..., as in the following example:

```
<TABLE>
  <FIELD name="FileName" datatype="char"/>
  <LINK href="http://server/getFile?${FileName}" />
  <DATA><TABLEDATA>
    <TR> <TD>photo/procyon.dat</TD> </TR>
    <TR> <TD>photo/vega.dat</TD> </TR>
  </TABLEDATA></DATA>
</TABLE>
```

  where the `<LINK>` indicates, via a substitution mechanism similar to shell scripting conventions, that the contents of the `FileName` column can be used to retrieve data related to each star described in a table row.
- the *Query Mechanism* in which a server may return as an answer a description of tables without actual data results, associated to a `<LINK>` action – in other words a server may answer with a form to be filled when questions are too generic or would generate huge volumes of data.

Despite of the usefulness of such conventions, it was felt that the more generic solutions which are currently emerging in the context of the Web service developments would probably bring "natural" solutions to this problem of how to describe the existing querying possibilities – and should moreover enable machine-processable data exchanges across different disciplines.

Other revisions and enhancements to the present VOTable specification will certainly be necessary in the future, and we will work to ensure that the evolution will continue to occur with full international participation and concurrence.

References

1. Blackburn, K., Lazzarini, A., Prince, T. and Williams, R. 1999, in HPCN'99, 513, Amsterdam, February (`http://citeseer.nj.nec.com/blackburn99xsil.html`)
2. Cotton, W.D., Tody, D. B., and Pence, W. D. 1995, A&AS **113**, 159
3. Derrière, S. *(this conference)*
4. Harten, R. H., Grosbøl, P., Greisen, E. W., and Wells, D. C. 1988, A&AS **73**, 365
5. Ochsenbein, F., Albrecht, M., Brighton, A., Fernique, P., Guillaume, D., Hanisch, R. J., Shaya, E. and Wicenec, A. 2000, ASP Conf. Ser. **216** (ADASS IX), 83

A C++ Parser for VOTables

Ajit Kembhavi[1], Hrishikesh Hegde[2], Sonali Kale[2], P.R. Krishnan[2], Vasudev Navelkar[2], and T.M. Vijayaraman[2]

[1] Inter-University Centre for Astronomy and Astrophysics, Ganeshkhind, Pune 411 007, India
[2] Persistent Systems Pvt. Ltd., "Bhageerath" 402, Senapati Bapat Road, Pune 411016, India

1 Introduction

This paper provides a brief description of C++ Parser for VOTables. The parser is developed by Persistent Systems in association with IUCAA as part of the Virtual Observatory India (`http://vo.iucaa.ernet.in/~voi`) initiative.

2 VOTable Basics

The VOTable format is a proposed XML (Extensible Markup Language) standard for exchange of tabular data, with particular emphasis on astronomical tables. A table is an unordered set of rows, each of a uniform format, as specified in the table metadata. Each row is a sequence of table cells, and each of these is either a primitive data type, or an array of such primitives.

There are three representations for data part in VOTable:

- FITS: VOTable can be used either to encapsulate FITS files, or to re-encode the metadata.
- BINARY: This format is supported for efficiency and ease of programming, no FITS library is required, and the streaming paradigm is supported.
- TABLEDATA: This contains pure XML format for small tables.

VOTable has built-in features for big-data and Grid computing. The underlying XML format allows applications to easily validate an input document, and facilitates transformations through XSLT engines. Here XSLT is the Extensible Stylesheet Language for Transformations that transforms a source XML tree into a result XML tree. It is based on template rules, which specify how XML documents should be processed.

3 Parser Features

The motivation behind the C++ VOTable parser is to provide a C++ library for API based access to VOTable files. The APIs can be used directly to develop VOTable applications without having to do raw VOTable processing. The VOTable Parser can be downloaded from
`http://vo.iucaa.ernet.in/~voi/html/iucaawriteup.htm`.

The parser is implemented as a wrapper over XALAN-C++, a robust implementation of the W3C recommendations for XSLT and the XML Path language. The user can retrieve a Table or a Resource within a VOTable directly using an XPath query. The parser can be easily used by users of CFITSIO library. Version 1.1 of this parser was released on May 31, 2002 and Version 1.2 was released on July 15, 2002. These versions support reading of VOTables and pure XML TABLEDATA. It runs on Windows NT 4.0, Windows 2000 and RedHat Linux 7.1.

4 Parser Implementation

The parser was developed on Windows NT 4.0 platform using VC++ and was ported to Redhat Linux 7.1 with zero effort. It consists of 18 classes representing various elements of the VOTable format. The VOTable Parser V1.1 has approximately 8500 lines of code.

5 Parser Design

The VOTable C++ Parser makes use of XALAN-C++ to read the VOTable file. The user specifies the <RESOURCE> or <TABLE> to be read using XPATH. An XPath expression contains one or more "location steps", separated by slashes to address part of an XML document. XPath syntax is similar to filesystem addressing. For example '/RESOURCE[1]/TABLE[1]' is the xpath to get the first Table from the first Resource.

XALAN builds the entire tree of the XML document in the memory. Once the entire tree is loaded in the memory, the VOTable Parser reads VOTable elements into class structures of its own. The class structure is explained in the next section. The whole tree is traversed and entire data is inserted into the classes representing each element. The tree is then destroyed.

6 Class Details

A brief description of the main classes is given below. Also see the figure.

- Resource: In memory representation of a single <RESOURCE> element from the VOTable. A Resource can contain other Resources as well as VTables. It also contains Coosys, Params and Links representing corresponding elements in the Resource.
- VTable: In memory representation of a single <TABLE> from the <RESOURCE> element. VTable contains meta data and data.
- TableMetaData: Contains meta data (Fields, Links and Table Description)
- TableData: Contains Rows of data.
- Field: Representation of <FIELD> from the Table. It describes the contents of a column.

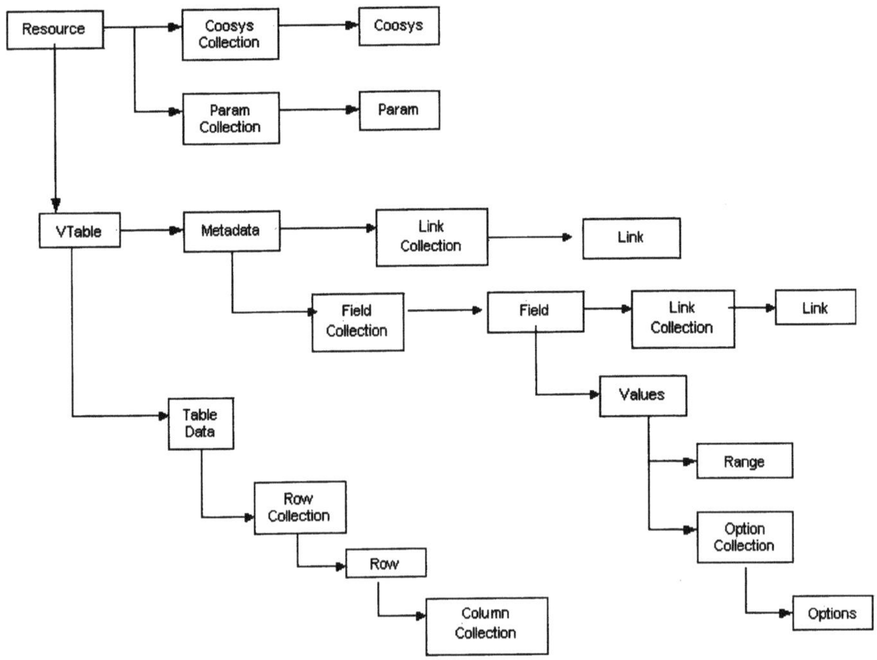

Fig. 1. Class diagram

- Row: Represents a single row of data i.e. a <TR> element. Rows contain columns.
- Column: Represents a column i.e. a <TD> element from VOTable. Column contains the actual data.

7 APIs – Typical Operations

The API mainly consists of

1. File Level I/O Routines that support basic file related operations on VOTable files. These include APIs to Open and Close VOTable files.
2. Resource and Table Access Routines to access meta data and data in Resource and VTable. These include APIs to
   - Get number of VTables.
   - Get number of Rows.
   - Get number of columns.
   - Get column (field) information (column name, column number, etc.)
   - Access table data.

8 Example Code

Given below is sample code to print the number of rows and columns from a VOTable.

```
int status;
VTable v;
char * str = NULL;TableMetaData tmd;TableData td;
// Open VOTable file 'SampleVOTable.xml' and read the first Table from first
// Resource
v.openFile("SampleVOTable.xml","/RESOURCE[1]/TABLE[1]", 0, &status);
// check the status of the operation
if (status != SUCCESS)
{
// Some error occured - exit with an error message.
cout << "Error : " << getVOTableErrorMessage(status);
cout << "VTABLE could not be opened properly." << endl;
exit(1);
}
// Read the name of the Table
v.getName(str, &status);if (status == SUCCESS && str != NULL)
{
cout << "VTable name is" << str endl;
delete[] str;
}
//Read meta data.
if (v.getMetaData(tmd, &status) == SUCCESS)
{
int numOfCols = 0;
//Get the number of fields from meta data.
tmd.getNumOfColumns(numOfCols, &status);
cout <<"Number of columns is" << numOfCols << endl ;
}
if (v.getData(td, &status) == SUCCESS)
{
//Read the number of rows in the table
int numOfRows = 0 ;
if (td.getNumOfRows(numOfRows, &status) == SUCCESS)
{
cout << "Number of rows in this table data are" << numOfRows <<
endl;
}
}
```

9 Future Work

The parser will be enhanced to include APIs for creating and writing VOTables (currently the APIs provide only read access to VOTable documents) and APIs for reading FITS and BINARY data. The VOTable format may be enhanced in the future to include query extension, etc. The parser can be enhanced to support these changes and any other changes that incorporated in the VOTable format. An important task is to develop APIs for streaming support for reading large VOTables. The parser may not be able to handle very large VOTables because it loads the complete VOTable in the memory. For supporting such large VOTables, the data needs to be read in parts or in streaming mode.

References

VOTable Details - `http://vizier.u-strasbg.fr/doc/VOTable/`
XML - `http://www.w3.org/XML/`
XALAN-C++ - `http://xml.apache.org/xalan-c/index.html`
XPath - `http://www.w3.org/TR/xpath`
XSLT - `http://www.w3.org/TR/xslt`
CFITSIO -
`http://heasarc.gsfc.nasa.gov/docs/software/fitsio/fitsio.html`

Appendix

About IUCAA - The Inter-University Centre for Astronomy and Astrophysics (IUCAA) is located in Pune, India. It has a vigorous programme of research in theoretical astrophysics, observational astronomy and instrumentation development.IUCAA has a well-developed data center and hosts mirror sites of various astronomical data bases and scientific literature archives.
Website - `http://www.iucaa.ernet.in/`

About Persistent - Persistent Systems Pvt. Ltd., located in Pune, India, is an 11-year old, 350-person, software development company. Persistent specializes in developing data infrastructure software. Persistent has developed data management components such as bit-vector indexes, query optimizers, ETL Tools, OLAP Tools etc. It has worked for companies like Microsoft, Hewlett-Packard, Agilent Technologies, Informix, i2 Technologies, Engage Technologies etc.
Website - `http://www.persistent.co.in`

ASTROVIRTEL: Tools and Operations

Francesco Pierfederici[1], Markus Dolensky[2], Alberto Micol[2], and Benoit Pirenne[1]

[1] ESO/ST-ECF, Karl-Schwarzschild-Str. 2, D-85748 Garching, Germany
[2] ESA/ST-ECF

1 Introduction

ASTROVIRTEL is a project supported by the European Commission and managed by the ST-ECF on behalf of ESA and ESO. It is aimed at enhancing the scientific return of the current ESO/ECF Archive as well as at being a test bed for future Virtual Observatories.

Astrovirtel is entering its third year of operation. So far, a total of 11 proposals have been accepted and supported (resulting in more than 15 publications). During these first years the ASTROVIRTEL team found itself in a privileged position. On one side, it has been listening to the needs of Astronomers interested in multi-archive research. On the other side, it has been trying to overcome the present limitations of astronomical information providers (archive/catalogue sites). In this the ASTROVIRTEL project managed to create a small-scale VO prototype for European users. This role enabled us to clearly understand what the technical prerequisites of the future Virtual Observatories are.

This paper will focus mainly on our recent software developments such as Querator while P. Benvenuti's paper (in this same volume) describes our experiences in the support of ASTROVIRTEL proposals.

2 Different Proposals, Different Needs

The eleven proposals we fully supported during the first two cycles had different scientific goals. Some of them were dealing with solar system objects such as Trojans of Saturn and Neptune, others were investigating photometric and kinematic properties of stars. A few others, instead, were focused on extragalactic astronomy. An example of the latter group are studies of the evolution of galaxy morphology with redshift (a complete list of ASTROVIRTEL accepted proposals is available from the general ASTROVIRTEL home page, see the Links section).

This diversity in science goals called, inevitably, for a number of different software solutions. In particular, the software needs of the different proposals fell into four main categories:

1. cross-correlation tools (catalogue-catalogue and catalogue-archive)
2. visualization tools and image preview generation
3. data reduction and analysis pipelines
4. sophisticated search engines and data-mining tools

3 Querator

Among the software products that ASTROVIRTEL has produced, Querator is the most advanced. It has been developed in order to address the needs of astronomers interested in multi-color imaging data. Querator has been used satisfactory in such fields as

1. Multicolor photometry
2. Photometric redshifts
3. Galaxy morphology
4. Luminosity function
5. Supernovae
6. Gravitational lensing and cosmic shear

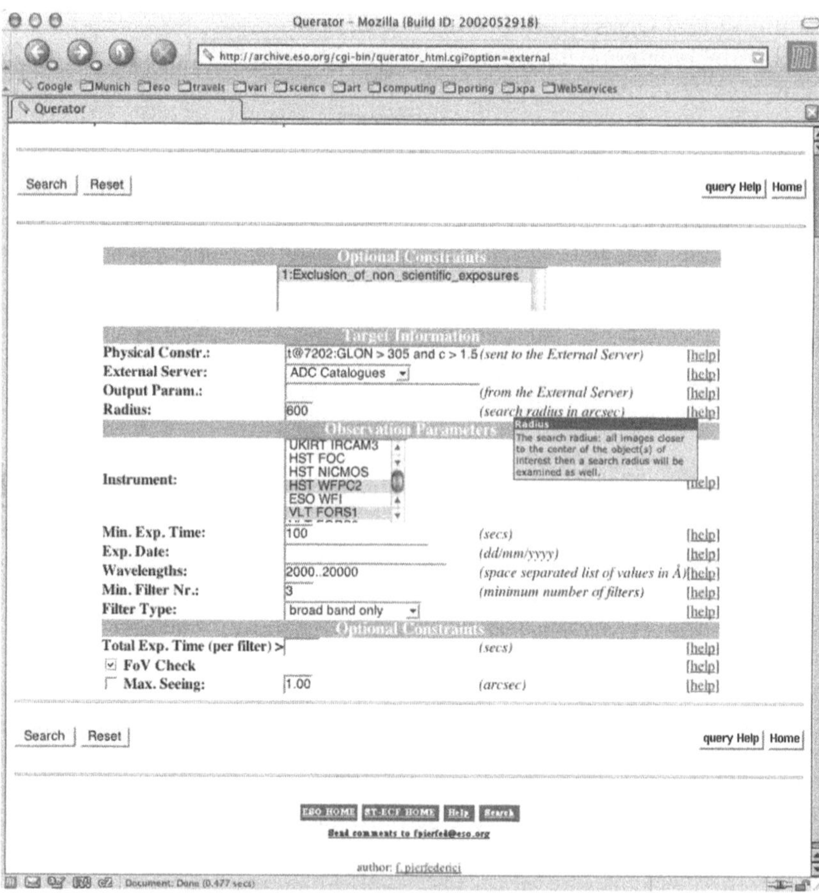

Fig. 1. Querator: Example of the "External Catalogue" query page

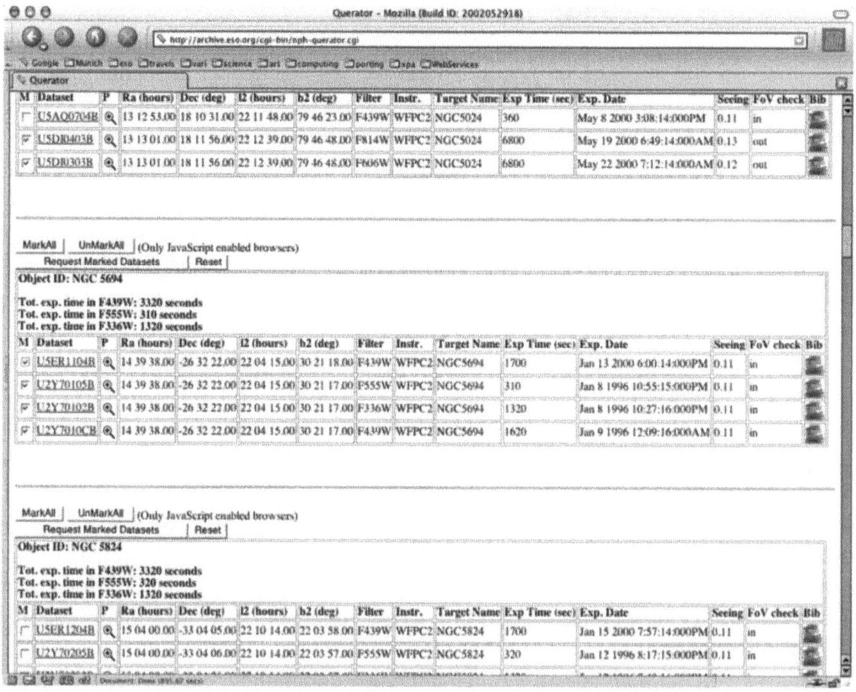

Fig. 2. Querator: Example of a query output page

The search engine itself has some quite unique features such as

1. Constraints on filters **using physical units**
2. Constraints on **colors**
3. Multi catalogue support
4. Multi archive/instrument support
5. cross-correlation of archives and catalogues
6. Field of view check, image seeing and previews

These features all together enable Querator to answer scientific questions like “Give me all the images of supernovae in spiral galaxies closer than 50 Mpc in three colors using the broad band filters U, B, V, R and I. I am only interested in images taken before the supernovae event and which actually contain the supernovae progenitor in their field of view.”

Querator is being used extensively not only by ASTROVIRTEL users, but also by the European astronomical community at large. Planned improvements include support for the ISO archive and major streamlining and optimization of the code.

4 Lessons Learned

The work on the support of ASTROVIRTEL proposals has shed light on some of the technical deficiencies of present-day archives and catalogues. First of all, and this is particularly relevant when doing cross-matching between archives and/or catalogues, there is a clear need for a common and standardized query language. This would allow users and search engines alike to make use of standard identifiers (UCDs) to access table contents. Another interesting issue is that of unit harmonization.Both are being addressed by the VOTable standard, as far as description of tabular data is concerned. What is still missing is an analogue for queries.

Since archives have, by design, a long 'memory', it is important to keep historical information on instruments, detectors, filters etc. Without those pieces of information, even a simple calibration of raw data might not be feasible. Just think about the difficulties of handling images for which filter transmission curves or detector characteristics are not known!

Another important issue is that of image quality. Ideally, as every physical measurement should be accompanied by its uncertainty, astronomical images should come with a quantification of their quality. This is an important indication of the suitability of a given image for a given analysis - and also for its use in a Virtual Observatory.

Clearly, the issue of image quality is closely coupled to that of the data-flow signature on the archived datasets. Users need to know whether a given image is, for instance, photometrically calibrated or not and, if calibrated, to which precision. It should always be possible to retrieve raw data and all the relevant calibration files from the archives and repeat all the reduction and calibration steps. Sometimes, this step is needed as our knowledge of instruments evolves with time. Unfortunately, most of the current archives do not yet provide image quality figures, not even a measure of their image seeing; let alone a description of the data-flow signatures.

Once resources are well described and their quality is well characterized, users still have to find them. And once they have been found, users still have to understand how they can access them and if they are of any interest. Resource discovery technologies are being developed in the industry world, especially in the context of Web Services applications. What is still unclear is whether emergent protocols like UDDI could be efficiently used in an astronomical context. A ranking system for resources is also essential. This system should be able to tell users whether a give resource has the characteristics relevant for a given analysis (spatial coverage, flux depth, statistical completeness etc.).

5 Links

In this section we present a list of internet links that were mentioned in our paper.

1. ASTROVIRTEL home page: `http://www.stecf.org/astrovirtel/`

2. Querator: `http://archive.eso.org/querator/`
3. ESO/ECF archive home page: `http://archive.eso.org/`
4. ST-ECF home page: `http://www.stecf.org/`
5. ESO home page: `http://www.eso.org/`
6. ESA home page: `http://www.esa.int/`
7. AVO: `http://www.euro-vo.org/`
8. NVO: `http://www.us-vo.org/`
9. IVOA: `http://www.ivoa.net/`
10. UCD at CDS: `http://vizier.u-strasbg.fr/doc/UCD.htx`
11. VOTable at CDS: `http://cdsweb.u-strasbg.fr/doc/VOTable/`
12. UDDI: `http://www.uddi.org/`

References

1. F. Pierfederici: 'Querator: an advanced multi-archive data mining tool'. In: *Proceedings of SPIE Vol. 4477*, 2001

Towards an AVO Interoperability Prototype

Mark Allen[1], Françoise Genova[1], Christophe Arviset[2], Sebastian Derriere[1], Pierre Didelon[3], Simon Garrington[4], Robert Mann[5], Alberto Micol[6], François Ochsenbein[1], Anita Richards[4], Guy Rixon[7], Alberto Salama[2], Andreas Wicenec[8], Christophe Benoit[9], and Jim Lewis[7]

[1] CDS, Observatoire astronomique de Strasbourg, UMR 7550, 11 rue de l'Université, F-67000 Strasbourg, France
[2] ESA
[3] SAP/Terapix
[4] University of Manchester, Jodrell Bank Observatory
[5] Royal Observatory Edinburgh
[6] ESO-ECF
[7] Cambridge University
[8] European Southern Observatories.
[9] OCA Nice

Abstract. A working interoperability prototype is an important goal for year 1 of the AVO interoperability Work Area. This prototype aims to federate a set of archives representative of a variety of techniques (space/ground, images/spectra, X-ray to radio wavelengths) into the CDS VizieR system. The target archives for federation are: VLT, NTT, EIS (ESO), HST/ECF, ISO, XMM (ESA), Wide field UK archives, Jodrell Bank, and Terapix. We describe the interoperability of some of the federated archives, highlighting new functionalities of the CDS tools. We also describe how case by case discussions with archive providers is helping to compile a set of recommendations for interoperability.

1 Introduction

The European *Astrophysical Virtual Observatory* (AVO) project [1] is a 3-year Phase A study in preparation for a future Phase B proposal to build the European Virtual Observatory. The projects comprises three work areas, "Science use cases and requirements", "Interoperability", and "Technology". Quinn [2] provides a description and status report project. This paper describes progress towards the goals of the interoperability work area.

An important first year milestone of the AVO project is to implement an interoperability prototype. The basic task of the interoperability prototype is to federate a set of key data archives into the CDS interoperability services. The aims for the prototype system are to:

- enable early science usage of Virtual Observatory like tools
- evaluate interoperability tools in terms of usability and function
- drive the specification and development of new functionalities
- provide a test bed for new and evolving standards (UCDs, VOTable)

We also aim to heighten awareness about interoperability among archive providers, and identify difficulties. Using experience obtained in the development of the prototype we are compiling a set of basic interoperability recommendations for data providers to assist in implementing their data within virtual observatory interoperability systems.

2 AVO Interoperability Prototype

The prototype is based on the existing interoperability tools at the CDS, namely the VizieR catalog browser, and the Aladin sky atlas image tool. VizieR [3] and Aladin [4] gather and maintain metadata about catalogs, data archives and image servers. Using this metadata, the interfaces provide access to catalogs, images and remote resources for browsing, downloading data and visualization of images and catalogs in a single environment.

Among the benefits of using the existing CDS systems are that they offer a working framework in which to test and develop high level metadata definitions and standards required for Virtual Observatories. Units and Uniform Content Descriptors [5] (UCDs) are already defined for all archives and catalogs in VizieR, providing a working testbed for development, complementary to parallel developments of distributed metadata systems and web-service type operations [6].

3 Target Archives

The target data archives for inclusion in the prototype consist of European data archives chosen to be representative of space and ground-based observations over a wide range of wavelength. The target archives include VLT, NTT, EIS (ESO), HST/ECF, ISO, XMM (ESA), Wide field UK archives, Jodrell Bank radio archives, and Terapix data.

The prototype system provides interoperability between these archives at the *data level*, meaning that coordinate systems of images from the different instruments and telescopes can be combined to create mosaics and overlays. This is done via the Aladin tool, which matches images by re-orientation and scaling. Recent enhancements facilitate display of radio images, and allow the possibility of making astrometric registration adjustments for image alignment based on common image features. Other improvements to Aladin, driven by the prototype include a facility to generate, display and manipulate contour plots of images, and to combine registered images into a RGB colour composite [7]. Some of these features are displayed in the Aladin interface shown in Fig. 1.

4 UCD Development

Significant future development is expected in the way metadata is recorded, manipulated and used. Emerging web service technologies describe topic maps,

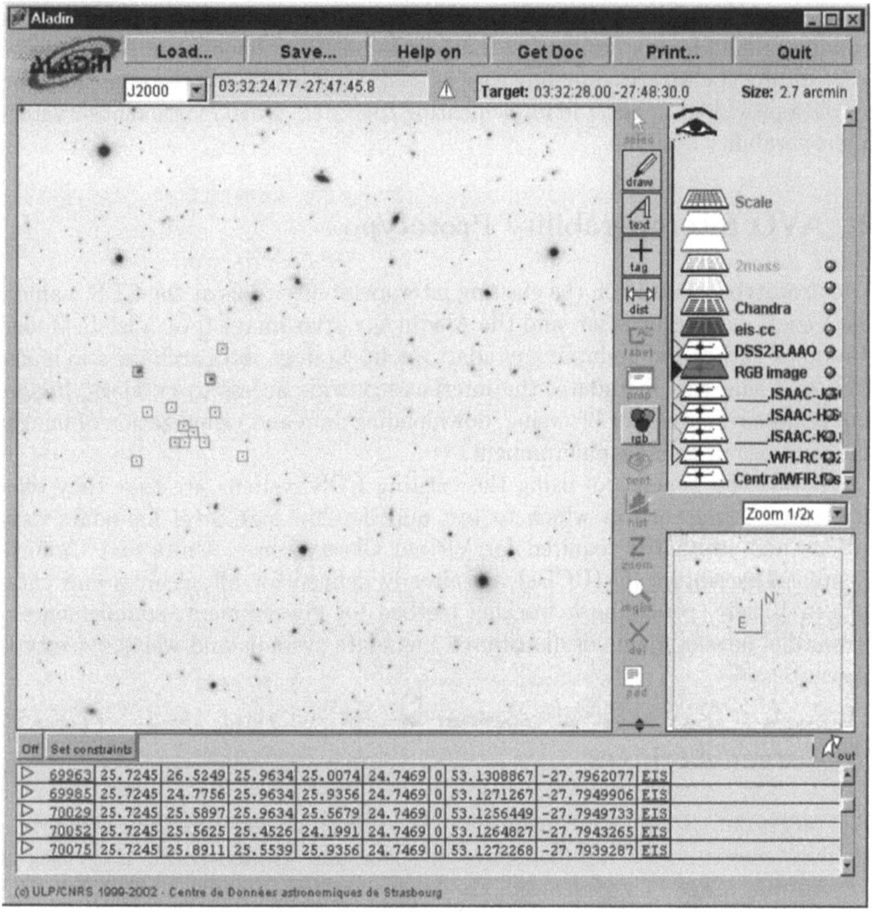

Fig. 1. Aladin Sky Atlas. This figure shows a number of the interoperability prototype target archive images and catalogs loaded into Aladin. The image is a colour composite of VLT ISAAC J, H and WFI-R frames of the central part of the Chandra Deep Field South overlaid with an EIS catalog. Also available in the image/catalog stack are locations of XMM and Chandra observations

ontologies and sophisticated schemes for embedding the meaning and function of data into the metadata. For the short term goals of the interoperability prototype we take a pragmatic approach and demonstrate the effectiveness of encoding semantic information into the CDS UCD system [5], using maps of UCDs onto physical quantities, and a framework for unit conversions.

A simple example of searching databases via wavelength has been implemented using the CDS UCDs that describe photometric wavebands. This has been developed to address common features of science scenario questions posed for Virtual Observatories, that often begin with: "Find B-band photometry ... in

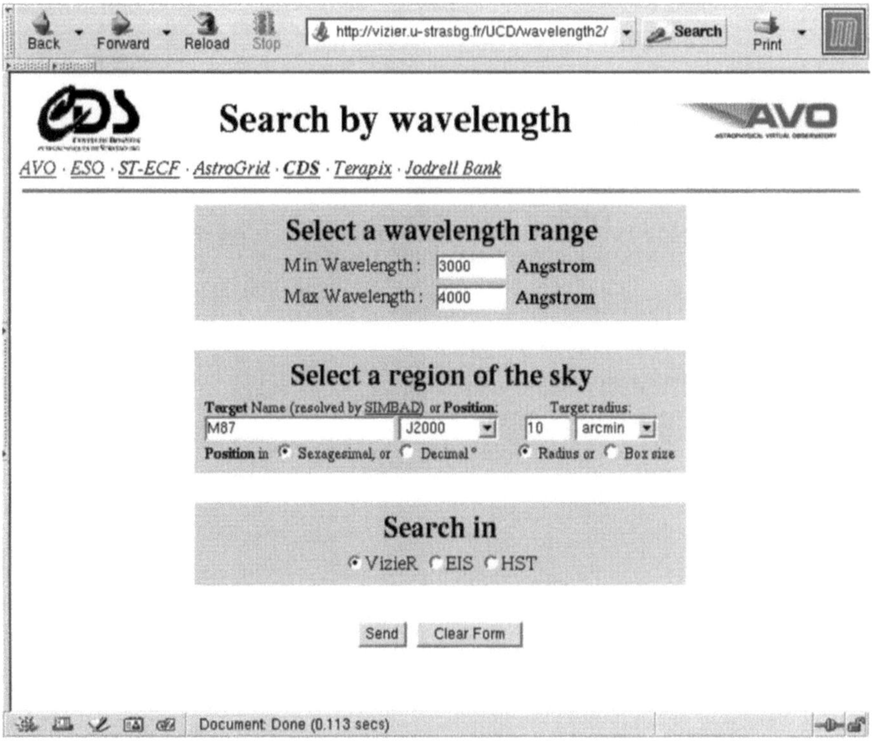

Fig. 2. VizieR Search by Wavelength facility allows searches of catalogs and archives by wavelength range, combined with an optional target name or region of the sky

a particular region of the sky", or "...of a particular type of object". The photometry UCDs (such as PHOT_JHN_V) have been mapped to a wavelength table which lists the wavelength coverage of the band. This map, in combination with a newly implemented *search by UCD* function in VizieR, allow searches on archives and catalogs using a wavelength range, with an option for also specifying an object name or region of the sky. Figure 2 shows the interface to this search, and Fig. 3 shows the result which gives access to tables and columns containing the selected UCDs. More details on the functions of UCDs are described in Derriere [5].

As part of the prototype, we also implement a framework for interoperability at the *units level* whereby measurements, listed in catalogs of extracted sources, can be identified, compared, and converted to the same units. This process is based on the units and UCDs embedded in the VizieR system. This is an important step towards combining quantitative information from different sources to make comparisons and table-joins. Figure 4 shows the test interface to the unit converter. Selected tables and catalogs can be compared against each other for columns with similar UCDs and convertible units.

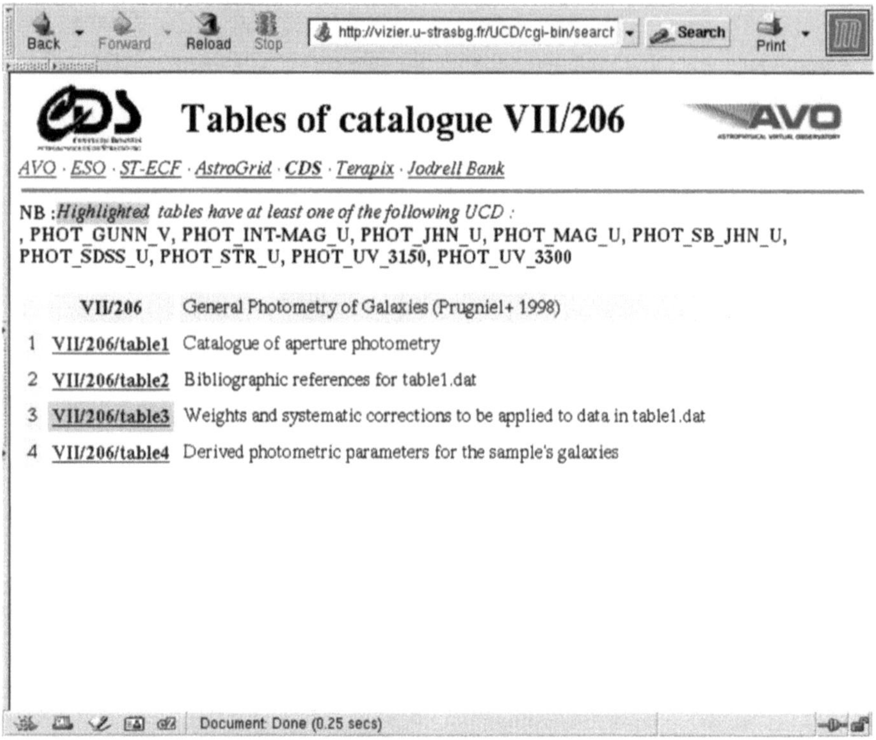

Fig. 3. Results window of searching by wavelength range. Note that the UCDs associated with the selected wavelength range are shown, and the tables matching those UCDs are highlighted

5 Practical Recommendations for Interoperability

Implementation of observatory observation log files into the interoperability system currently requires significant interaction with the archive holders. This process identifies the required metadata definitions to properly describe a remote resource. In some cases this process stimulates archive holders to review their archive exposure mechanisms. Lessons learned from archive implementation are being systematically reviewed to produce a set of interoperability recommendations to archive providers. The basic requirements for implementation are the availability of direct http access to archive imaged, and that astrometry information is supplied in the FITS World Coordinate System. For "living" archives, such as a growing observation log, a simple mechanism is required for updating the metadata. Emphasis is also placed on including uncertainties and errors on measurements where possible, and on the inclusion of auxillary information such as filter transmission curves. Data rights and access are also key questions to be addressed by archive holders prior to interoperability implementation.

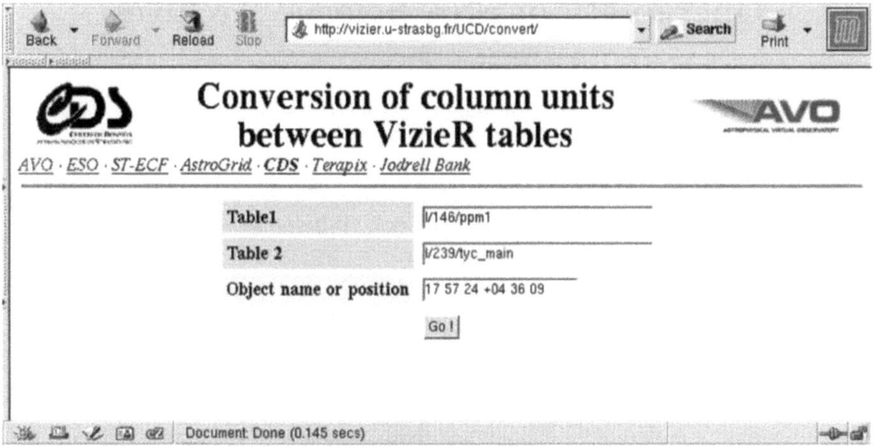

Fig. 4. Interface window for Unit conversion between VizieR tables

6 Conclusions

The AVO interoperability prototype is implementing a increasing set of European data archives into the CDS VizieR and Aladin system, making these data available for early science usage. The prototype provides a test bed for development of metadata descriptions and standards, and of new functionalities for working with multi-wavelength data. Early examples of searching by UCD and by wavelength are already available with test interfaces. The requirements of the prototype have driven the developments of new functionalities such as RGB and contour displays, and additional developments and customizations are also being driven by the requirements of the AVO Early Science Demonstration foreseen for early 2003.

References

1. `http://www.eso.org/avo`
2. P. Quinn (these proceedings)
3. F. Ochsenbein, P. Bauer, J. Marcout, Aston. Astrophys. Suppl 143, pp. 23–32, (2000)
4. F. Bonnarel, P. Fernique, O. Bienayme, D. Egret, F. Genova, M. Louys, F. Ochsenbein, M. Wenger, and J.G. Bartlett, Astron. Astrophys. Suppl 143, pp 33–40, (2000)
5. S. Derriere et al. (these proceedings)
6. A. Szalay, T. Budavári, T. Malik, J. Gray, A. Thakar, *Web Services for the Virtual Observatory*, Microsoft Research Technical Report MSR-TR-2002-85 (2002)
7. F. Bonnarel et al. (these proceedings)

Inter-Operability of ESA Science Archives

Christophe Arviset[1], Matteo Guainazzi[1], Alberto Salama[1], John Dowson[2], José Hernández[3], Pedro Osuna[4], and Aurèle Venet[5]

[1] ESA, European Space Agency, Research and Scientific Support Department, Science Operations and Data Systems Division, Villafranca del Castillo, P.O. Box 50727, 28080 Madrid, Spain
[2] VEGA, Madrid, Spain
[3] AURORA, Madrid, Spain
[4] SERCO, Madrid, Spain
[5] DEIMOS SPACE, Madrid, Spain

Abstract. ESA Science Archives for ISO and XMM-Newton have been developed by the Science Operations and Data Systems Division in Villafranca, Spain. By using an open 3-tier architecture (Data Products and Database, Business Logic, User Interface) together with Java and XML technology, inter-operability has been achieved from these archives to external archives (NED/ SIMBAD, ADS, IRAS). Furthermore, that has allowed external archives (CDS, ADS, IRSA, HEASARC) to directly access ISO and XMM-Newton data without going through their standard user interfaces.

1 IDA and XSA Quick Overview

1.1 ISO Data Archive

The European Space Agency's (ESA) Infrared Space Observatory (ISO) is an astronomical satellite that was operational between November 1995 and May 1998. It operated at wavelengths from 2.5 to 240 microns, in the infrared range of the electromagnetic spectrum. Exciting new results on the cool universe came in practically all astronomical fields from comets to cosmology, with detecting water throughout the cosmos being one of the highlights. The satellite essentially consists of: a large liquid-helium cryostat; a telescope with a 60-cm diameter primary mirror; four scientific instruments and the service module.

The ISO Data Archive (IDA) has been developed at the ESA ISO Data Centre at Villafranca, Spain. It opened to the community in December 1998 and offers a state-of-the art archive facility to access ISO data products and auxiliary files. It is accessible at `http://www.iso.vilspa.esa.es/ida/`. It contains all the ISO raw and fully processed, science and calibration data as well as all ancillary data (engineering, uplink and downlink data) for a total of about 400 GBytes stored on magnetic disks. Through a powerful and user friendly Java User Interface, over 1200 registered users have already downloaded the archive content more than five times since it was opened to the public. The IDA main characteristics can be summarized as:

- powerful and complex queries against the observations catalogue
- configurable results display, including product visualization tools

- customizable product retrieval via a shopping basket
- selection of product level for product retrieval via FTP
- inter-operability with other archives and applications

It has been designed in an open and flexible way, which allows configurability and re-usability for other archive projects.

1.2 XMM-Newton Science Archive

XMM-Newton, ESA's X-ray Multi-Mirror satellite, is the most powerful X-ray telescope ever placed in orbit. It was launched by an Ariane 5 rocket from ESA's spaceport in Kourou, French Guiana, on 10 December 1999. With its unprecedented sensitivity it observes the X-ray sky, helping to solve many cosmic mysteries, ranging from extremely violent and exotic processes, such as enigmatic black holes, to the formation of galaxies. XMM-Newton also observes celestial objects within our Solar System, such as comets and planets.

The XMM-Newton Science Archive (XSA) has been developed at the ESA XMM-Newton Science Operations Centre at Villafranca, Spain. It has been available to the scientific community since mid April 2002 at `http://xmm.vilspa.esa.es/xsa/`. It contains all the XMM-Newton raw and fully processed, science and calibration data as well as some ancillary data for an expected total of about 1 TBytes stored on magnetic disks. Based on the IDA architecture and code, it offers similar data query and retrieval (FTP and CDROM) facility through a user friendly User Interface. Proprietary data is accessible to the observation owner only while public data has unrestricted access.

Inter-operability of the XSA with other archives is currently being developed along the same lines as it is already with the IDA.

2 Open and Flexible 3-Tier Architecture

The IDA and the XSA were both built -by a common team- using the open 3-tier architecture described in Fig. 1. The main goal of this architecture is to separate the data from the presentation, which allows a more modular and flexible development.

As the data volume is not that big, data are saved on magnetic disks for fast access as a normal UNIX file system. From the data products, metadata is extracted and put in a Relational Data Base, SYBASE. Note that the data ingestion from the data producer and the metadata extraction are separate processes to allow new metadata data extraction when user requirements evolve.

The middle tier, also called the Business Logic, provides a transparent access to the data products and to the metadata. This key layer has been developed in Java and XML and resides on the archive server.

On the client side, several types of applications can be found. The standard IDA and XSA User Interface is a Java applet downloaded by the end user to access the archive content.

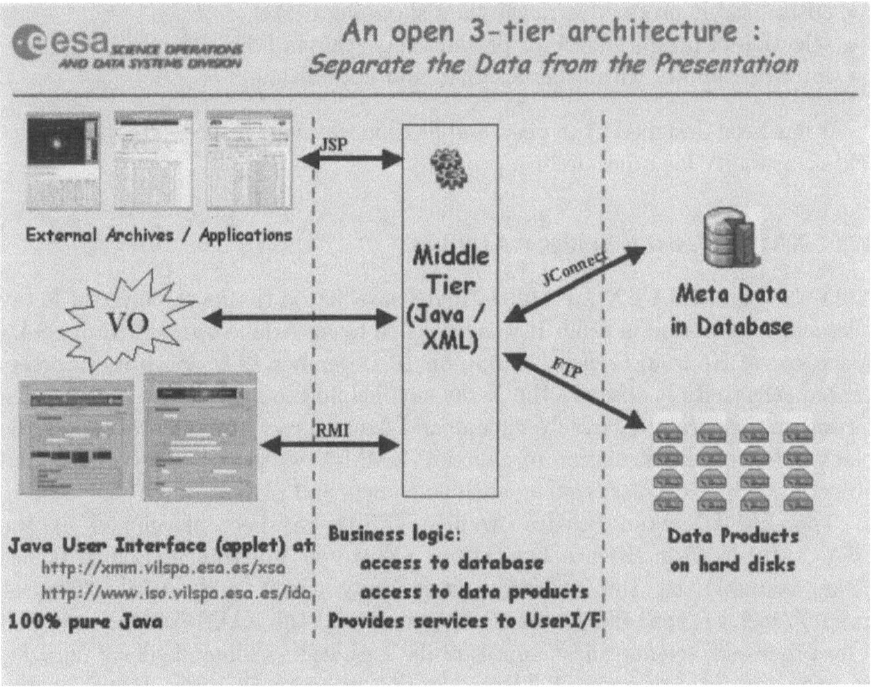

Fig. 1. IDA and XSA 3-tier Architecture

Remote applications and other archives can also have access to the data and the metadata, bypassing the standard User Interface, by speaking to the Business Logic that will provide them with the required services via Java Server Pages.

This architecture is especially powerful in the context of the Virtual Observatory initiatives worldwide where archives will all have to inter-operate together in a transparent manner for the end user.

3 From ESA Archives to External Archives

3.1 Name Resolution with NED or SIMBAD

On the IDA Query Panel, one can query against a target name (see Fig. 2). By entering a target name and choosing the name resolver (NED or SIMBAD), the IDA will make contact with the IPAC or CDS server to resolve the target name into coordinates and then search the ISO observations catalog against these coordinates. This is done completely transparently to the user.

For SIMBAD and NED, target name is resolved into coordinates, calling the CDS server at Strasbourg, France or the NED one at IPAC, USA respectively via a specific TCP/IP socket.

Same mechanisms are already in place for the XSA.

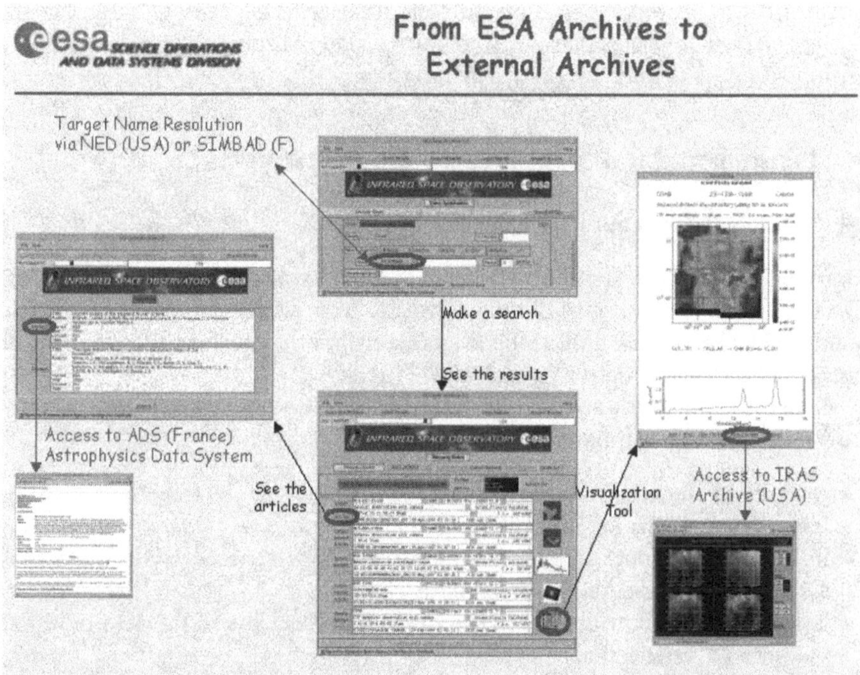

Fig. 2. Access from IDA to External Archives

3.2 Access to Electronic Articles

In cooperation with the NED project, all major astronomical journals are scanned for finding reference to ISO observations. Later on, the ISO Data Centre finds the link between the article and the used ISO observations. From the IDA Latest Results Panel, the button "Articles" indicates if there are known publications linked to the observation (see Fig. 2). By clicking on the button, one gets an extra window showing the Title, Authors, Journal, etc. By clicking the "Abstract" button, the applet will launch a browser window with the ADS WWW mirror at Strasbourg, France with the abstract of the article associated with the selected ISO observation. The call is made through a standard URL / cgi-bin script as defined by the ADS interface.

Same mechanism is being developed for the XSA.

3.3 Access to IRAS Data

From the IDA Latest Results Panel, one can see the small icons giving a quick overview of what the observations is about. By clicking on one of them, a bigger window is launched with the postcard giving more information on the ISO observation (see Fig. 2). By clicking the button "Access to IRAS", a browser window will open, from the InfraRed Science Archive (IRSA) WWW page located at

IPAC, USA. The window will contain the data covering the region of the sky of the selected ISO observation. The call is made through a standard URL / cgi-bin script as defined by the IRSA interface.

4 From External Archives to ESA Archives

4.1 General Concepts

Apart from using the standard IDA and XSA user interfaces, ISO and XMM-Newton products can also be obtained directly from external archives or applications. That allows such data to be available from other well established archive where multi-missions data can be found.

Generally, the direct access to ISO and XMM-Newton data, outside the standard user interface can be made in four steps:

- delivery of the archive observation / exposure log to the remote archive
- integration of this log into the remote archive
- link from the remote archive to the *Postcard Server*: access to the quick look GIF image representation of the data
- link from the remote archive to the *Product Server*: link to the data products via an FTP connection

4.2 The Observation / Exposure Log

The Observation log consists of a file (generally ASCII) containing the list of ISO or XMM-Newton observations with associated parameters (observation id, PI name, coordinates, time, quality flag, release date, ...). In the case of XMM-Newton, each observations contains several instrument exposures with also relevant specific parameters.

According to the requirements of the remote archive, the log can contain more or less parameters depending of the use they want to be make of it.

The standard format is ASCII (fixed width, tabular or character separated). This log is just a view of the content of the IDA or XSA database and can be easily exported into other formats such as XML, HTML, VOTable, etc... that will make it easier to ingest in remote archive databases.

The delivery mean can be via electronic mail or also with automatic updates available on our FTP server.

4.3 ISO and XMM-Newton Postcard Server

Through calling a URL / Java Server Page (JSP) containing the ISO or XMM-Newton observation identifier, the Postcard Server returns the ISO or XMM-Newton postcard (GIF image) of this observation and ancillary quality information embedded into an HTML page. Within this HTML page, links to relevant documentation as well as the data quality flag can be found.

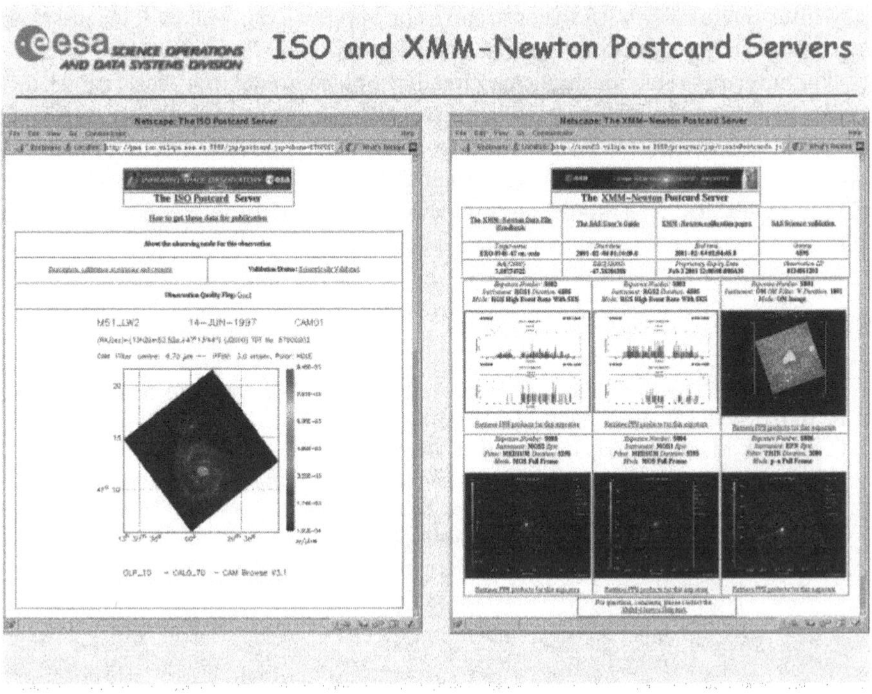

Fig. 3. ISO and XMM-Newton Postcard Servers

This service is already available for ISO in CDS, IRSA, ADS and HEASARC as shown in Fig. 3.

Same service is being developed for XMM-Newton as shown in and will be released soon in similar remote archives.

4.4 ISO and XMM-Newton Product Server

The Product Server uses similar concepts as the Postcard Server. Nonetheless, instead of returning an HTML page, it returns directly the ISO or XMM-Newton data products by initiating a FTP session.

The Postcard Server helps the users to decide which dataset he or she wants to retrieve and the Product Server allows him or her to immediately download them onto his or her machine.

Both ISO and XMM-Newton Product Servers have been available since fall 2002 from the remote archives where the Postcard Server was already in place.

5 Conclusions

The ESA science archives, in particular the ISO Data Archive and the XMM-Newton Science Archive are based on an open 3-tier architecture which allows easy inter-operability with other archives or applications. This will ensure their easy and fast integration in the VO architecture.

The inter-operable mechanisms already in place are flexible, fast, direct and secure both from the ESA archives to external archives and in the other direction.

Currently, dedicated interfaces have been defined with the remote archives inter-operating with the IDA and the XSA. But having common standards (XML, VOTables, ...) would help integration of services in the context of the VO.

References

1. C. Arviset, T. Prusti: 'The ISO Data Archive'. In: *ESA Bulletin, Number 98*, ed. B. Battrick (ESA, June 1999) pp. 133–140
2. C. Arviset, M. Guainazzi, J. Dowson, J. Hernández, P. Osuna, A. Venet: 'The XMM-Newton Science Archive'. In: *New Visions of the X-ray Universe in the XMM-Newton and Chandra era Conference at ESTEC, The Netherlands, 26-30 November 2001*, ed. F. Jansen (ESA SP-488, August 2002)

The AstroGrid Pilot Programme

R.G.Mann[1], A.C. Davenhall[1], C.G. Page[2], M.G. Watson[2], A.M.S. Richards[3], S.T. Garrington[3], A.J. Holloway[3], R.D. Bentley[4], C.D. Pike[5], C.H. Perry[5], and R. Stamper[5]

[1] Institute for Astronomy, University of Edinburgh, Edinburgh, UK
[2] Department of Physics and Astronomy, University of Leicester, Leicester, UK
[3] Jodrell Bank Observatory, University of Manchester, Macclesfield, UK
[4] Mullard Space Science Laboratory, University College London, Dorking, UK
[5] Rutherford Appleton Laboratory, Chilton, UK

Abstract. We describe the AstroGrid Pilot Programme, a set of testbeds conducted during AstroGrid's one-year Phase A study, and intended to complement the science requirements analysis and technology evaluations comprising the remainder of the Phase A programme. We selected a pilot to address an issue of particular relevance to each of AstroGrid's five broad scientific areas, to allow empirical study of many aspects of the general database federation problem and to engage, from the outset, all parts of our user community. The lessons learnt during the course of the Pilot Programme will help define the direction of AstroGrid's Phase B implementation.

1 Introduction

The AstroGrid programme (e.g. [1]) consists of a one-year Phase A study and a two-year Phase B implementation, leading to the delivery of a working data-grid for UK astronomy in the autumn of 2004. The Pilot Programme was designed to complement the Phase A technology assessment workpackages (e.g. [2], [3]). While the latter were intended to evaluate some of the technologies likely to be deployed to help meet AstroGrid's science requirements [4], the Pilot Programme was designed to elaborate those requirements, by developing prototype systems to deliver small portions of AstroGrid's desired functionality. An important goal from the outset was that all pilots should proceed to the point of delivering software that could be used by test users to do real science, but, equally, it was stressed that no undue effort should be expended in producing pretty user interfaces, etc, as the lessons to be learnt concerned the delivery of new functionality, not its presentation to users.

A set of five pilot topics was selected, one each for the five broad fields covered by AstroGrid, namely

- optical/near-IR astronomy: *Large object catalogues*
- X-ray astronomy: *Association methods*
- radio astronomy: *Fourier data*
- solar physics: *Data selection from summary information*
- solar/terrestrial physics (STP): *Time-series data*

The five pilots are described in turn below: a much fuller report on the Pilot Programme can be found on the AstroGrid WWW site [5].

2 The Optical/Near-IR Pilot – *Large Object Catalogues*

The optical/near-IR object catalogues to be included in the AstroGrid system greatly exceed all its other datasets in size, so there was a clear need for a pilot that addressed the practical problems of federating large object catalogues. Two approaches to this issue were planned, for eventual comparison. Firstly, data from the SuperCOSMOS Sky Survey(SSS [6]) and Sloan Digital Sky Survey (SDSS [7]) data would be federated, within the region of the latter's Early Data Release (EDR), and using a version of the SDSS science archive software (called SX [8]), modified for use with the SSS data model. Secondly, catalogues of objects derived from INT imaging with the Wide Field Camera (optical) and CIRSI (near-infrared) would be federated by making them both accessible via the VizieR [9]) system developed by the CDS: in the event, the VizieR evaluation was performed by the Astrophysical Virtual Observatory (AVO [10]) *Interoperability* Work Area (to be reported elsewhere). Delays with the installation of the SDSS-EDR database in Edinburgh, coupled with the SDSS consortium's decision to switch from SX (based on the object-oriented database Objectivity/DB [11]) to the SkyServer [12] system (based on Microsoft's SQL Server[13] relational database), led to the suspension of this pilot midway through Phase A. However, with the advent of a new SSS database, also built on SQL Server, work has restarted, and it is intended that the original goals will all be met by the end of December 2002, when the Phase A extension ends.

3 The X-Ray Pilot – *Association Methods*

Astronomical data has a natural indexing - spatial location in the sky - aiding the association of entries in different databases identified as being observations of the same astronomical object. Association by spatial proximity alone is, however, often inadequate, and a probabilistic method, encoding prior knowledge of the properties of the populations under study, must be employed. Astronomers already use several such probabilistic prescriptions, but their extension to the VO era poses a challenge. Not only does the size of the datasets to be made available by the VO mean that associations could be sought between databases containing millions of objects each, raising concerns about the scalability of the association techniques commonly used, but there are also issues relating to how the method employed to make a set of associations can be recorded so that a later user can judge whether s/he can employ them with confidence, rather than recomputing them all anew. This pilot was designed to start addressing these important issues, through seeking optical counterparts for sources in the first XMM-Newton catalogue, produced by the XMM-Newton Survey Science Centre [14]. This was delayed, so the X-ray pilot was suspended, too, but not before much useful preparatory work had been performed – such as the development of the PCODE algorithm [2] for performing the "fuzzy join" spatial matches which are likely to be first step in any VO association procedure – and work on this pilot can be easily restarted once the XMM-Newton source catalogue is available, leading to its planned completion by the end of December 2002.

4 The Radio Pilot – *Fourier Data*

High resolution radio interferometer arrays produce sets of "visibility data", which are sparse samples of the Fourier transform of the radio sky and which can be processed, via non-linear deconvolution, to produce images in a number of different ways, depending on the astronomical requirements. It is therefore efficient and productive to maintain access to the data in the Fourier plane and produce data products on demand, according to the user's specifications.

This pilot was designed to produce a prototype of such a system, using visibility data from the Hubble Deep Field, accessed by a cgi wrapper to AIPS [15], enabling users to extract maps at any position within the field of view from calibrated visibility data. For the HDF, these data are several GB in size, making it impractical to supply them to off-site astronomers even if they could use AIPS locally; thus a Virtual Observatory service like the one developed here is not merely a convenience but a necessity for such data sets. Building on this work, the MERLIN Archive [16] now supports simple queries (returning text and ready-made plots and FITS images) either via its web page or via CDS. User feedback on this prototype system was positive, confirming that it produces (in an automated fashion) data of the same quality as that produced by an expert reducing the same data, although it was noted that the system ran slower than might be expected. Several performance bottlenecks have been identified, and possible enhancements are currently being investigated.

The remaining part of this pilot has seen work on deriving interferometry metadata standards proceed in conjunction with the AVO Interoperability Work Area. A library of terms describing radio interferometry observations has been drawn up and these have been translated into CDS Unified Column Descriptors (UCDs [17]). Work is ongoing to expand these where they are not at present sufficient: consultations are underway with experts on higher frequency Fourier data to expand the scope of these metadata to ensure that it covers the requirements of interferometry in general, not just those of radio astronomy.

5 The Solar Pilot – *Data Selection from Summary Information*

In solar physics, it is not typical for all data from a particular experiment to be automatically pipeline-reduced all the way to science-quality products. What is more common is that summary data products are made available, together with catalogues listing observational parameters, which the user may then interrogate, to select interesting datasets for further study. The main goal of the solar pilot was to improve this selection mechanism, as this is currently a major stumbling block to doing science with the data.

The central design issue for this pilot was the definition of the minimal set of parameters required for solar observing catalogues. This was constructed by collating the set of parameters that a user might need to query in order to select a particular dataset, but bearing in mind that the set defined must work within the context of the standard SolarSoft [18] software package ubiquitous within the solar physics community. Another design issue was the format for storing the solar observing catalogue data. Currently, these data are usually held within an IDL [19] database, for ease of manipulation using SolarSoft (which is written in IDL), but there is a desire to remove the reliance on IDL, to ease the integration of solar physics software and Grid middleware. Another possibility would be to store these data in an XML repository, such as Xindice [20], possibly using VOTable [21] documents for each entry. This would require the addition of further UCDs relevant to solar physics: this avenue is being explored within the context of EGSO [22], and it was decided that the time constraints on the delivery of working prototype software within this pilot would necessitate the use of an IDL-based system for this pilot. Similar concerns led to the decision to use only data stored on-line at MSSL, so the prototype system offered users access to Yohkoh-SXT [23] and TRACE [24] data selected using GOES [25] solar X-ray activity data.

User comments towards the solar pilot User Interface were generally favourable, but people wanted more than had been included in the Pilot once they started to realize what might be possible. The variety of suggestions confirmed the initial suspicion that a single User Interface would never suit all users and that some means of configuring the interface on a user-by-user basis may be required. It was also clear that commonly-used data must be prepared ahead of time. For example, by producing versions of the Yohkoh-SXT observing catalogues that include pointing information, so this does not have to added at run-time. It also became clear that there is a need for a tool to predict the output size of the requested dataset, to ensure that unreasonable requests can be intercepted and refinements requested. Another welcome addition would be a way of storing a request so that the user can return to it at a later time. This allows the user to re-run the request (e.g. if problems were encountered), add additional data, or refine the existing selection; during the testing phase, it was found very irritating to have to re-enter complex requests into the UI repeatedly.

6 The STP Pilot – *Time-Series Data*

Solar Terrestrial Physics (STP) data sets include a large number of *in situ*, multi-point and remote sensing measurements made across a wide range of scales in both time and space, and this pilot was designed to study the metadata needed for their federation. Its short timescale necessitated that the STP pilot make fullest use possible of existing software, notably the Solar Terrestrial Physics Data Facility (STPDF [26]) system, and use data sources selected from those of the Cluster Coordinated Data Handling Facility (CHDF [27]) and the World Data Centre for Solar-Terrestrial Physics (WDC [28]), already on line at RAL and comprising about 35 million time-series records in total, of several types.

The development of the pilot's metadata translation layer started with an assessment of a number of existing XML formats that might be used. A report on this may be found on AstroGrid WWW site [29], and it may be summarised as follows: XSIL [30] is too simple; XDF [31] is too complicated; CDFML [32] is too STP-specific; and VOTable looks usable. Despite the fact that VOTable was considered usable for STP work, it was decided to use an *ad hoc* XML format for the pilot work, for simplicity; both to avoid having to implement the full VOTable DTD/schema and to avoid having to define the additional UCDs that would have been needed.

Queries to the pilot system are constructed using a simple UI which allows selection of system or user datasets (i.e. the results of previous operations, thereby allowing compound queries to be built). The UI is dynamically updated to show available fields, and the number of records for each selected dataset. Four basic operations are supported: select/output fields; query data set; find/select time interval; and time series join (nearest neighbour). Finally, a Quicklook UI (based on the CHDF UI) was produced to view the selected dataset(s), allowing the user to zoom in on a time range subset of the data, select parameters to plot and obtain an ASCII dump of the data.

7 Summary and Conclusions

Many useful lessons have been learnt from the AstroGrid Pilot Programme. At the most general level, test users responded positively to the additional functionality they were offered, but quickly wanted the ability to do more. This reassuringly confirms that the VO enterprise is worthwhile, but also that it will be difficult to meet expectations, and that considerable flexibility will have to be designed into VO systems to help them meet the range of user requirements. A number of more specific results emerged, too, for example that the VOTable prescription for presenting tabular data in XML may well be useful well beyond its originally intended domain, if appropriate UCDs (e.g. for interferometric, solar physics and STP data) can be defined. It was also clear that the VO must provide the means of estimating the resource implications (e.g. result dataset volume, time taken to run, etc) of proposed operations, lest users grind the system to a halt and/or generate much more data than they can handle.

A questionnaire circulated by the teams running the STP and solar pilot, in conjunction with ESA's SpaceGRID [33] initiative, and addressed to a wide cross-section of the international solar system research community, produced some quite explicit performance requirements, for example: the system should provide feedback on an action with 30s; simple, online tasks should be completed within an average time of a minute; and complex, offline tasks should be completed within an average time of 24 hours. Interestingly, this survey also identified some Intellectual Property Rights issues not discussed much within the VO community to date, as some respondents thought that there should be a possibility of keeping workflows and query results private within the VO.

Perhaps the most valuable lesson learnt from the AstroGrid Pilot Programme is the importance of keeping (at least some) users closely engaged in the VO development process. New technologies, and the vast wealth of astronomical data now available, mean that the possible directions that the VO can take greatly exceed what can possibly be delivered, given the finite funding for the various VO projects, and it is essential that the course of VO development is decided by what users most need to do their science and not what the technology can deliver most readily.

References

1. A. Lawrence, these proceedings
2. C.G. Page, these proceedings
3. G.T. Rixon, these proceedings
4. N.A. Walton, these proceedings
5. AstroGrid Pilot Programme Final Report: `http://wiki.astrogrid.org/bin/view/Astrogrid/WPA5FinalReport`
6. SuperCOSMOS Sky Survey: `http://www-wfau.roe.ac.uk/sss`
7. Sloan Digital Sky Survey: `http://www.sdss.org`
8. A.R. Thakar, P.Z. Kunszt, A.S. Szalay: 'Case study of handling scientific queries on very large datasets: the SDSS Science Archive'. In: *Mining the Sky, MPA/ESO/MPE workshop at Garching, Germany, July 31–August 4, 2000*, ed. by A.J. Banday, S. Zaroubi, M. Bartelmann (Springer, Heidelberg 2001) pp. 624–630
9. VizieR: `http://vizier.u-strasbg.fr/viz-bin/VizieR`
10. Astrophysical Virtual Observatory: `http://www.eso.org/avo`
11. Objectivity/DB: `http://www.objectivity.com`
12. SkyServer: `http://skyserver.fnal.gov`
13. Microsoft SQL Server: `http://www.microsoft.com/sql`
14. XMM-Newton Survey Science Centre: `http://xmmssc-www.star.le.ac.uk/`
15. Astronomical Image Processing System: `http://www.aoc.nrao.edu/aips`
16. MERLIN Archive: `http://www.merlin.ac.uk/archive`
17. Unified Column Descriptors: `http://cdsweb.u-strasbg.fr/doc/UCD.htx`
18. SolarSoft: `http://surfwww.mssl.ucl.ac.uk/surf/surf_software.html`
19. Interactive Data Language: `http://www.rsinc.com/idl`
20. Xindice: `http://xml.apache.org/xindice`
21. VOTable: `http://cdsweb.u-strasbg.fr/doc/VOTable`
22. European Grid of Solar Observations: `http://www.mssl.ucl.ac.uk/grid/egso`
23. Yohkoh: `http://www.lmsal.com/SXT/`
24. TRACE: `http://vestige.lmsal.com/TRACE/`
25. GOES: `http://www.goes.noaa.gov`
26. Solar Terrestrial Physics Data Facility: `http://www.ssd.rl.ac.uk/stpdf`
27. Cluster Coordinated Data Handling Facility: `http://www.cluster.rl.ac.uk/cdhf.htm`
28. World Data Centre for Solar-Terrestrial Physics: `http://wdcc1.bnsc.rl.ac.uk`
29. Report on "XML for STP data": `http://wiki.astrogrid.org/pub/Astrogrid/PilotDocs/agstp-0002.pdf`
30. Extensible Scientific Interchange Language (XSIL): `http://www.cacr.caltech.edu/SDA/xsil`
31. eXtensible Data Format (XDF): `http://xml.gsfc.nasa.gov/XDF/XDF_home.html`
32. CDF Markup Language (CDFML): `http://nssdc.gsfc.nasa.gov/cdf/html/cdf_xml.html`
33. SpaceGRID: `http://spacegrid.esa.int`

Making Ground-Based Optical/Infrared Imaging Surveys VO-Compatible

Luiz da Costa

European Southern Observatory, Karl-Schwarzschild Str 2, Garching bei München, D-85748, Germany

Abstract. To cope with the rapidly increasing volume and complexity of data being generated by ESO's public optical/infrared imaging surveys a high-performance, integrated, end-to-end survey software system has been developed. Its main aim is to provide a robust framework to efficiently transform raw data into science-grade survey products. The system consists of several tasks wrapped together into an integrated framework. These tasks include: the un-supervised reduction of optical/infrared images generated by different imagers, the astrometric and photometric calibration of the data, the creation of image stacks and mosaics, the preparation of catalogs and the selection of different targets of potential interest for spectroscopic follow-ups. Since the data is meant to be public, the system also provides an extensive description of the products and the required information for users to assess their quality. While originally designed for handling survey data, the system can also be used as a general front-end to ESO's raw data archive, and as such serve as a site-specific interface to the general VO infra-structure. In this contribution the system is briefly described.

1 Introduction

In recent years the number of large digital sky surveys has dramatically increased. This increase has been driven by the advent of large format CCD arrays, dedicated imaging telescopes and the need to supply data sets to 8-m class telescopes, and it has led to an unprecedented growth in the rate data are being accumulated in digital archives. Furthermore, in addition to survey data, large quantities of data gathered by private programs, publicly available after a finite proprietary period, are also being stored in on-line archives kept by most major astronomical observatories.

These recent developments have created a glut of data and currently the main challenge is on how to cope with the large increase in data volume available and allow the scientific exploitation of the ever growing number of multi-wavelength, multi-resolution federated archives. In particular, in order to fully exploit the wealth of information that these archives provide it is imperative to develop a system enabling individual users to efficiently transform raw data into science-grade products as well as more advanced analysis tools allowing one to collect and combine information provided by different archives. This is the purported goal of the ongoing Virtual Observatory (VO) effort which aims at providing an integrated framework for data reduction and analysis, introducing new technologies to efficiently handle large volumes of data and to interconnect different archives and databases.

It is important to emphasize that the problem of building the VO infra-structure is, in many respects, akin to that faced by individual teams conducting large surveys for public use. First, because a survey software system developed for a particular instrument or set of instruments can eventually be interfaced to the VO infra-structure as long as they are "VO-compatible", fulfilling some basic requirements such as providing a full description of the products they generate and quality control parameters from which users can assess their quality and suitability for different applications. Second, because survey systems need VO-like tools to carry out data quality checks of the science-grade products. Finally, because archival data will rapidly overtake those currently available in data centers, turning site- or survey-specific systems into the main content providers of the VO. Therefore, considerable attention should be paid to the products generated by these systems if one is to ensure that only well-described data are used in VO-based research. All of this means that there should be a close collaboration between groups building survey system and VO software.

As a concrete example of a broadly-speaking "VO-compatible" system, in this contribution the survey software system developed by the ESO Imaging Survey (EIS) project is presented. It is the result of the experience accumulated over the past few years from the ongoing series of ESO-coordinated public optical/infrared imaging surveys. Even though moderate in size, these surveys have served to proto-type a broad cross-section of scientific goals and observing strategies, Equally important have been the lessons learned from several data releases and the feedback from users. All these have been essential in defining the design and the required functionalities of the system.

In section 2 the EIS project is briefly described, emphasizing the requirements set by its public and long-term nature which differs considerably from large but private programs. An overview of the EIS survey software system is presented in section 3, followed in section 4 by a short description of the individual processes. The main characteristics of the system and its possible use in the VO paradigm are summarized in section 5.

2 Real and Virtual Surveys

The ESO Imaging Survey (EIS) project started in 1997 [6]. Its primary goal was to conduct public, relatively modest, optical/infrared imaging surveys and provide suitable data sets to the ESO community at large before the start of regular operation of the VLT. The project was also seen as a good opportunity to garner experience for more ambitious surveys to be carried out using dedicated imaging telescopes planned for La Silla and Paranal (*e.g.*, WFI@2.2m MPG/ESO telescope, VST, VISTA). Most of the original work was conducted using available software (*e.g.*, IRAF, Eclipse), which were assembled together using simple scripting languages. Besides the observations and data reduction, the project also required the delivery of a host of survey products which included astrometrically and photometrically calibrated pixels maps, single and multi-passband catalogs, image and catalog mosaics, and lists of candidate clusters of galaxies and of color-selected targets.

The continuing increase in volume and complexity of the surveys made the development of a proper end-to-end survey system necessary, which involved: 1) the development of an administrative wrapper, written in Python, to control processes, and manage data and information; 2) the design and implementation of user-friendly graphic user interfaces (GUIs) using the Tcl/Tk library, to provide access to the various processes, to the data and to a suite of tools for examining products and system generated logs; 3) the integration of a supporting database (Sybase); 4) the development and implementation of a new C-based image processing system to reduce optical/infrared images in a common environment (MVM library); 5) the consolidation of procedures to produce and assess the quality of science-grade products (images and catalogs); 6) the consolidation of administration tools to monitor the observations, pipeline and surveys; 7) the development and implementation of a new XML-based Web service. The end-product of this development is a user-friendly, high-throughput, fully integrated and configurable system, capable of reducing optical/infrared images in a unsupervised 24/7 operation cycle. It is also capable of creating image stacks and mosaics, extracting catalogs, producing color catalogs, classifying objects based on their spectral properties and producing comprehensive logs describing all the processing steps, input parameters and quality checks carried out automatically by the system.

Since the system is interfaced to the ESO Science Archive, handles nearly all the imagers currently available at La Silla and Paranal, and by and large fulfils the requirements of being "VO-compatible", it can be used not only to reduce survey data but also any other data set, as long as these follow standard observing strategies. Therefore, the EIS survey system can, with a proper interface, be also used to conduct archive-based virtual surveys, defined from high-level queries to ESO's Science Archive. This is a core functionality envisioned for the VO infra-structure.

3 The EIS Survey System: Overview

The end-to-end EIS survey system consists of nine distinct modules, described in more detail below, each representing a possible entry point into the pipeline framework. These modules can be assembled into two distinct groups.

The first, representing the front-end of the system, is responsible for requesting data from the archive, processing images, and astrometrically and photometrically calibrating the derived final images on a nightly basis. These various processes can be executed interactively one-by-one or automatically. Furthermore, in regular operation, these modules are automatically executed in sequence with one process triggering the other. After reduction, the final products are transferred to an image repository and the raw data are deleted from disk. For survey operations this sequence of processes is triggered by the so-called archive-scanner, which constantly searches the ESO Science Archive for survey data.

The second group of modules forms the back-end of the system from where the results of different observations are combined into final image stacks or mosaics, and from where catalogs are extracted, targets selected and survey products released.

To each module corresponds a graphic user interface (GUI) which allows the user to interact with the system in the interactive mode, or to monitor the progress of the processes in the automatic mode. The individual panels can be called from a single widget, displayed at log in. The widget not only provides the access to the various panels but also reports the CVS version of the system being used and the date the system was last updated. In automatic mode, at the end of each process the corresponding panel is automatically iconized, while the next in the sequence opens up, thus preventing overcrowding of the workspace. When a new panel is launched, temporary disk directories are created to store all the files generated by the process. These are automatically deleted when the panel is closed.

Each module has an interface to the system's supporting database (DB) and to a data access layer (DAL) which is fed by a search engine. The latter supports generic queries to locate the data suitable to the specific process being considered. The supported queries are combinations of survey, instrument, filter/passband and sky region (pre-defined in the case of surveys). The results of a given query provide a list of entries (*e.g.*, runs, images, catalogs) followed by information describing them, from where the user may select any number of entries. For entries with more than one version the user may select either the most recent or a default version which can also be set by the user.

Figure 1 shows the image pipeline panel. The figure illustrates the general layout of the panels, as adopted throughout the survey system, with small variations depending on the specific process. The top part of the panel is split into three sections. The first, lists all the sub-processes that can be executed. It also allows the user to set the mode of operation (automatic/interactive) and, in some cases, the type of process to be executed (*e.g.*, catalog production). In the banner of the panel one finds the name of the process, the execution mode, the revision of the code and the user.

From the second section one can access the configuration file and the search-engine/DAL combination, described above. The configuration is presented as an HTML form consisting of a combination of field boxes, radial and pull-down buttons depending on the nature of the information to be provided. The configuration file displayed depends on the mode of operation selected. In interactive mode only parameters referring to the specific process being called are shown. In automatic mode, the configuration of all processes being executed in sequence are shown in distinct, superposed HTML forms. In the banner of the browser, information about the configuration file is displayed including user name, creation date and the type (*e.g.*, last used, user's default, system's default). When a process is completed and saved, the configuration file used is ingested into the process table, and included into the process log, thus providing a link be-

Fig. 1. Example of the GUI used throughout the EIS system. The one shown is that of the image pipeline panel

tween process, configuration, input data and product, which is at the core of the versioning mechanism of the system.

Finally, on each panel a set of keys are available under the administration section. With the exception of that labelled panel tools, all others are common to all panels and provide short-cuts to a variety of administration tools for the pipeline, survey, Web and database. Under panel tools there is a large variety of tools specific to each panel. The last key, labelled save is used at the end of the process to ingest the required information into the EIS DB and move final products to appropriate directories.

The panels also include a TTY, where some of the more relevant information about the process being executed is reported. In addition, on top of the TTY one has: a status button, reporting the name of the sub-process being executed; a progress bar, reporting the progress of the sub-process; a data rate meter reporting the mean or the instantaneous data rate, whenever possible; and a clock which reports the elapsed time for each sub-process. In the process log both the mean data rate and time fraction spent on each sub-process is reported so as to enable the monitoring of the performance of the system over time.

The lower part of the panel consists of listboxes where the results of most sub-processes of the panel are listed. Next to each listbox there are keys that call tasks that allow the user to examine intermediate and final products in various ways depending on their nature. These keys are associated with tasks to display images and catalogs, to show other listboxes providing more details about each entry, and to convert XMLs into HTMLs and display process and product logs. In the example shown in Fig. 1, there are four listboxes listing different runs (upper-left) and raw images (upper-right), groups (lower-left) and RBs (lower-right) for the selected run, respectively.

4 The EIS Survey System: Individual Modules

4.1 Observations

From this panel information required to fully describe a new survey or the continuation of an ongoing survey is entered into the system and used as input to other tasks required in the preparation of observations and to compute quantities required by the data reduction process. For instance, to define a new survey the user is prompted for a program-id, number of strategies, number of regions and area to be covered, filters to be used, desired limiting magnitudes and estimated integration times (set either by the user or using the exposure time calculator for the specific instrument). The id assigned is then used to automatically search ESO's database of accepted programs and retrieve additional information about the observations such as the mode of observations (service, visitor) and allocated telescope time (nights, hours), which are entered into the DB. These data are critical for administration purposes, including the creation of Web pages. The use of automatic procedures minimizes the possible occurrence of typos leading to mismatches between the contents of ESO's and EIS' databases and prevents wrong information from being published on the Web.

The information entered by the user is then used to: define mosaics of pointings, if required; extract DSS cutouts; create observation blocks (OBs) and finding charts as required by the ESO's phase-2 proposal preparation (P2PP); extract lists of stars from astrometric reference catalogs (*e.g.*, USNO, GSC2); extract lists of objects from all-sky surveys (*e.g.*, 2MASS); compute model predictions of stellar number counts, color-color, color-magnitude and color distributions for the same filters and regions as the survey. This information is stored in the EIS DB and in disk repositories managed by the system for future use. The prepared OBs are also ingested into the EIS DB, serving as an internal OB repository which is used to update the status and quality of each observed (reduced) OB and to prepare new observations. The status of an OB is updated automatically by the system following the completion of an observing run and later after reduction.

The process associated to this panel also triggers the creation of HTML pages which are published in the external Web and rely exclusively on the contents of the EIS DB. These pages are then regularly and automatically updated by the archive-scanner mentioned above, thereby minimizing the time lag between observations and update of the Web, as required for public surveys. The task also creates run and night summaries as well as host of other information describing the observations and the progress of ongoing surveys.

4.2 Data Request

The data request panel has two main functions: one to launch and monitor the archive-scanner and the other to retrieve raw data from the ESO's Archive. The archive-scanner is used to locate exposures according to a set of conditions based on dates, instrument and program-ids. Its most common use is to detect the ingestion into the ESO Science Archive of new survey data, based on program-ids read from a table in the EIS DB. In normal operation, the archive-scanner runs constantly in background and once new images are located it triggers the ingestion of information about the exposures and OBs into the EIS DB. Depending on configurable parameters it can also trigger the retrieval of the new data automatically, as is the case for normal survey operation.

There are several possible ways of retrieving data from the archive. Currently, the system supports the following type of requests: survey runs; a survey or user-specified sky region; and more generically by a user-specified program-id. In the first case, the user can specify the survey and instrument by using the search-engine/DAL mechanism. The system not only identifies and requests all the relevant science frames but also all calibration or standard star frames for the relevant nights, instrument, mode, and filters, regardless of their program-id. This ensures that all calibration frames prepared by the telescope team are available during reduction.

For a given selection of parameters, the datasets fulfilling the query are displayed in the DAL from which the user can accept one or more. Before launching the request to the Archive, the user can specify the type of process desired and the machine/disk combination to be used in the reduction. This is done using the configuration file associated with the panel. Several options are available, but the most common in regular operation is to execute the entire sequence consisting of request, reduction, calibration, ingestion and deletion of the raw data and to set the choice of machine/disk to automatic, allowing the system to decide which machine to use depending on the workload already assigned and free disk space.

Once a dataset is selected the information available for each exposure in the ESO Archive is displayed, an estimate of the seeing for each exposure is computed, by interpolating values sampled by the seeing monitor, the moon distance calculated and a category assigned (science, standard, calibration). The exposures are then assigned to nights, and for each night relevant information such as sunset, sunrise, moon-set, moon-rise, and moon illumination fraction are computed and stored in the database for future reference. The exposures are also grouped into their original observation blocks so that the completeness of the OBs can be assessed and the internal OB repository updated accordingly.

Once a request is submitted to the Archive, data are retrieved from the medium storing them, uncompressed (gzipped), transferred by ftp to target machine/disk set by the system, and ingested into the EIS DB, if not yet done so by the scanner. The speed of these various tasks are also saved for future reference, In automatic mode, after these sub-processes have been executed the panel is iconized and the image pipeline panel, described below, is triggered.

4.3 Image Pipeline, Data Calibration and Photometric Pipeline

The image pipeline process consists of seven distinct steps: data preparation, data grouping, RB preparation, data reduction, data calibration, preparation of final products and verification. It can be started either by the user or launched automatically by the system after the retrieval of data from the Archive.

The data preparation sub-process inspects the files retrieved to determine if they are valid FITS files, rejecting and reporting those of zero-length or corrupted. It also checks for the existence and integrity of mandatory FITS header keywords, taken from an instrument-dependent list retrieved from the EIS DB. This list is prepared at the time a new instrument is defined. Next, the entries in the EIS DB, corresponding to each exposure, as given by the ESO Archive database, are compared to the values given in the header. If different, the EIS DB entries are updated accordingly. Finally, multi-extension FITS files from multi-chip instruments are split into different FITS images and stored in separate directories to be processed individually.

In data grouping the information available for each exposure is used to sort them in time and to group them by night, category, filter and position on the sky, in the case of science and standard exposures. Once groups of exposures are formed, the properties of these groups are computed and written to a XML file. These groups are used to prepare reduction blocks (RBs) which are the input to the image processing library. In general, RBs are created to match the original OBs, using the OB name and the sequence number of the exposure in the OB, if available. Additional constraints may be imposed such as airmass and time interval. In addition, individual frames may be rejected by imposing conditions on the seeing and/or the background counts. Each RB is then assigned a reduction mode which depends on their category, instrument, filter and type of observation. The reduction mode defines the reductions steps and optimal parameters for the reduction of a given RB. A more detailed description of the image processing software can be found elsewhere [7]. For science RBs the user may also define a reference grid onto which reduced images should be mapped by specifying the reference position, pixel scale, projection and orientation. For instance, in survey work these are taken to be the position of the center of a region, the pixel scale of the instrument used, TAN and north up and east left, respectively.

Once the RBs are prepared and a reduction mode and reference grid are assigned to them, the system launches the C-based image reduction software. This package takes control of process, reducing all the images and producing a comprehensive history of the process in the form of XML files. After the reduction is finished some of the most relevant information is read by the Python wrapper and appended to the process log. Next, the image pipeline panel is temporarily iconized, followed by the opening of the data calibration panel.

SExtractor is used to extract catalogs for all reduced images associated with standard stars and the positions and flux of the detected objects are cross-matched with those of photometric standards available in the literature and stored in the EIS database. The matching of standards is first carried out by position. If a match is found within a search radius, then the fluxes of the measured star and its counterpart in the database are compared. An estimate of the measured flux is computed using the median values of the zero-point and extinction of all previous solutions for that instrument and filter. If the magnitude difference is less than a user-specified value the match is accepted. If not the search continues up to a maximum value of the search radius. The SExtractor flags for the matched star are also examined to check for problems such as saturated pixels. In the case of optical observations, besides primary standards, the system also attempts to find matches of secondary standards, key for dealing with multi-chip instruments. After all matches are found the panel is iconized, the photometric pipeline panel is launched and the information about the measurements of the matched-objects is passed to the new process.

For each data set solutions, ranging from a 3- to a 1-parameter fit, are computed for each night, chip, and filter combination. The best solution, that with smallest rms and deviation from the median values, is then used to calibrate the science frames for the corresponding nights. Measurements, growth curves, fits and diagnostic plots are produced and written to a XML log file stored together with the photometric solution in the database. For non-photometric nights the zero-point is set to the median value of the trend analysis and the error is set to -1.

After solutions are found for all nights in the relevant filters, the photometric pipeline is iconized and the control of the process is returned to the image pipeline panel. The derived zero-points are used to calibrate all science frames.

For multi-chip instruments the individual reduced images for each CCD are calibrated using chip-based zero-points, whenever possible, mapped to the same reference grid and combined. Next, image headers are re-written and compared to a template to ensure uniformity of the products, and each final image is combined with its weight and moved to the image repository. Finally, the main attributes of the final images such as the seeing, limiting magnitude and parameters characterizing the PSF are computed. In addition, any detection of fast moving objects (satellite tracks) on an image is recorded together with a postage stamp of the frame on which it was detected. All this information, together with plots, are included into a XML file associated with each product. These product logs are ingested together with the corresponding images and can be accessed at a later time.

The final step of the image pipeline is to automatically verify the quality of the products. This is done by either using previously reduced data, over the same field and passband, or external data already available in the database. These data are used to compare positions and fluxes to ascertain the quality of the astrometric and photometric calibration of the images produced. As a last resort, objects from all-sky catalogs, created at the start of the process, are also used as a crude check. From these comparisons a set of quality control parameters are produced and a quality assessment is made according to configurable acceptable ranges of these parameters. In addition, a final *a posteriori* by-eye verification is also possible. All the results from the verification step are also appended to the product logs.

4.4 Image Products

This panel is used for the generation of final survey products such as image stacks (from images associated with RBs) and of mosaics (from stacks). The first step of the process groups images in stacking blocks according to their filter and position on the sky and a user-specified minimum overlap area between images. Once these stacks are formed they are validated. The validation includes checking that all images are unique, share the same reference grid and astrometric reference. After validation the final stacking blocks are prepared according to user-defined constraints on, for instance, the seeing, image distortion, surface brightness and/or airmass. The images are then stacked and flux-scaled using a weighted co-addition, with the weights given by their respective weight maps.

Essentially the same procedure is applied to create a mosaic. A special validation is required to group only connected images satisfying a minimum condition of overlap so that the relative photometry for the final mosaic can be calculated. Using the mean difference in magnitudes of common objects in the overlap region the differences in zero-point from frame to frame are computed and used to derive a common zero-point of the mosaic either iteratively or by a least-square method.

As in the case of images produced by the image pipeline, described above, for each final image produced a series of checks (astrometry, photometry and header) are carried out and stored in the product XML logs which are ingested into the database.

4.5 Catalog Products

From the catalog production panel different types of catalogs (single-passband, color, mosaics) can be created. Single passband catalogs can be extracted using different source extraction algorithms such as SExtractor, DAOPHOT, PSFex and KSB, depending on the intended application. Similarly, color catalogs can be produced by association of several single passband ones or by using a reference image, either a science image in a given filter or a χ^2 image. Finally, mosaic catalogs are produced by stitching together overlapping single passband catalogs, using objects in the overlap areas to compute a uniform zero-point, as discussed above for image mosaics.

The creation of science-grade, single-passband catalogs is carried out in two steps. First, sources are extracted using a low S/N detection threshold to create a parent catalog, which is normally used in the production of color catalogs by association, in order to preserve objects that may dropout in one of the passbands. During the creation of a catalog, bright objects are masked and objects within the masks properly flagged. For galaxy catalogs, the magnitudes are corrected for galactic extinction and Vega magnitudes are transformed into AB magnitudes for the appropriate pointing and filter, respectively. Note that the value for the AB correction is computed and stored in the EIS DB at the time a filter is first introduced into the system, using its transmission curve. In addition, the original image is used to generate different mock images from which the completeness and the fraction of spurious objects in the original catalog, as a function of magnitude, are computed as well as different limiting magnitudes characterizing these curves.

Next, depending on user-selected settings, a science-grade catalog is produced by culling from the parent catalog low S/N objects or objects fainter than a given limiting magnitude. In the preparation of the final catalog, sources are also assigned a rough morphological classification. This is commonly done using a given value of the stellarity index parameter of SExtractor to separate detections into point or extended sources. An alternative way, also available, is to adopt a magnitude-dependent statistical criteria proposed by Groenewegen *et al.* [3]. Finally, the spatial distribution of objects, the associated smooth projected density distribution, number counts of galaxies and stars and the two-point angular correlation function for galaxies at different limiting magnitudes are computed for the final catalog. These results provide different ways of visualizing and characterizing the catalog produced and are part of quality control procedure.

To assess their quality, the catalogs produced are compared with those previously produced by the pipeline or by other authors, if available in the EIS DB, to check the astrometric and photometric calibration. The EIS DB also stores published results of number counts and the variation of the amplitude of the angular correlation function at a given scale as a function of magnitude for different passbands, which are used to compare with the results derived from the catalogs generated. Finally, the stellar counts are compared with predictions of a population synthesis model created using the same filter and region as the data [2]. As mentioned earlier, this model is computed at the time of the definition of a survey and stored in the EIS DB. For color catalogs, comparisons with model predictions of color distributions, color-magnitude and color-color diagrams are also made. An interface with XGOBI also allows the user to visualize and cross-reference different projections of the multi-dimensional data.

All the results generated during the catalog production are included into a XML file which also provides information about all the steps of the catalog production including the original settings of the source extraction algorithm. At the end of the process these logs are also ingested into the EIS DB for future reference.

4.6 Analysis

Even though several tests are carried out during catalog production, undoubtedly the best way of identifying undesirable features in the products is to use them in different scientific applications that depend on different characteristics of the data. For this purpose, this panel is available to integrate different scientific application software into the survey system framework. So far the following applications are available: photometric redshifts [1]; spectral classification of sources [4]; simple mass reconstruction methods; and identification of clusters of galaxies using the matched-filter method [5]. Some of these were developed by the EIS team, while others are adaptations of algorithms provided by the original authors. The available framework makes it relatively easy to insert other plug-ins, thus allowing the scientific library to grow in time as needed. This panel is clearly open-ended and many of the functionalities required are similar to those envisioned by the VO. In the future, as VO tools mature it may be possible to expand and improve the functionalities available.

Integrating these algorithms into the survey system framework allows the user to benefit from the infra-structure provided by the system such as the database, the search engine/DAL mechanism, history, versioning and visualization tools. In addition, from this panel targets of potential interest for spectroscopic follow-up can be created, stored and published on the Web using the tools available for other survey products. The stored lists are also used to drive the cutout/co-addition server which is used to produce postage stamps of objects of interest from images of all available instruments and passbands. These tools are complemented by others which allow the visualization of SEDs, color-magnitude and color-color plots customized for the inspection of individual objects. Finally, these tools are used for the verification of the classification and photometric redshifts assigned automatically by the system.

4.7 Data Release

This panel provides the required infra-structure to publicly release survey products. It allows one to select and examine final products, in the form of pixel maps, catalogs or lists of targets. Once the products to be released are selected, several automatic procedures are triggered that update the EIS home page announcing the release, create a new entry in the tables listing data releases, assign a version number based on information available in the database, establish links between these tables and HTML pages containing tables listing the products being made available, and move the released products to a repository accessible by the ESO Science Archive for the preparation of packages for distribution. The HTML pages created by the system list information about each product, a link to the product log produced by the pipeline and a check-box that allows the user to select the products to be retrieved. The system also triggers an entry in Remedy and notifies via email all of those that need to be informed of the release, streamlining the entire process. It is also envisioned that the main differences between versions of a product will also be available. This will report the differences in the configuration file and the results of the comparison between versions, which will depend on the type of product. The system is also currently being updated to provide the mechanisms necessary to monitor the data requests and produce suitable reports.

5 Summary

In this paper the EIS survey system developed to carry out public surveys has been described. It is a high-performance, fully integrated, end-to-end system capable of sustainable and un-supervised operations over long stretches of time. It supports the reduction of image data from different instruments covering both the optical and infrared domains, providing well-defined survey products, with complete creation history and quality assessment, intended for general distribution. The system has been built in such a way as to minimize as much as possible the need of human intervention. From the EIS survey system it is possible to prepare observations, retrieve raw data, convert these to science-grade products in the form of fully calibrated pixel maps, catalogs and other derived products, and to monitor the status of the observations and reductions. Judging by its current throughput, the system will be able to meet the requirements of the next generation of public surveys to be conducted with dedicated imaging telescopes such as VST and VISTA.

Even though developed to support long-term public surveys, the EIS survey system can be easily adapted to operate as a front-end to ESO's raw data archive to produce well-described science-grade products for any data set. Therefore, it can also be seen as one of the building blocks of the VO infra-structure which, among many other requirements, needs site-specific tools to provide the desirable connection between raw data archives and high-level VO tools.

Acknowledgements

I would like to thank all present and former EIS team members for their hard work in putting together the system described here.

References

1. Arnouts, S., 2001, *private communication*
2. Girardi, L. *et al.* , 2002, A&A, 391, 195
3. Groenewegen, M. *et al.* , 2002, A&A, 392, 741
4. Hatziminaoglou, E *et al.* , 2002, A&A, 384, 81
5. Olsen, L.F., *et al.* , 1999, A&A, 345, 681
6. Renzini, A. & da Costa, L., 1997, The Messenger, 87, 23
7. Vandame, B., 2002, SPIE conference "Astronomical Telescopes and Instrumentation, (astro-ph/0208230)

e-Star: Telescopes and Databases as a Single Information Grid

Tim Naylor[1], Iain Steele[2], Dave Carter[2], Alasdair Allan[1], Jason Etherton[2], and Chris Mottram[2]

[1] School of Physics, University of Exeter, Stocker Road, Exeter EX4 4QL, U.K
[2] Astrophysics Research Group, Liverpool John Moores University, Twelve Quays House, Egerton Wharf, Birkenhead CH41 1LD, U.K

Abstract. We present a sketch of a system which consists of a Grid of databases and telescopes within which intelligent agents assess the information available to them, and as a result request the telescopes take new data. These systems allow programmes to be carried out much more efficiently than is currently possible, and their speed of reaction enables novel programmes that are unavailable to current technology. We then present eStar, a currently functioning technology prover for such systems which use computational grids software (Globus) to create an information Grid.

1 Introduction

Not so long in the future, in a galaxy not far, far away astronomers had built a Virtual Observatory. But a small, determined rebel group continued to make new observations! "What has the mighty federation of databases ever done for us?" they cried.

I began my talk at the AVO conference with this spoof of the Star Wars opening sequence (and with a minor dig from Monty Python's Life of Brian) to make a very serious point about the aims of the Virtual Observatory initiatives. Whilst the end result may be a seamless federation of databases, there will be a significant number of astronomers (possibly the majority), still involved in data acquisition. That data acquisition process often requires the use of databases, and increased use of databases will clearly benefit the process. What I hope to persuade you in this contribution is that making a seamless interface between telescopes and databases, indeed making telescopes look like databases will bring enormous benefits. But, there is functionality telescopes need that databases don't, so a degree of planning is needed at this stage to ensure convergence in the paradigms used for grid enabled telescopes and databases.

What can be achieved is best seen by considering a couple of possible science drivers. So, imagine we have built a system which not only has unified access to databases and telescopes, but also to the astronomical literature. In addition the system has intelligent agents which can interpret the results. One can view this system as a unified information grid, within which these intelligent agents live.

2 Driver 1: The Space Density of Dwarf Novae

Dwarf novae are semi-detached interacting binaries where a Roche-lobe filling low-mass star loses mass to a white dwarf. There is a thermally unstable accretion disc around the white dwarf. When disc is in outburst it is almost a standard candle, which outshines all the other sources of light in the system. Hence acquiring a sample which is magnitude limited when in outburst is almost identical to obtaining the volume limited sample required for measuring the space density. The problem is that dwarf novae spend only $\sim$10% of their time in outburst, the remaining time they are $2-5$ astronomical magnitudes fainter. Thus conventional, one epoch sky surveys are not a good way to tackle this problem.

Instead we could imagine a programme where an intelligent agent (IA) compares every CCD field taken in the world with the SuperCosmos scanning of Schmidt plates. The IA could then carry out the following tasks.
1) Objects which brighten above some fixed magnitude (say 16th) are cross-correlated with the SIMBAD stellar database.
2) Those which SIMBAD shows are known dwarf novae are noted, and those which are known not to be dwarf novae discarded.
3) For the previously unknown objects, archive data are searched to identify their nature from their lightcurves.
4) Where the archive data are too sparse, the IA requests new observations, building up a lightcurve, ensuring that new observations are placed at critical times to define whether or not the object is a dwarf nova.
5) The IA catalogues objects it is certain are dwarf novae, and requests a parallax measurement from HST to confirm the distance.

After this entire process is complete, the astronomer returns from a (presumably quite extended) lunch break, and is ready to write a paper.

3 Driver 2: What Was That?

The above example is a science problem we could already solve, but the system proposed here could do it faster and cheaper than more traditional methods. However, the proposed system offers a speed of action which not currently possible, opening new possibilities. Consider the following example, with timings which I will justify later.

02:11:03UT. The shutter closes on a wide-angle CCD image of Centaurus taken using a large camera lens.

02:12:30UT. The data have been processed, and a list of magnitudes and positions is made available on the astronomical data grid.

02:12:45UT. An astronomer's intelligent agent has compared the source list with archive data, finding a bright object.

02:13:00UT. In response to the IA's request for confirmation a small telescope slews to acquire another image. Whilst waiting for this the IA queries SIMBAD and discovers there is no known variable at this position.

02:15:06UT. The new image confirms the object, so the IA requests spectroscopy from the Liverpool Telescope. During this waiting period the IA pulls all the other available data and papers.

02:22:34UT. The IA recognises that the spectrum is odd. There hasn't been a γ-ray burst, but VISTA shows a very faint red object mentioned in a paper last year.

02:22:50UT. Less than 11 minutes after the initial exposure ended, the IA has run to the end of its pre-programmed "understanding", so it wakes up an astronomer.

That astronomer already has a unique dataset, and a great deal of the preliminary work of searching databases and literature has been done. But now some creative thinking is required. The majority of the time delay has been introduced by exposure time; I have assumed that the data pipelines can reduce the data quickly, and will discuss below how rapidly the IAs can react.

4 The eStar Project

So how close are we to achieving a system like this? The eStar project (e-Science Telescopes for Astronomical Research) is being funded by the UK's Department of Trade and Industry, initially as an e-Science demonstrator project. We use Meade LX200 and ETX telescopes with Santa Barbara Instruments Group (SBIG) or Apogee CCD cameras. These are chosen mainly because they have a simple programming interface, allowing us time to concentrate on the novel issues of the problem. At the moment the system works across the network as a daytime demonstration, slewing the telescopes and taking CCD images, but then sending fake data to the IAs. We plan complete tests taking real data later in 2002.

Our main design criterion was to have something that works in an analogous way to a computational grid, with no overall supervisor. Each telescope has its own scheduler, which can talk to the IAs using the Robotic Telescope Markup Language (RTML – see F.V. Hessman's paper in these proceedings) and transported over Globus. The IAs also have to communicate with databases such as SIMBAD and the Astrophysical Data Service (ADS), and datasets such as the US Naval Observatory's A2 astrometric catalogue (USNO-A2), and the Digital Sky Survey (DSS). This is done using WWW interfaces. Such a system should be eminently scalable, with many telescopes (we have already tried four) and many intelligent agents. The intelligent agent we see as carrying out a specific scientific programme, and probably resides on the astronomer's desktop computer.

A typical sequence begins with the IA carrying out a Globus resource discovery (LDAP), finding each telescope and its associated instrument suite. Next it asks which telescopes could carry out a particular observation, each node returning a score, which encodes the predicted quality the observation. The IA then selects a telescope it wishes to carry out the observation, and then requests it, with the scheduler placing it in the observing queue. Once the observation is complete the raw and reduced data are made available to the IA. Our current

IA can compare the new observations with USNO-A2, finding variable stars, and carrying out a search at these co-ordinates in SIMBAD. The IA will then grab a DSS image of the area. If the variable star turns out to be a dwarf nova, then the IA will request new observations, in addition to acquiring a complete set of references for the object. Having created this system, we were pleasantly surprised to discover how fast the cross-correlation runs, normally about six seconds, including recovering data from three web sites. This shows that the IAs would not be the limiting speed factor in a rapid reaction system.

If a system similar to this is to become widespread, the idea of requesting observations is important. It allows telescopes to join the system with outside intelligent agents being, perhaps, only a small part of the observations the scheduler has to deal with. Furthermore, telescopes can reject requests from an IA if, for example, the astronomer has not got time allocated on the telescope. In the longer term we hope to use the authentication and authorisation of users of grid resources, as well as accounting for the resources used. Observatories will want to be sure that only authorised people are using their resources. Equally important, though, is that such a system will be easier to create if databases and telescopes have a common interface. Databases are only special cases of telescopes, that can return data instantly, rather than after a time delay.

Massive Variability Search and Monitoring by OGLE and ASAS

Bohdan Paczyński

Princeton University, Princeton, NJ 08544, USA

Abstract. OGLE and ASAS are long term observing projects operated by the Warsaw University Observatory at the Las Campanas site in Chile. OGLE is currently monitoring almost 200 million stars in the Galactic Bulge and the Magellanic Clouds, and has detected so far almost 1,000 events of gravitational microlensing with the dedicated 1.3-meter telescope. ASAS uses several very small instruments to monitor all southern sky for variability down to approximately 14 magnitude. A total of almost 300 thousand variable stars were discovered so far by the two projects, and all photometric data is available on the WWW. Both projects aim at real time recognition and verification of all new phenomena in the sky. OGLE is likely to discover planets by 2003 and stellar mass black holes by 2004-2005. All OGLE and ASAS data is made public domain as soon as possible, and may be used by a Virtual Observatory.

1 OGLE – Past and Present

OGLE (Optical Gravitational Lensing Experiment, the leader: Andrzej Udalski) is a long term project, with a gradual expansion of its capability. The dedicated 1.3 meter telescope is located at the Las Campanas Observatory in Chile, a site owned and operated by the Carnegie Institution of Washington. OGLE is operated by the Warsaw University Observatory. Information about it may be found at:

OGLE: `http://sirius.astrouw.edu.pl/~ogle/`
OGLE: `http://bulge.princeton.edu/~ogle/`

OGLE-III is the current, third stage of the project. The observations are done with a mosaic $8K \times 8K$ CCD camera built by A. Udalski. OGLE-III begun in May 2001. In the 2002 Galactic Bulge season almost 400 candidate microlensing events were discovered:

OGLE-EWS: `http://sirius.astrouw.edu.pl/~ogle/ogle3/ews/ews.html`

A search for planetary transits was conducted on 32 nights during a 45 day interval in 2001. A total of 59 stars were identified, for which low depth flat bottom transits with orbital periods shorter than 10 days were found (Udalski et al. 2002a,b). Some of these are likely caused by 'hot Jupiters', others by brown and red dwarfs.

The analysis of OGLE-II stage (Udalski, Kubiak & Szymański 1997), which covered years 1997-2000, is advanced, and almost all data is in public domain. This includes catalogs of I, V, B magnitudes and positions for a total of almost

40 million stars in the Galactic Bulge, and in the Magellanic Clouds (Udalski et al. 2002c, 2000b, 1998a), catalogs of over 500 microlensing events (Udalski et al. 2000a, Woźniak et al. 2001), catalogs of over 270 thousand variable stars (Zebruń et al. 2001, Woźniak et al. 2002), and in particular thousands of eclipsing binaries in the SMC (Udalski et al. 1998a) and Cepheids in the LMC and SMC (Udalski et al. 1999a,b). In addition OGLE astrometry provided proper motion measurements of thousands of stars (Soszyński et al. 2002, Sumi et al. 2002), leading to the discovery of streaming motion (rotation) in the Galactic Bar.

Of particular interest are special microlensing events. OGLE-2000-BUL-43 was found to have a spectacular parallax effect (Soszyński et al. 2001). Even more dramatic was OGLE-1999-BUL-19, the first event with multiple peaks in its apparent brightness caused by the Earth's orbital motion and a very long time scale: $t_E = R_E/V = 372$ days (Smith et al. 2002). Even longer time scales were found for OGLE-1999-BUL-32 (Mao et al. 2002, $t_E = 641$ days and OGLE SC_5 2859 (Smith 2002, $t_E = 551$ days). The last two events are likely caused by massive lenses, probably stellar mass black holes.

2 ASAS – Past and Present

ASAS (All Sky Automated Survey, the leader: Grzegorz Pojmański) is a long term project, with gradual expansion of its capabilities. Currently it has 4 small instruments, with apertures of 2 cm, 7 cm, 7 cm, and 20 cm, all located at the Las Campanas Observatory in Chile, at a site owned and operated by the Carnegie Institution of Washington. ASAS is operated by the Warsaw University Observatory. Information about it may be found at:

ASAS: `http://www.astrouw.edu.pl/~gp/asas/asas.html`
ASAS: `http://archive.princeton.edu/~asas/`

All four instruments use as detectors Apogee $2K \times 2K$ CCD cameras. The two instruments with 7 cm aperture cover all sky every two nights in standard V and I filters down to about 14 magnitude. A total of over 6 thousand variables was discovered so far by ASAS (Pojmański 1998, 2000, 2002), and all photometric data are on the WWW.

All instruments are fully robotic, with some support provided by the OGLE observers.

3 Future of OGLE and ASAS

Both projects aim at real time data processing, and in particular at real time recognition and verification of any new events in the sky, among them microlensing events, supernovae, novae, dwarf novae, stellar flares, GRB afterglows, etc.

Two specific goals of OGLE are: a firm detection of microlensing by planets, and by stellar mass black holes. So far stellar mass black holes were suggested as lensing masses for several long microlensing events (Bennett et al. 2002, Mao et al. 2002, Smith 2002), but evidence is not conclusive. No definite planetary microlensing event has been discovered so far (cf. Bennett et al. 1999, Albrow et al. 2000, Gaudi et al. 2002).

Inspecting almost 400 microlensing events reported by OGLE-III EWS (Early Warning System) in 2002 Jaroszyński & Paczyński (2002) noticed that the event OGLE-2002-BLG-055 has a single data point deviating from otherwise smooth microlensing light curve by 0.6 magnitudes, while the photometric errors were about 0.01 magnitude. Dr. Udalski kindly examined the CCD image and found that there was nothing wrong with it, i.e. the bright point appears to be real. As nearby data points show no obvious departure from a smooth light curve a plausible interpretation of the phenomenon is in terms of a binary lens with a very extreme mass ratio. While a unique value of the mass ratio cannot be determined from so sparse time coverage, a good fit to all data was obtained for the mass ratio of 0.01 and 0.001, indicating a planetary mass companion to a stellar mass lens. Obviously, it is not possible to make a strong claim based on a single data point, but a modest modification of the future OGLE observing procedure will provide a far better coverage of future planetary events. There are typically only several hundred stellar microlensing events unfolding at any given time, compared to over 150 million stars monitored on a given night. Small CCD sub-frames covering known events can be processed within minutes of data acquisition. When an anomalous data point is noticed the observation of the field will be repeated. If the anomaly is confirmed the field will be observed every 30 or 60 minutes, to provide a good coverage of the rapidly changing brightness, and allowing a unique determination of the mass ratio. We expect that OGLE-III EWS system will have this capability by the spring of 2003 (A. Udalski, private communication), and it is likely that the first definite planetary microlensing event will be detected in 2003.

At the opposite end of the lens mass spectrum are the very long duration events, which are plausible candidates for stellar mass black holes. Currently two independent lens parameters can be determined for long events with the OGLE-III photometry: the event time scale t_E and the magnitude of the parallax effect. One additional parameter: the angular separation between the two images, or the astrometric shift in the combined light centroid is needed to determine uniquely the lens mass. It will be possible to make such measurements with the future VLT Interferometer (Delplancke, Górski & Richichi 2001), or with the existing HST. A massive lens like the one associated with OGLE-1999-BUL-32 had the two images separated by several milli arc seconds (Mao et al. 2002, eq. 9), leading to the centroid motion of a comparable amount. This astrometric effect is within easy reach of the HST (e.g. Benedict et al. 2002). The long events last several years, and with the OGLE-III data rate there will be several of them unfolding at any given time within several degrees of the Galactic center. Very likely the HST observations may be scheduled in advanced, with no need for the TOO (Target of Opportunity) mode of operation. Little is known about the number and the distribution of stellar mass black holes in the Galaxy, and the range of their masses. OGLE is likely to lead to the first definite mass determination in 2004–2005.

We expect that OGLE as well as ASAS will expand their capability in the future. There are no definite and specific plans for the expansion, but there is a general idea how to proceed. A natural step to OGLE-IV would be another 1.3 - 1.8 meter telescope with a CCD camera with a total of ~ 1 giga pixels, i.e. an instrument with the data rate 16 times higher than current OGLE-III. The data rate is the single most important parameter for the search and/or monitoring projects like OGLE and ASAS. For ASAS we expect more frequent all sky coverage, with a more gradual increase in its depth. A virtue of bright transients is that they may be followed up for a longer time, in greater detail and with more instruments than faint transients can. Obviously, bright transients are rare, hence the need for all sky coverage and a frequent time sampling.

It is a pleasure to acknowledge the support by NSF grants AST 9820314 and AST 0204908, and NASA grant NAG5-12212.

References

1. Albrow, M.D. et al. (PLANET) ApJ, 534, 894 (2000)
2. Bennett, D.P. et al. Nature, 402, 57 (1999)
3. Bennett, D.P. et al. (MACHO) ApJ, 579, 639 (2002)
4. Benedict, G. et al. AJ, 124, 1695 (2002)
5. Delplancke, F., Górski, K.M., & Richichi, A. A&A, 375, 701 (2001)
6. Gaudi, B.S. et al. (PLANET) ApJ, 566, 463 (2002)
7. Jaroszyński, J., & Paczyński, B. in preparation (2002)
8. Mao, S. et al. (OGLE) MNRAS, 329, 349 = astro-ph/0108312 (2002)
9. Pojmański, G. (ASAS) AcA, 48, 35 = astro-ph/9802330 (1998)
10. Pojmański, G. (ASAS) AcA, 50, 177 = astro-ph/0005236 (2000)
11. Pojmański, G. astro-ph/0210283 (2002)
12. Smith, M.C. in preparation (2002)
13. Smith, M.C. et al. (OGLE) MNRAS 336, 670 = astro-ph/0206503 (2002)
14. Soszyński, I. et al. (OGLE) ApJ, 552, 731 = astro-ph/0012144 (2001)
15. Soszyński, I. et al. (OGLE) AcA 52, 143 = astro-ph/0205289 (2002)
16. Sumi, T., Eyer, L., & Woźniak, P.R.: astro-ph/0210381 (2002)
17. Udalski, A., Kubiak, M., & Szymański, M.: AcA 47, 319 = astro-ph/9710091 (1997)
18. Udalski, A. et al. (OGLE): AcA 48, 147 = astro-ph/9806313 (1998a)
19. Udalski, A. et al. (OGLE): AcA 48, 563 = astro-ph/9812348 (1998b)
20. Udalski, A. et al. (OGLE): AcA 49, 223 = astro-ph/9908317 (1999a)
21. Udalski, A. et al. (OGLE): AcA 49, 437 = astro-ph/9912096 (1999b)
22. Udalski, A. et al. (OGLE): AcA 50, 1 = astro-ph/0002418 (2000a)
23. Udalski, A. et al. (OGLE): AcA 50, 307 = astro-ph/0010150 (2000b)
24. Udalski, A. et al. (OGLE): AcA 52, 1 = astro-ph/0202320 (2002a)
25. Udalski, A. et al. (OGLE): AcA 52, 115 = astro-ph/0207133 (2002b)
26. Udalski, A. et al. (OGLE): AcA 52, 217 = astro-ph/0210278 (2002c)
27. Woźniak, P.R. et al. (OGLE): AcA 51, 175 = astro-ph/0106474 (2001)
28. Woźniak, P.R. et al. (OGLE): AcA 52, 129 = astro-ph/0201377 (2002)
29. Zebruń, K. et al. (OGLE): AcA 51, 317 = astro-ph/0110623 (2001)

Statistical Analysis of Observed and Simulated Surveys

Andrei Doroshkevich

TAC, Juliane Marie vej. 30, DK 2100, Copenhagen Ø, Denmark

Abstract. The statistical analysis of large astronomical datasets requires also application of powerful fast codes and corresponding basic models. Some examples of such statistical analysis are presented. They are the Minimal Spanning Tree approach for investigation of large observed and simulated 3 dim surveys and 6 dim cluster analysis in arbitrary functional space. Both methods are tested with 20 000 000 simulated points in 3 dim space and 5 000 points in 6 dim space, respectively.

1 Introduction

With the advent of the Las Campanas Redshift Survey (LCRS, [1]) with $\sim$ 20 000 galaxies, the galaxy distribution on scales up to $\sim$300 h^{-1} Mpc could be studied. Now these investigations are extended using the public data sets from the Sloan Digital Sky Survey Early Data Release (SDSS EDR, [2]), which contains $\approx$ 30 000 galaxies in two slices for distances $D \leq 600h^{-1}$Mpc. It can be expected that soon this sample will be extended up to $\sim$ 1 000 000 galaxies. Such growth of statistic promises the corresponding growth of information about the Large Scale Structure in the galaxy distribution.

Analysis of these observations can be essentially complemented by measurements of high resolution absorption spectra of farthest quasars which allow to test the main characteristics of the Large Scale Structure formation and evolution at high redshifts. The already available Keck and VLT high resolution observations of the Ly-α forest provide a reasonable database with $\sim$ 5 000 lines that allows to apply statistical methods for such investigations (see, e.g., [3]). It can be expected that this database will also be essentially extended. Furthermore, high progress achieved now in observations of CMB and in X-ray astronomy can also significantly extend the available information.

All these observations are interconnected and allow us to reconstruct the evolutionary history of the Universe.

These observations can be successfully compared with the simulations of formation and evolution of the Large Scale Structure in the matter distribution. Now such simulations operate already with $\sim$20 000 000–1 000 000 000 particles that allows to improve the available resolution and to test the possible impact of many physical factors. Comparing with observations these simulations allow us to test the set of the most promising scenarios of the evolution of the Universe.

Appearance of so large datasets poses the hard problem – how is possible to extract the information of interest, to obtain reliable characteristics of the Large

Scale Structure and to describe its evolution. The first step of such investigation is the preparation of effective codes allowing to obtain the general description of a dataset.

2 Minimal Spanning Tree Approach

One of the most effective methods for the preliminary analysis of large catalogues is the Minimal Spanning Tree approach which allow to characterize the *general network* of objects. Now we have in TAC and Fermilab an effective code allowing to build the MST for both observed and simulated samples. Thus, for 30 000 galaxies in the SDSS EDR the MST is builded during $\sim 3 - 4$ minutes and for $\sim 20\,000\,000$ points in simulations the required time is ~ 2.5 hours only.

The MST is a construct from graph theory, originally introduced by Kruskal [4] and Prim [5], which has been widely applied in telecommunications and similar fields. It is a *unique network* associated with a given point sample and connects all points of the sample to a *tree* in a special and unique manner which minimizes the full length of the tree. One of its earliest uses in the study of the Large Scale Structure was that of Bhavsar & Ling [6], who successfully applied it to extract filamentary structures from the original CfA Redshift Survey. Later on this approach was applied to simulations [7], the LCRS [8] and the SDSS EDR [9].

Further definitions, examples, and applications of this approach are discussed in Barrow, Bhavsar, & Sonoda [10] and van de Weygaert [11]. More references to the mathematical results can also be found in [11].

Here we will restrict the presentation to the *probability distribution function of the MST edge lengths* $W_{MST}(l)$ (the PDF MST method) and to the morphological description of individual clusters. The potential of the MST approach is not, however, exhausted by these applications.

Basically, the MST for a object distribution contains within it all 'friends-of-friends' cluster catalogues for all linking lengths; if the MST is separated for a given linking length, one then has the 'friends-of-friends' cluster catalogue for that linking length. This strengths of MSTs approach can be successfully used for fast selection of clusters at different linking lengths and to find the variations of cluster characteristics with the linking length.

The PDF MST techniques can be used to examine the geometry of a galaxy distribution. In the general case $W_{MST}(l)$ depends on the correlation functions (or cumulants) of all orders. For larger point separations, however, where some correlations become small and the cumulants tend to constants, the appearance of a Poisson-like point distribution can be expected (White [12]; Sahni & Coles [13]; Borgani [14]). In such a case the PDF MST techniques can characterize the geometry of a point distribution.

For example, consider a random point distribution situated along randomly distributed straight lines and flat planes. Even for such a distribution of points, correlations do not vanish in 3D space and for the suitable set of such lines and planes the two-point correlation function is consistent with the observed power law, $\xi(r) \propto r^{-1.8}$ [15]. The PDF MST techniques are able to reveal the geometry of such a point distribution.

For the 1 dim and 2 dim Poissonian distributions analytical expressions for the PDF MST [16] are:

$$W_{MST}^{(1)}(l) = \frac{1}{\langle l \rangle} e^{-l/\langle l \rangle}, \tag{1}$$

$$W_{MST}^{(2)}(l) = 2\frac{l}{\langle l^2 \rangle} e^{-(l^2/\langle l^2 \rangle)}. \tag{2}$$

These relations remain valid also for curve filaments and walls when their curvatures and thicknesses are small as compared with the point separations.

For 3 dim Poissonian point distribution the analytical expression is unknown. For comparison with simulated PDF we can use the standard expression for the distance up to the closest neighbor:

$$W_{MST}^{(3)}(l) = 3\frac{l^2}{\langle l^3 \rangle} e^{-(l^3/\langle l^3 \rangle)}. \tag{3}$$

The simulated PDFs MST for 1D, 2D and 3D Poissonian samples are plotted in Fig. 1 together with fits (1, 2 & 3). For 3D Poissonian point distribution the cutoff of the PDF MST at $l \sim 2\langle l \rangle$ describes the percolation process.

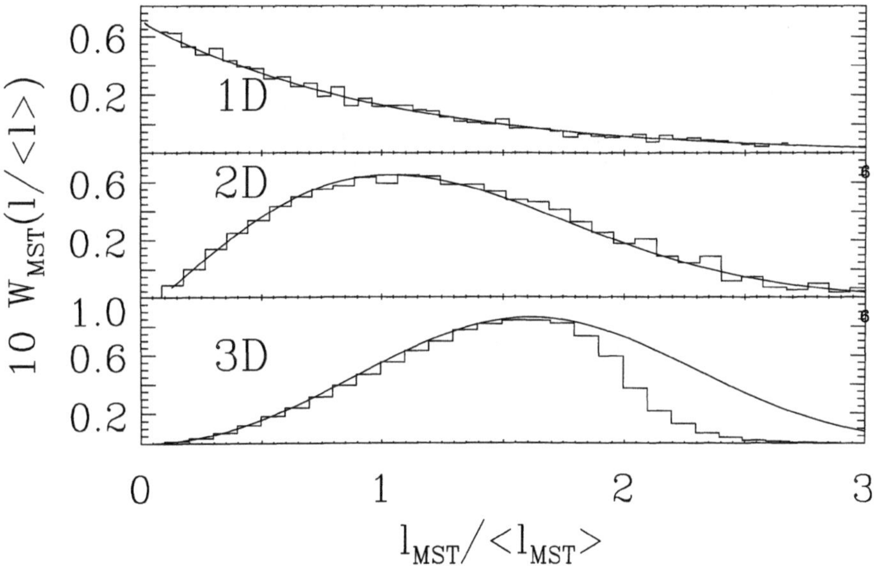

Fig. 1. Simulated PDFs MST for 1 dim, 2 dim and 3 dim Poissonian distribution together with fits (1, 2 & 3)

3 PDF MST for the SDSS EDR

The PDFs MST plotted in Fig. 2 for the SDSS EDR catalogue is well fitted by the superposition of Rayleigh and exponential functions. This fact indicates that this distribution can be roughly described as a network of 1D filaments and 2D sheets (or walls) occupying the sparser and denser part of tree, respectively.

Of course, the complex internal structure of both filaments and walls makes this description very approximate. The more detailed investigation shows that walls accumulate some number of filaments, clusters and groups of galaxies etc. In turn, galaxy groups are clearly seen within filaments. This internal structure distorts the measured PDF MST and makes interpretation unreliable. However, using the PDFs MST for the full sample we can separate several subsamples dominated by walls and filaments.

3.1 Separation of Subsamples of Walls and Filaments

The methods for an approximate statistical decomposition of sample into subsamples of wall–like structures and filaments were proposed and tested in our previous publications [7–9]. Such decomposition can be easy performed for a given overdensity contour bounding the clusters and a given threshold richness of individual structure elements. In particular, to characterize the overdensity, δ_{thr}, we can use the threshold linking length, r_{lnk}, and a relation familiar from

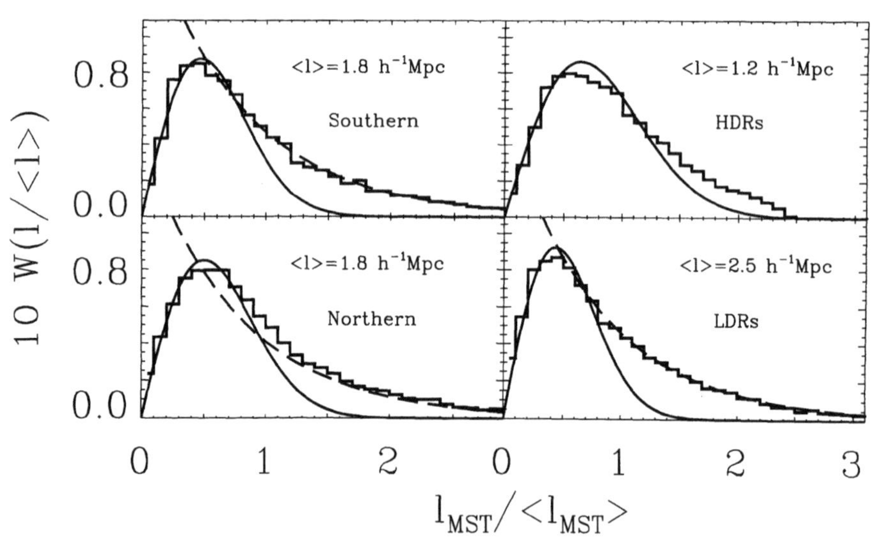

Fig. 2. The PDFs MST for south (left top panel) and north (left bottom panel) samples of the SDSS. For the high and low density regions the PDFs MST are plotted in right panels. Exponential and Rayleigh fits (1, & 2) are plotted by solid and dashed lines

friends-of-friends algorithms [17]:

$$\delta_{thr} = 3/[4\pi\langle n_{gal}\rangle r_{lnk}^3] \,. \tag{4}$$

As is shown in [9], for the SDSS EDR the wall–like high density regions are suitably identified with clusters found for a threshold richness $N_{thr} = 50$ and a threshold overdensity contour bounding the cluster equal to the mean density, $\delta_{thr} = 1$. The sample contains 45–50% of all galaxies. The samples of low density regions, which are occupied mainly by filaments and poor groups of galaxies, are complementary to the wall–like high density sample and it was prepared by removal the wall–like sample from the original total sample.

The PDFs MST for both subsamples are plotted in Fig. 2 together with fits (1 & 2). This figure demonstrates that for the wall-like subsample the PDF MST is quite similar to the Rayleigh function that verifies the separation procedure. For the low density subsample the PDF MST demonstrates that, for less separation, the PDF is again similar to the Rayleigh function. This means that of about half of galaxies are concentrated within galaxy groups. However, for larger separations the exponential function fits quite well the PDF.

3.2 The Morphology of the Structure Elements

With the MST technique we can extract from the wall-like high density and filamentary-like low density regions subsamples of structure elements for various threshold overdensities. We can then suitably characterize the morphology of each structure element by comparing the sum all edge lengths within its full tree, L_{sum}, with the sum of all edge lengths within the tree's trunk, L_{tr}, which is the longest path that can be traced along the tree without re-tracing any steps:

$$\epsilon = L_{tr}/L_{sum} \,. \tag{5}$$

For filaments, we can expect that the lengths of the full tree and of the trunk are similar to each other, whereas for clouds and walls these lengths are certainly very different. This approach takes into account the internal structure of each element rather than the shape of the isodensity contour bounding it, and, in this respect, it is complementary to the Minkowski Functional technique (see, e.g., [18]).

However, even this method cannot discriminate between the wall–like and elliptical clouds and those rich filaments having many long branches for which again $\epsilon \leq 1$. This means that the PDF of this ratio, $W(\epsilon)$, is continuous function and the morphology of structure elements can be more suitably characterized by the *degree of filamentarity and wall-ness.* This also means that we can only hope to distinguish statistical differences between the morphologies of structure elements.

For the SDSS EDR the distribution function of the ratio, $W(\epsilon)$, is found to be close to Gaussian with $\langle\epsilon\rangle \approx 0.5$ & 0.70 for wall–like and filamentary–like regions, respectively. The mass functions, $f_m(\epsilon)$, are shifted to left (for wall–like) and to right (for filamentary–like) regions in respect to the middle point.

These results verify the objective nature of the differences in the structure morphologies in the wall–like denser and filamentary–like sparser regions.

4 Conclusion

The short presentation of the MST approach demonstrates that it can be successfully used for the detailed description of rich sets of observed and simulated point distribution. This approach allows to obtain general characteristics of the sample, to separate set of subsamples with different properties and to characterize individual structure elements separated with various isodensity contours. More details can be found in [7], [8] and [9].

The high efficiency of the method noted above makes it a suitable tool for many investigations of larger observed and simulated catalogues. Now this approach is generalized that allows to investigate the point distribution in 6 dim space.

References

1. S.A. Shectman, S.D Landy, A. Oemler, D.L. Tucker, H. Lin, R.P. Kirshner, P.L. Schechter, ApJ, 470, 172, (1996)
2. S. Stoughton et al. AJ, 123, 485; 576, (2002)
3. M. Demiański, A. Doroshkevich, & V.I. Turchaninov, MNRAS., submitted, astro-ph/0207136, (2002)
4. J.B. Kruskal, Proc. Amer. math. Soc., 7, 48, (1956)
5. R.C. Prim, Bell. System Tech. J., 36, 1389, (1957)
6. S.P. Bhavsar, E.N. Ling,ApJ, 331, L63, (1988)
7. M. Demiański, A. Doroshkevich, V. Müller, & V.I. Turchaninov, MNRAS., 318, 1177, (2000)
8. A. Doroshkevich, D. Tucker, R. Fong, V. Turchaninov, H. Lin, MNRAS, 322, 369, (2001)
9. A. Doroshkevich, D.L. Tucker, S. Allam MNRAS, submitted, astro-ph/0206301, (2002)
10. J. Barrow, S. Bhavsar, D. Sonoda, MNRAS, 216, 17, (1985)
11. R. van de Weygaert, Voids and the geometry of large scale structure, Ph.D. Thesis, University of Leiden, Leiden, (1991)
12. S.D.M. White, MNRAS, 186, 145, (1979)
13. V. Sahni, P. Coles, Physics Report, 262, 1, (1995)
14. S. Borgani, Dark Matter in the Universe, Varenna 1995, eds. S. Bonometto, J. Primack, A. Provenzale, World Scientific Press, p. 126, (1996)
15. O. Buryak, A. Doroshkevich, A&A, 306, 1, (1996)
16. M. Kendall, P. Moran, Geometrical Probability, (London: Griffin), (1963)
17. J.P. Huchra, M.J. Geller, ApJ, 257, 423, (1982)
18. J. Schmalzing, S. Gottlöber, A. Klypin, A. Kravtsov, MNRAS, 309, 1007 (1999)

Catalogue Intersection: Beyond Basic Cross Correlations

Patricio F. Ortiz[1,2]

[1] **AstroGrid**, Department of Physics and Astronomy
University of Leicester, LE1 7RH UK
[2] INAF, Osservatorio Astronomico di Capodimonte, Naples, Italy

Abstract. Matching objects listed in different catalogues by position is a primary need for VO users: we want to find counterparts (or absence of counterparts) in one or more catalogues. However, the VO concept opens the door to questions like *which areas of the sky have been most intensely observed in optical, radio, X-ray and IR?* To answer these questions it is not necessary, nor desirable, to perform a cross correlation at an element by element basis between the most relevant catalogues in each wavelength. If multi-million source catalogues are involved in the question, eg, USNO, the cross correlation becomes impossible to perform in a reasonable time. Sky indexation methods are used to accelerate matching catalogues. I discuss how the "sky grids" generated from the indexation can then be used to answer this and other questions in a very short time (seconds) regardless of catalogue size and the number of catalogues involved.

1 Introduction

During the last few years data-centres around the world have put a big effort to solve the question of positional cross correlation between two catalogues and cone searches with user provided positions. The reason is simple: identifying the same object in a multi-wavelength search is basic to get the *grand picture* of the physical properties of an object. Detecting time variations has the same basic requirements: one needs to first identify the same object in different catalogues to then produce a light curve or to detect positional shifts.

Solving this problem has no intrinsic difficulty in principle: just look for objects which lie in the same area of the sky to find the matches; it can even be done by hand if the number of targets is small, with the caveat that different angular resolutions and flux limits make the match a non trivial issue. As the data volume increases alongside users expectations for quick procedures, several methods have been used to accelerate the computing time these queries require. Sky indexation has become very relevant, as one would like to look for matches only in the areas of the sky where they are likely to occur. Different methods have been used over the years to achieve this objective with different degrees of efficiency and *popularity*: HEALPix [1], Hierarchical Triangular Mesh (HTM) [2,3], Quadrilateralized Spherical Cube [4], iso-declination tiles [5], etc. Some of these methods have been designed for different purposes but could be easily implemented to tackle this problem.

The natural extension to positional cross correlation is to take into consideration additional constraints, like *find the galaxies which are closer than 30′ and whose redshift differ at most by 0.2 from the reference object.* This is an issue currently addressed by some services and which will surely be an important part of VO.

However, there is another class of positionally oriented questions which can be asked, and which make more and more sense as the catalogues increase in size and number. This class of questions deal with properties of the distribution of objects in the sky. The locations of the Hubble Deep Fields were defined by this class of questions: *select areas of the sky where the density of objects is minimal in several wavelength regimes.*

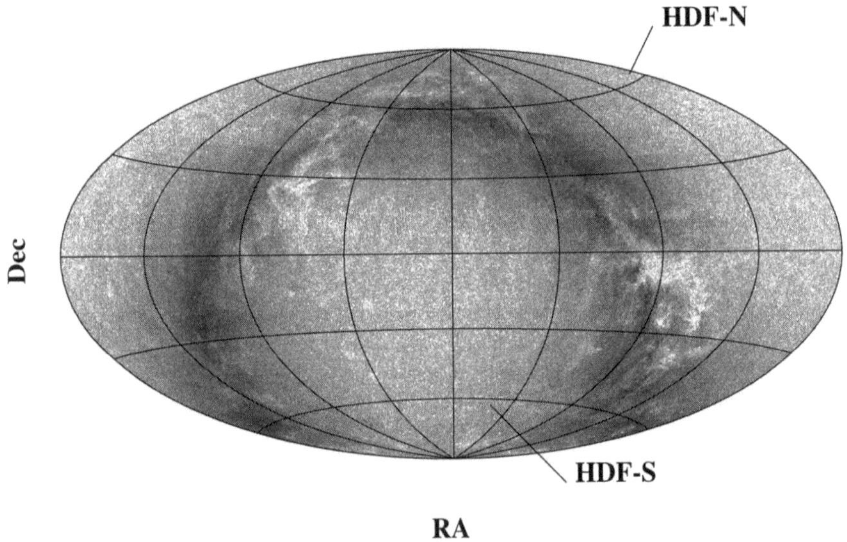

Fig. 1. Tycho 2, ESA. Distribution of the 2.5×10^6 brightest stars in the Sky

Figure 1 shows the distribution of stars in the Tycho 2 catalogue and the position of HDF North and South. Several other considerations were taken besides stellar density to choose these fields. The opposite situation may be of interest as well: *locate areas of the sky rich in galaxies down to R=26 to be selected as target areas in a supernova search.*

The last question would require, in principle, the analysis of one catalogue while the HDF selection involved much more information than a stellar catalogue. Therefore, the family of questions involving global properties of the "object" distribution in the sky is quite a valid one. Notice that "object" is used without meaning a physical source in the sky. Digging in the observing log of HST provides pointings; locating frequent pointings to study variability or to form other deep field areas are valid problems not involving specific sources.

And looking for the most observed areas of the sky between HST, VLA, and XMM (and/or other instruments) to find "multi-wavelength deep fields" may prove to be a very important scientific question once the archives grow with more and more observations. Time studies would benefit from the archives enormously, provided that identification of frequently observed areas is at the fingertips. Areas like cosmology, stellar variability, galactic dynamics and astrometry could make use of these resources.

2 Requirements

As mentioned above, the very existence of sky-positional indexation schemes applied to many catalogues makes the problem much easier to solve. It would be enough to keep the information on how many "objects" exist per sky element (tile) to determine areas of interest based on sky density. It should be up to each data-centre to select the indexation scheme to use, as long as the above-mentioned operations are possible.

It is worth noting that this scheme will most likely reside outside the database system used to handle the catalogues, and in fact, it could be considered as yet another piece of meta-information for each catalogue, which combined with those of similar catalogues can lead to answer quite interesting scientific questions.

3 Benefits and Possible Applications

One important benefit of using the described technique is that it permits to pre-test the properties of a cross correlation without accessing the involved catalogues. If the catalogues are small, the difference in execution time is not marked, but in cases like USNO and other large databases, the benefit is enormous.

On the other hand, providing users with an "image" of the sky coverage is a very effective method of describing the global properties of a catalogue, an issue not always easy to solve by describing the sky coverage with statistical means. The risk of "blind data-mining" can be minimized if a user can have a snapshot of the distribution of sources in the sky, preventing the "temptation" to run statistical tests which require a uniform degree of completeness. Figure 2 illustrates the case of a relatively populated catalogue ($\sim$ 25000 objects), covering the whole sky with a very inhomogeneous density distribution.

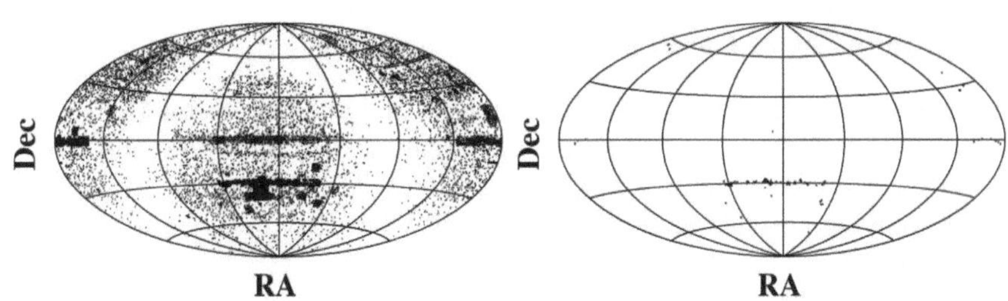

Fig. 2. Veron et al. 2001. Distribution of AGNs in the Sky. (a) All QSOs listed in the catalogue (b) QSOs located in areas of the sky with densities greater than 50 sources per square degree

3.1 Selecting Areas According to Sky Density

A relevant aspect is the capacity to select areas of the sky and/or objects located in areas selected according to the 2-D number density. Applications like locating HDF are possible, as it is to locate areas of the sky where the QSO density is high enough to use these object as astrometric references instead of stars. This case is illustrated by Fig. 2, where all the sources are shown in **(a)**. It is evident that some areas show a much higher density than others. The galactic plane is shown as an under-density, while "popular search areas" like the Equator and the -30° zone show a higher density. **(b)** shows only the QSOs located in regions where the density is greater than 50 sources per square degree. The figures were created using datoz2k [6], a system which allows querying by sky density.

3.2 Determining the Intersection of Catalogues

Given a number of catalogues, for which one has an indexation grid, determining the areas where intersection occurs is straightforward. It would be enough to AND the two grids and look for the non-zero elements. A direct application would be to determine if it is worth matching a list of targets against a given catalogue. Although this operation may appear trivial or unnecessary, in the case where a user wants to access a long list of catalogues determining that there exists a empty/non-empty intersection does save a significant amount of time.

3.3 Combining Different Density Criteria for Different Catalogues

This operation may imply computing whether the source density satisfies certain conditions for each catalogue, a 0 or a 1 could be assigned and the result for each area should be ANDed. Only the areas with a 1 shall survive. The applications are various, from looking for areas of low stellar density and high galaxy density, to areas of high QSO density with low density of radio sources. At the observing log level, look for frequently observed areas in the optical which have been observed at least once by any given space mission.

3.4 Search for Absolute or Relative Extrema from a Set of Catalogues

Particularly in the multi-wavelength regime one might be interested to find areas where density enhancements or depletions occur, as defined by the total number of sources in the area compared to adjacent areas. These areas can not be determined applying absolute criteria to the density, but they must be studied after the information of one or more catalogues requested by a user is combined. These density enhancement or depletions may be the result of physical phenomena or observational biases (or a combination of both). No automatic tool is able to handle this question up to now, but when built it should take into consideration the size of the sky areas. Clusters of galaxies or any other kind of objects or even pointings accumulations should be made evident with this method.

4 Applications to Archives and VO

Target-less positionally-oriented questions may provide the preliminary steps toward deciding which catalogues to use in the context of the Virtual Observatory, as well as for selecting areas of the sky which satisfy a number of criteria and which could represent a particular physical situation. The examples given above are just the tip of the iceberg of exploring the data space in this fashion.

In order for these methods to be an integral part of VO, standards should be adopted for the exchange of positional information among data centres, which may choose to use their own approach or not to use them at all. Depending on the patchiness of a catalogue, different methods to transmit the data could be used. If only a minor part of the sky is covered, then passing the information could be trivial: just point the tiles used and the sky density in each area. If most of the areas were used, then perhaps a binary transmission is due.

Within VO, some operation may imply moving data from one site to another to perform correlations. Time can be saved if the number of objects listed in a catalogue is larger than the number of indexation tiles and comparisons can be made using logical operands. A catalogue intersection method could tell us if it is worth moving the data from one point to another saving valuable bandwidth.

Archives, where the amount of data is an increasing function of time, could provide a tool to their users to explore the most/least explored areas of the sky. This could: **a)** avoid repeating observations – searching by object name may not always be effective, as if the original PI was interested in a particular galaxy (which happens to be close to a Planetary Nebula), any query searching for the name of the PN will give no results. **b)** help finding heavily observed areas suitable for variability studies as well as the formation of deep images with a very long exposure time.

5 Conclusions

In a certain way, the operations described in this paper represent the *exploration phase* leading to the mining points. Few of these operation will lead to results at a source level, which one could consider the real mining places, but they point to areas of the data space which could be worth mining in a variety of domains. Users may know ahead of time where they want to start "digging", or they may engage in an exploration mode to find the material, in the form of images, spectra, or catalogues suitable for their purposes.

If sky source distribution is considered as meta-data, other operations involving meta-data can fall into the same category "beyond basic cross correlation". Recovering data for proper motion studies could require finding areas with a large number of observations AND that a significant number of observations were made with a time spacing of about a year (to rule out other known dynamic effects as parallax and aberration).

It would be good to start developing the ideas and the standards for data exchange which the issues mentioned in this paper involve in the context of the

Virtual Observatory. Some of the methods mentioned here have been described before in a somewhat different context – sky pixelization to create images, yet the highly useful applications coming from exploring this "sector" of the data domain makes it worth to work on and re-work these ideas while the first examples of a VO are being built, as well as identifying any other piece of meta-information needed to achieve these goals.

References

1. Górski, K. M.; Hivon, E.; Wandelt, B. D. 1998: *Proceedings of the MPA/ESO Cosmology Conference Evolution of Large-Scale Structure.* eds. A.J. Banday, R.S. Sheth and L. Da Costa.
2. Kunszt, P.; Szalay, A. S. Thakar, A. R. 'The hierarchical triangular mesh'. In *Mining the Sky, Proceedings of the MPA/ESO/MPE Workshop held at Garching, Germany, 31 July-4 August, 2000.*, edited by A. J. Banday, S. Zaroubi, and M. Bartelmann. Heidelberg: Springer-Verlag, 2001., p.631
3. O'Mullane, W.; Banday, A. J.; Górski, K. M.; Kunszt, P.; Szalay, A. S. 'Splitting the Sky - HTM and HEALPix'. In *Mining the Sky, Proceedings of the MPA/ESO/MPE Workshop held at Garching, Germany, 31 July-4 August, 2000.*, edited by A. J. Banday, S. Zaroubi, and M. Bartelmann. Heidelberg: Springer-Verlag, 2001., p.638
4. White, Richard A.; Stemwedel, Sally W. 'The Quadrilateralized Spherical Cube and Quad-Tree For All Sky Data'. In *Astronomical Data Analysis Software and Systems I, A.S.P. Conference Series, Vol. 25, 1992*, edited by Diana M. Worrall, Chris Biemesderfer, and Jeannette Barnes, p. 379
5. Ortiz, P. F.; Ochsenbein, F. 'Mining the CDS collection: A learning experience'. In *Mining the Sky, Proceedings of the MPA/ESO/MPE Workshop held at Garching, Germany, 31 July-4 August, 2000.*, edited by A. J. Banday, S. Zaroubi, and M. Bartelmann. Heidelberg: Springer-Verlag, 2001., p.677
6. Ortiz, Patricio F.: 'Accessing astronomical Data over the WWW using datOZ'. In *Astronomical Data Analysis Software and Systems VII, A.S.P. Conference Series, Vol. 145, 1998, R. Albrecht, R.N. Hook and H.A. Bushouse, eds., p.371*

Meeting the User Science Challenge for a Virtual Universe

Nicholas Walton

Institute of Astronomy, University of Cambridge, Madingley Road, Cambridge, CB3 0HA, UK

On behalf of the AstroGrid Consortium

`http://www.astrogrid.org`

Abstract. The AstroGrid project, a core UK eScience programme, is rapidly implementing the UK's contribution to the global drive towards the Virtual Observatory. An important aspect in the definition of AstroGrid's capabilities is to encourage and enable community interaction in specifying the science and use case drivers. From these, priorities can be derived, and the project architecture shaped to ensure that the relevant science requirements are met.

AstroGrid is aiming to support a broad spectrum of astronomical activity, with an initial emphasis on meeting the needs expressed by the UK community. This paper discusses how AstroGrid is balancing the scientific requirements of the community (which includes the Solar and STP communities). Note is made of the use of interactive mechanisms provided by the project in capturing user requirements.

1 The Virtual Observatory: An Introduction

The Virtual Observatory (VO) will be a set of co-operating and interoperable software systems that will allow users to interrogate multiple data centres in a seamless and transparent way. The VO will provide powerful new analysis and visualisation tools and give data centres a standard framework for publishing and delivering services using their data.

The long term vision of the VO is not one of a single software package, but rather of a framework which enables data centres to provide competing and co-operating data services, and software providers to offer compatible analysis and visualisation tools. The VO aims to make all archives speak the same language, with all being accessible, and manipulatable by a common set of tools through a uniform access method.

The first priority of AstroGrid, along with the other VO projects worldwide, is to develop this standardised framework to allow such creative diversity.

2 The AstroGrid Project

AstroGrid is one of three major funded world-wide projects (along with the European Astrophysical Virtual Observatory (AVO) [14] and US National Virtual Observatory (NVO) [6]) projects which are aiming to create an astronomical Virtual Observatory.

AstroGrid will develop a working implementation of immediate use to astronomers. As a consortium of UK data centres and software providers, AstroGrid aims to pool common resources, including key UK databases, storage, and compute facilities. As a UK e-Science project, its architecture is firmly based on a data-grid approach: making use of grid components produced by other projects, with AstroGrid developed components being made freely available to the community. The scientific aims of AstroGrid are:

- to improve the quality, efficiency, ease, speed, and cost-effectiveness of on-line astronomical research
- to make comparison and integration of data from diverse sources seamless and transparent
- to remove data analysis barriers to interdisciplinary research
- to make science involving manipulation of large datasets as easy and as powerful as possible.

Other papers in these proceedings describe the AstroGrid project in more detail. An overview of the project is given in [8]. The current AstroGrid approach to supporting data mining is described in [13]. The identity, authentication and authorisation model being developed by AstroGrid is detailed in [15]. The AstroGrid, year one pilot programme, is described in [9].

3 AstroGrid Development Activities

In its first year AstroGrid has been undertaking activities in a number of areas, with the primary emphasis being to capture the science requirements required to scope the project, together with the development of a project architecture within which the required capabilities can be developed.

Subsidiary activities have included engagement with the other international VO projects, closely with the AVO [3] where AstroGrid is a formal partner, and informally with projects such as the NVO [10]. Important activities such as agreeing low level interoperability standards have been undertaken within the context of the International Virtual Observatory Alliance [7] (see in [14]). AstroGrid has also engaged with other Grid and eScience projects, especially in the areas of grid technology and database access through the OGSA-DAI project [12].

Astrogrid technology demonstrations have been initiated (e.g. see in [15]) and the year one pilot programme has focused on learning from current practical problems encountered in federating multi-wavelength datasets (see in [9]).

4 The AstroGrid On-Line Collaboration Tools

AstroGrid has deployed a set of ground-breaking collaboration tools which have enabled team members to share experiences, seek opinions and expert advice, and create an extensive library of documents pertaining to all aspects of the project. These tools are:

- **Portal:** [http://www.astrogrid.org] – These pages give an overview of the project and provide entrance links into the collaborative pages.
- **News:** [http://news.astrogrid.org] – The news site is to announce issues of interest, including details of AstroGrid product releases. This is a public access site.
- **Forum:** [http://forum.astrogrid.org] – The forum is an area where discussions related to the development of the project can be initiated. This is a public access site.
- **Wiki:** [http://wiki.astrogrid.org] – The wiki is where all AstroGrid project documentation is not only stored but also developed. The wiki is an area where documents can be edited by public users, with version control enabling differences between versions to be tracked. The wiki has proved itself to be extremely useful in the development of the science and user requirements.

The great majority[1] of AstroGrid project development is therefore available for public perusal on the AstroGrid web sites. The Wiki and Forum sites also allow active participation and input from any user. These collaborative and open sites have proved to be essential for enabling the operation of the distributed development teams that make up the personnel of the project. These tools are now being deployed on other project web sites, for instance by the AVO [4].

5 Capturing the Science Requirements

The AstroGrid consortium members generated and gathered science cases reflecting the scientific interests of their research groups. Because the consortium contains representatives involved in a wide range of UK astronomy, solar and STP research activity, a broad range of science cases stressing areas such as radio astronomy, solar physics and solar/terrestrial physics resulted. This was a major and early task of the Project.

In the first instance a number of specific use cases were formulated. These were often concerned with how a part of science problem might be approached, for instance, running a query on a database to locate and show the positions of known QSO's. Other cases are aimed at easing the process of acquiring sufficient data to address a particular problem, e.g. returning the colours of galaxies and their bulges as a function of redshift to study alternative theories of galaxy evolution. The distinction between 'science cases' and more generic 'use cases' rapidly became apparent. Emphasis was placed on capturing and formalising the science case.

Further input for the science cases was sought from a number of areas. The project scientist was responsible for assessing current major scientific research strands with a view to identifying those areas likely to benefit from the promise of access to the distributed data and processing capabilities to be opened up by a virtual observatory. Areas here included those science areas serviced by large

[1] A small number of documents are held on an internal intranet server

scale multi-wavelength data survey's or those requiring access to large scale computational facilities. Note was made of the similar requirements survey undertaken by the Science Definition Team [11] of the NVO. Further requirements were captured via user input to the requirements area of the wiki.

The AstroGrid Phase-A Pilot Programme [9] provided a number of inputs into the Science Requirements of the project. In a similar vein, five science topics were fed into the pilot programme as the basis for the pilots.

5.1 Scoping the AstroGrid Science Drivers

Through internal meetings and contact with external scientists, we developed an extensive list of requirements, documented as science problems. The complete list is held online [2]. This selection of science drivers will be continually expanded upon throughout the project lifetime. These drivers will form an important resource for input into partner VO initiatives, especially the AVO. Partner projects such as EGSO [5] are also likely to draw from them. These 'VO' science cases, together with those being highlighted by other virtual observatory projects, will be used in shaping the evolution of longer term VO initiatives.

With the collation of the VO science drivers, the project recognised that AstroGrid would not be able to produce a virtual observatory capable of meeting the demands of all of these cases. Thus a formal and rigorous process was undertaken in order to select a well defined set of science drivers, the AstroGrid Ten, which would be used to shape the AstroGrid deliverables. The project scientist and scientific members of the team analysed the science drivers over a number of review meetings. Based on selection criteria the science problems were distilled to produce the key set of 'ten drivers', chosen to:

- Represent a cross section of currently topical Astronomy, Solar and STP research areas with functionalities covering a wide spread of technical areas.
- Be achievable within the AstroGrid Phase-B project span both in terms of technical complexity of solution, but also in terms of availability of input science data sets.
- Have a well defined user base who would benefit from capabilities generated by the project
- But at the same time, the tools generated to satisfy the science project would be of use across a wide range of problem areas.

The AstroGrid Ten drivers will be under formal version control from the beginning of AstroGrid's development Phase B.

5.2 The AstroGrid 'Top Ten' Science Drivers

The AstroGrid 'Ten' science cases have been selected as they demand the delivery of a set of capabilities that will be relevant to aiding scientific output in these topical science areas.

- Discovery of High Redshift Quasars (Astronomy: involves optical and near-IR data sets)

- Locating galaxy clusters at a range of redshifts. This will be linked to predictions from model data. (Astronomy: Models)
- Brown Dwarf selections (Astronomy: involves optical and near-IR data sets)
- Deep Field Surveys (Astronomy: involves linking radio with other data on deep fields such as the Hubble Deep Fields).
- Low Surface Brightness galaxy discovery (Astronomy: optical, near-IR, radio, X-ray)
- Supernova galaxy environments (Astronomy: Optical)
- Solar Stellar Flare Comparison (Astronomy, Solar: linking understanding of solar flares with those seen in stellar events)
- Solar Coronal Waves (Solar: unravelling this phenomenon)
- STP Solar Event Coincidence (STP and Solar, links solar data with STP data in understanding space weather)
- Magnetic storm onsets (STP, federation of in-situ and remote sensing data)

These science topics encompass a wide cross section of areas, and include drivers aiming to bridge subject area gaps; thus there is a driver bringing together Astronomy and Solar studies, whilst another links the Solar and STP domains. Importantly, the capabilities demanded by these drivers are seen as being of use more generally in the support of other areas of astronomy. The ten key science drivers have been selected to enable the project to be scoped to ensure that a Virtual Observatory capability is generated by the end of the AstroGrid Phase-B period. The ten science cases are being collated into the overall 'AstroGrid Science Requirements' document, a constituent part of the AstroGrid Phase-A report [1].

6 Architecture Development

For each science driver a typical flow of events was constructed, decomposing the tasks required to complete that process. Sequence Diagrams were thus generated for each of the science cases, and for the generic technical use cases (these covering activities such as negotiating access to jobs, logon to the system etc). The sum of these tasks represents the components of the system that are required to form the AstroGrid Phase-B product, to be developed within the framework laid down by the project architecture.

The capabilities derived for the AstroGrid Ten will have a usefulness to a wider scientific audience. Any science problem with a similar workflow to one of the AstroGrid Ten will be supported. For instance, searching for AGB candidates would benefit from the system developed to support searches of high redshift Quasars – the difference being one of types of input catalogues, and constraints on the discovery space.

6.1 From Science Driver to AstroGrid Product

The AstroGrid Ten science drivers are used to define the scope of the AstroGrid *deliverables. Each use case* was analysed and decomposed into a work flow, with the tasks required by the science cases being identified.

- **Generic Use Cases:** A number of use cases were captured which have generic utility. These cases are those that would be needed in any baseline virtual observatory system, and are largely infrastructural in nature.
- **Specific Use Cases:** Analysis of each science use case revealed that in addition to the need for the generic use cases, each would require more specialised use cases. For instance, the HiZQuasars case requires use cases such as ones to allow the determinations of redshifts of objects in the field.
- **AstroGrid's Use Case Set:** Analysis of the Ten science cases revealed the minimum set of use cases that would be required to enable the construction of the capabilities require to meet the demands of these science drivers.

In parallel to the scientific requirements process the architectural shape of AstroGrid was formulated. The use cases demanded by the science drivers, together with any of those indicated by the outline system architecture thus represent the reduced set of use cases that need to be developed by the AstroGrid project. A full description of the architectural development is given in [1].

7 Concluding Remarks

AstroGrid will provide tools and capabilities to help the researcher in producing solutions to these science topics. AstroGrid will not in itself provide the answers, the astronomer will be presented with new capabilities to aid their research. This will be especially so in the areas of data discovery, transformation of data into information via access to processing facilities, and management of the processing flow of events. The researcher will be able to devote more time to the understanding of the astrophysics revealed by their results, more time can be given to the important step of transforming information into knowledge.

References

1. Phase-A Report:
   `http://wiki.astrogrid.org/bin/view/Astrogrid/PhaseAReport`
2. AstroGrid Science Problems:
   `http://wiki.astrogrid.org/bin/view/VO/ScienceProblemList`
3. Astrophysical Virtual Observatory: `http://www.euro-vo.org`
4. Astrophysical Virtual Observatory Wiki: `http://www.euro-vo.org/intranet`
5. European Grid of Solar Observations: `http://www.mssl.ucl.ac.uk/grid/egso/`
6. R. Hanisch, 2002, these proceedings
7. International Virtual Observatory Alliance: `http://www.ivoa.net`
8. A. Lawrence, 2002, these proceedings
9. R. G. Mann et al, 2002, these proceedings
10. National Virtual Observatory: `http://www.us-vo.org`
11. National Virtual Observatory Science Definition Team: `http://www.nvosdt.org`
12. OGSA-DAI Project: `http://www.nesc.ac.uk/NeSC/general/projects/OGSA_DAI`
13. C. G. Page, 2002, these proceedings
14. P. J. Quinn, 2002, these proceedings
15. G. T. Rixon, 2002, these proceedings

Theoretical Astrophysics and the US-NVO Initiative

David De Young

National Optical Astronomy Observatory, 950 N. Cherry Av., Tucson, AZ 85719 USA

1 Introduction

Early discussions of the Virtual Observatory concept revolved almost exclusively around the use of large scale datasets derived from major observing programs. Subsequent development of the VO initiative has also been largely focussed on the problems and prospects of observational astronomy. However, it became clear even in the early stages of the VO development that theoretical astrophysics could play an important role in the Virtual Observatory. This paper provides some background about the development of trends in theoretical astrophysics in recent years and shows how these trends have made the contributions of theory an essential part of any viable VO model. A specific implementation of these ideas comes from an examination of current developments in the US vis a vis the incorporation of theoretical astrophysics into the US-NVO initiative.

2 Theory Then

Theoretical astrophysics is almost as old as observational astronomy, and some of the greatest minds in the history of thought have turned to contemplation of problems in astrophysics. With this grand history and tradition, it is no surprise that many of the past contributions to theoretical astrophysics have been characterized by brilliance, elegance, and a high level of mathematical sophistication. In the past the search has usually been for an "exact solution" to the problem in which a complete analytical description, without approximations, is found. Such solutions are often very general and of widespread applicability. They are also rather rare. The complexity of all but the most simple of astronomical systems has, in the past, often prevented the discovery of exact solutions, and thus in most cases some degree of approximation has been employed. Often such approximate methods themselves were sophisticated and mathematically complex, and again these approximate solutions attracted the attention of some of the most talented theorists of the time. More recently, the trend in approximations for analytic solutions has become ever more simplistic; these are often characterized by order of magnitude estimates or mathematical relations that are guessed by recourse to dimensional arguments. These treatments often result in very simple algebraic relations that may or may not have some applicability to the actual astrophysical problem being considered. This trend in analytic treatments may be due in part to the inexact nature of the observational data for some of the more exotic and interesting areas of current research; such uncertainty often provides constraints only to within an order of magnitude and so does not merit theoretical efforts with greater refinement. Employment of order of magnitude analytic estimates also does not require to mathematical prowess that was often required of past participants in theoretical astrophysics.

Another character of past theoretical work, even in the case of exact solutions, was the difficulty in making significant comparisons with observations. The analytic solutions usually employed idealized geometries and boundary conditions rarely encountered in nature. In addition such solutions often did not (and often still do not) include many important processes simply because construction of a solution with such processes included would be extremely difficult. Finally, such solutions usually do not calculate the radiative signature that would be expected from the proposed solution process. This makes the comparison with data and hence the verification of the theory quite difficult, especially for complex solutions.

A particularly irksome barrier for theoretical astrophysics in the past was the inability to treat non-linear phenomena with any degree of precision. In particular, all hydrodynamic phenomena that involve supersonic flows and/or turbulence are resistant to analytic treatment. Thus important phenomena such as star formation, protogalaxy evolution, stellar structure, stellar winds, mass transfer in binary systems, accretion disks and active galactic nuclei could only be treated in the most simple approximations. This difficulty arises not out of ignorance or lack of talent but is due instead to the non-existence of analytic solutions to the hyperbolic differential equations that govern such flows. The only recourse is numerical integration of these equations, and the very earliest techniques, such as the method of characteristics, were extremely painful. More recently (i.e., three decades ago) early numerical integration schemes using numerical techniques of limited spatial resolution and/or limited degrees of freedom provided results that were actually misleading. A clear example of this is the use of two-dimensional axisymmetry in modeling the propagation of supersonic astrophysical jets, particularly in the case of MHD jets.

3 Theory Now

The discovery of exact solutions in theoretical astrophysics is as difficult and rare now as it was in the past, and the satisfaction gained from finding such solutions has not diminished. However, the situation regarding the treatment of non-linear problems in theory has undergone a dramatic revolution. During the last ten years the computational capability available to theoretical astrophysicists has grown by orders of magnitude. As a result, high resolution, time dependent, three dimensional numerical simulations of realistic astrophysical problems can now be carried out in a reasonable time. These calculations can include the effects of gravity, magnetic fields, radiation and radiative transfer, as well as the effects of special and general relativity. Complex physical processes such as nuclear reactions can be calculated. Geometries are no longer constrained to simple symmetries, and realistic initial and boundary conditions can also be included. This increased computational capability also means that the radiation resulting from the models being calculated can be determined with high accuracy, and this enables for the first time serious and stringent comparisons between theoretical models and observations. This level of detail in modeling comes at just the right time, for it is only now that arcsecond and subarcsecond resolution is available from observations spanning the range from x-ray through optical and infrared into radio frequencies.

Thus theory and observation have both reached a pivotal point at the same time, and a new era of confrontation between theory and observation is now possible. For the first time one will be able to perform truly significant tests of models, and conversely, the models are now becoming detailed enough to provide meaningful guides to new observational programs. This fortuitous conjunction of theoretical and observational prowess now makes possible a new paradigm for scientific inquiry, and that is the discovery of new scientific results through data mining of a truly virtual sky, coupled with the mining of data from the Universe itself. This process will be enabled by the Virtual Observatory. Before discussing those aspects of the VO that will make this possible, it is useful to consider a few examples of data from this new era of computational astrophysics.

3.1 Some Examples of Current Theory Programs

Three kinds of observational data that will be common in the datasets incorporated into the VO are imaging data, tabular data, and spectroscopic data. Theoretical calculations can now produce large datasets in these three formats as well, thus making possible the direct and immediate comparison between theory and observation.

1. Imaging Data: 3-D MHD Jet Propagation These complex calculations have been recently carried out by Tregillis, Jones and Ryu [1], and they include solving the full three dimensional MHD equations for a supersonically propagating jet. In addition, the calculations include particle reacceleration at shock fronts within the jet, which is a difficult calculation by itself, together with the calculation of radiation losses due to both synchrotron losses from the relativistic particles in the magnetic field and inverse Compton losses due to electron collisions with photons both from the synchrotron radiation (synchrotron self-Compton or SSC) and from the cosmic microwave background. Thus this is an extremely complex calculation. Figures 1 and 2 show a comparison of the radiation emitted in x-rays by one of the model jets when the emission mechanism is SSC and when the x-rays arise from inverse Compton scattering of the CMB.

It can readily be seen that the two radiative signatures are very different, and thus for the first time there is a real possibility of determining which of these two processes is actually at work. Figures 3 and 4 show the radiation emitted from a model jet after it has been convolved with the VLA beam for the radio radiation and also after convolutions with the 1.2 keV Chandra response in the x-ray regime.

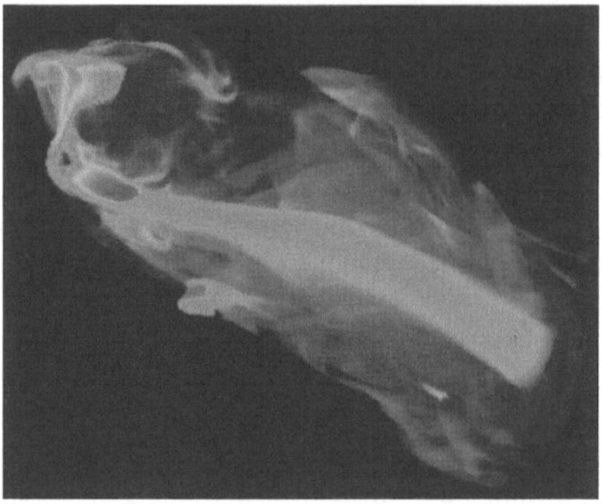

Fig. 1. X-ray radiation produced by the synchrotron-self Compton mechanism in a relativistic jet

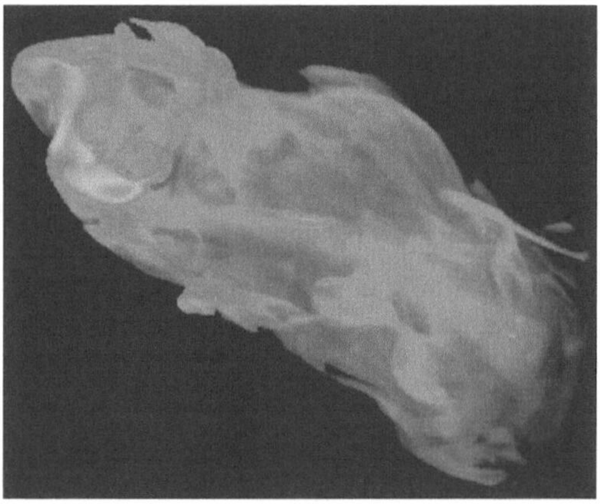

Fig. 2. X-ray radiation produced by Compton scattering of the microwave background in a relativistic jet

Hence the calculations produce results that allow a direct and immediate comparison with the observational data. Thus it is clear that a library of such calculations that cover a significant range in the multidimensional parameter space of jet speed, energy, age, particle distribution, environment, etc. can provide a very valuable (and very large) dataset for comparing models with observations.

2. Tabular Data: The Evolution of Globular Clusters Use of dedicated hardware components to solve the gravitational potential for N-body systems now allows the determination of the evolution of globular clusters. These amazing calculations actually follow the dynamical and evolutionary history of *every* star

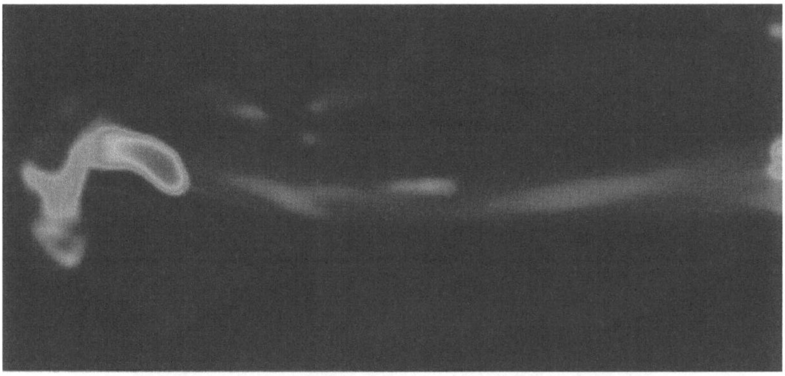

Fig. 3. Radio emission from a relativistic jet as seen by the VLA at 5.4 GHz

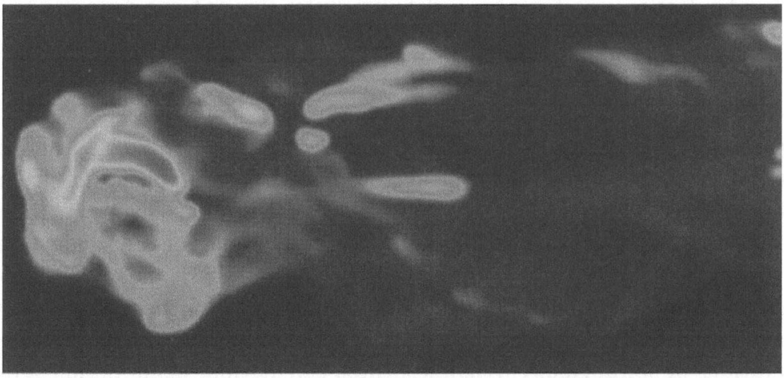

Fig. 4. X-ray radiation from the relativistic jet in Fig. 2 as seen by the Chandra observatory at 1.2 keV

in a globular cluster. This enormous calculation thus enables an investigation of many different problems in many different areas of interest. For example, stellar dynamical issues such as the occurrence and evolution of binary systems, the roles of stellar collisions and coalescence, the formation of a central core and core collapse can all be addressed. In addition the evolution of the stellar populations, stellar mass loss, injection of metals into the ISM and the effects on galactic metallicity can also be investigated. The roles of dark matter and the evolution of black holes is another area of inquiry to which these calculations can be applied, and Fig. 5 shows some results on the evolution of a globular cluster near the Galactic center.

The very large datasets resulting from calculations such as these form a valuable resource for data mining to address a wide variety of astrophysical problems.

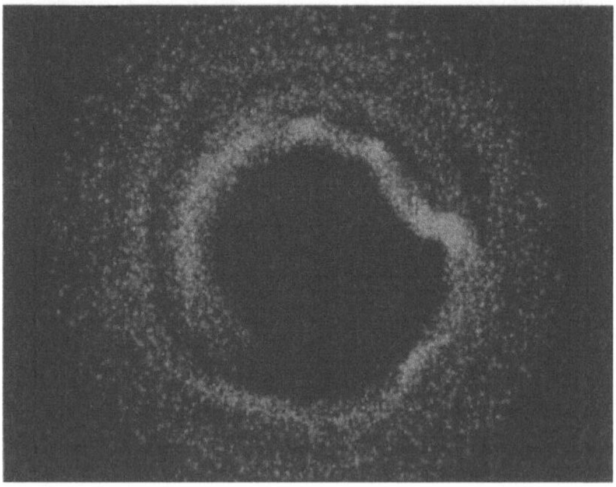

Fig. 5. The distribution of stars in a globular cluster that is being destroyed by tidal forces near the Galactic center. Only the stars in the globular cluster are shown. From Portegies Zwart et al. [2]

3. Spectral Data: Simulation of Galaxy Dynamics A third class of dataset produced by contemporary theoretical astrophysics calculations comes from the numerical simulation of galaxy dynamics. These calculations involve inclusion of the evolution of the gravitational potential, the dynamics of the stellar population, and the hydrodynamic evolution of the gaseous ISM under the influence of this changing potential. The spectral line emission from the ISM, CO in particular, is then calculated. Since CO is a commonly used diagnostic line for observations of spiral galaxy dynamics, these calculations again permit a direct comparison between the various theoretical models and the observational data. Figure 6 shows an example of the kinds of results that can be obtained for CO observations of barred spiral galaxies.

4 Theory and the Virtual Observatory

The three examples shown above illustrate the very different nature of theoretical astrophysics calculations in the current era of high speed computing capability. They also illustrate the vastly increased ability to compare large scale and complex theory calculations with contemporary high resolution and high sensitivity observations over a wide range of wavelengths. In many cases the sizes of the datasets resulting from complex calculations covering a large range in parameter space are comparable to those resulting from large survey programs - from 10 GB to 10 TB. Moreover, as shown above, the results of these calculations are very general and can be applied to many different specific astrophysical inquiries. Hence the the datasets arising from present day theory calculations are ideal candidates for inclusion in the Virtual Observatory. They permit "data mining" from both observations and theory, they can provide motivation for new observational programs, and they can reveal "buried physics" not previously seen or

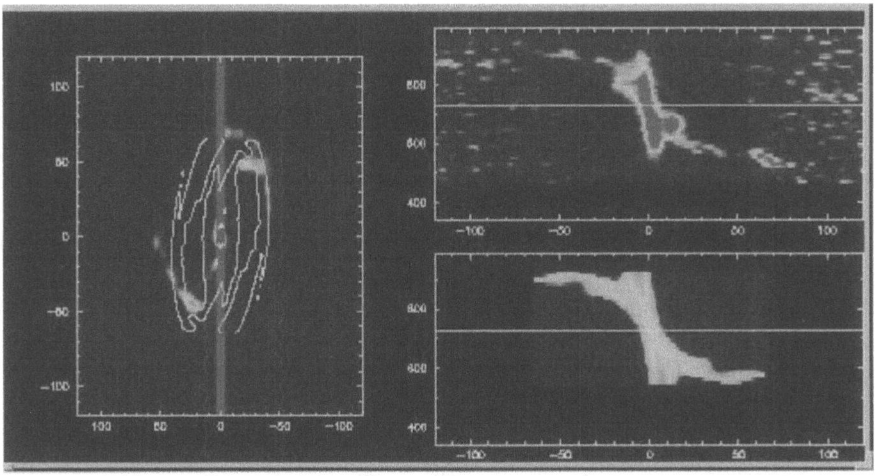

Fig. 6. Results of simulation of barred galaxy evolution. The left panel shows the distribution stars (contours) and gas (greyscale) in the galaxy from the simulation, together with the position of a spectrograph slit (vertical line). The top right panel shows the observed CO data, while the bottom right panel shows the velocity- position diagram obtained from the "slit" on the right panel. From Piner, Stone & Teuben [3]

understood. In addition, the data acquisition and analysis tools used to examine the results of these calculations are often similar to or identical with those already in use for observational datasets in the VO environment. Finally, the theory datasets are readily adapted to the growing VO infrastructure, such as the use of grid technology for remote computing and web access to the data via VO portals.

Another very important contribution to the the VO effort from the current theory efforts lies in the area of education and outreach (EO). EO is a central part of all VO programs, and theory calculations, with their dramatic graphics, clear animations, and detailed portrayal of the underlying physics provide ideal examples and tools to be used in formal and informal education and public outreach. The availability of such useful graphical results is a recent phenomenon, and it permits for the first time access to the understanding of fundamental astrophysical processes by educators and by the general public. This use of the results of theory calculations promises to be one of the most powerful tools in the education and outreach programs of the Virtual Observatory.

5 Theory and the US-NVO

Soon after the concept of a VO began to be discussed in the US astronomical community, it was recognized by many theorists that the VO idea included functions that had been under discussion for some time in the theory community. In particular, the idea of creating "libraries" of calculations covering a range in parameter space was already being examined by some N-body theorists. In addition, the dissemination of codes or subroutines among working groups in subspecialities was also taking place, and thus the VO appeared to be a natural extension of these ideas. The ability to quickly and easily compare theory models with observations also struck a natural resonance. Hence by late 2000 discussions were taking place among an interested group of theorists in the US, and the formation of a "Theory Virtual Observatory" or TVO was gaining support.

One year later, an interim "TVO Site" had been formed, and this is illustrated in Fig. 7.

In addition, a group of theorists is now included in the Science Working Group of the US NSF/ITR NVO program. Theoretical astrophysics is also represented on the Executive Committee of the US NVO management team. Input from the

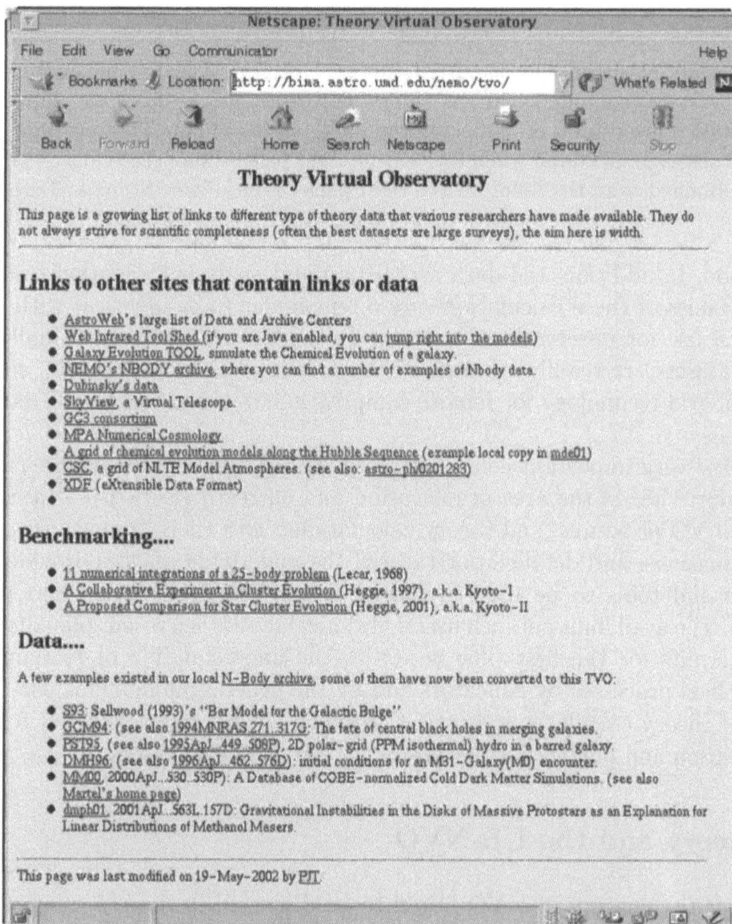

Fig. 7. Interim TVO site for the US-NVO

US theoretical community is also being sought during the development of the science demonstrations for the US-NVO. These demonstrations are planned to be shown at the next meeting of the American Astronomical Society in January 2003.

Hence there is in the US-NVO initiative a recognition that the contributions from theoretical astrophysics are an essential part of any VO program, and steps are being taken to ensure inclusion of large scale theory datasets in the NVO archives and analysis toolsets. It seems clear that theory must play a well defined role in any VO program, since the results of contemporary theory calculations can enrich current and future observational programs, can provide further stimulus to other theory initiatives, and can make major contributions to the educational and public outreach programs of every VO initiative.

I thank Dr. Nigel Sharp for valuable discussion and consultation.

References

1. I. Tregillis, T. Jones & D. Ryu: ApJ. **557**, 475 (2001)
2. S. Portegies Zwart, J. Makino, S. McMillan & P. Hut: ApJ., **565**, 265 (2002)
3. B. Piner, J. Stone & P. Teuben: ApJ., **449**, 508 (1995)

Artificial Intelligence Tools for Data Mining in Large Astronomical Databases

Giuseppe Longo[1,2], Ciro Donalek[2], Giancarlo Raiconi[3,4], Antonino Staiano[3], Roberto Tagliaferri[3,4], Fabio Pasian[5], Salvatore Sessa[6], Riccardo Smareglia[5], and Alfredo Volpicelli[3]

[1] Department of Physical Sciences, University Federico II, via Cinthia, Napoli, Italy
[2] INAF - Osservatorio Astronomico di Capodimonte, via Moiariello 16, Napoli, Italy
[3] Department of Mathematics and Informatics - DMI, University of Salerno, Baronissi, Italy
[4] INFM - Section of Salerno, via S. Allende, Baronissi, Italy
[5] INAF - Osservatorio Astrofisico di Trieste, via Tiepolo 13, Trieste, Italy
[6] Dipartimento di Matematica Applicata, Facoltà di Architettura, Università Federico II, Napoli, Italy

Abstract. The federation of heterogeneous large astronomical databases foreseen in the framework of the AVO and NVO projects will pose unprecedented data mining and visualization problems which may find a rather natural and user friendly answer in artificial intelligence (A.I.) tools based on neural networks, fuzzy-C sets or genetic algorithms. We shortly describe some tools implemented by the AstroNeural collaboration (Napoli-Salerno) aimed to perform complex tasks such as, for instance, unsupervised and supervised clustering and time series analysis. Two very different applications to the analysis of photometric redshifts of galaxies in the Sloan Early Data Release and to the telemetry of the TNG (telescopio nazionale Galileo) are also discussed as template cases.

1 Introduction

Data mining has been defined as "the extraction of implicit, previously unknown and potentially useful information from the data (Witten & Frank, 2000)". This definition fits well the expectations raised by the ongoing implementation of the International Virtual Observatory (or IVO) as a natural evolution of the European AVO (Astrophysical Virtual Observatory, `http://www.eso.org/avo/`) and of the American NVO (National Virtual Observatory, `http://nvosdt.org`). The scientific exploitation of heterogeneous, distributed and large databases (multiwavelength, multiepoch, multiformat, etc.; [1–3]) will in fact require the solution - in an user friendly and distributed environment - of old problems such as the implementation of complex queries, advanced visualization, accurate statistics, pattern recognition, etc. which are at the core of modern data mining techniques. The experience of other scientific (and non scientific) communities shows that some tasks can be effectively (id est in terms of accuracy and computing time) addressed by those Artificial Intelligence (A.I.) tools which are usually grouped under the label "machine learning". Among these tools particularly relevant are neural networks, genetic algorithms, fuzzy C-sets, etc. [4,5]. A rather complete review of ongoing applications of neural networks (NNs)to the fields of astronomy, geology and environmental science can be found in [6].

In this paper we shall address only some topics which were tackled in the framework of the AstroMining collaboration and, furthermore, we shall focus on tools which work in the "catalogue space" of processed data (as opposed to the "pixel space" represented by the raw data) which, as stressed by [2] is a multiparametric space with a maximum dimensionality defined by the whole set of measured attributes for each given object.

2 Methodological Background

The main aim of the AstroMining collaboration (started in 1999 at the Universities of Napoli and Salerno[1]) is the implementation of a set of tools to be integrated within the data reduction and analysis pipeline of the VLT Survey Telescope (VST): a wide field 2.6 m telescope which will be installed, by the end of year 2003, on Cerro Paranal, next to the four units of the Very Large Telescope (VLT). VST will be equipped with a large format 16k×16k CCD camera built by the OmegaCam Dutch-Italian-German Consortium and the expected data flow is of more than 100 GB of raw data per observing night. Due to the planned scientific goals, these data will often need to be handled, processed, mined and analysed on a very short time scale (for some applications, less than 8-12 hours).

Before proceeding at describing some of the main tools implemented in AstroMining, it is useful to stress a few points. All the above described tasks may be reduced to clustering or pattern recognition problems, *id est* to the search of statistical (or otherwise defined) similarities among the elements (data) of what we shall call the input space. As already mentioned above, NN's may work either in supervised or in unsupervised mode, where supervised means that the NN learns how to recognize patterns or how to cluster the data with respect to some parameters, by means of a rather large subset of data for which there is an *a priori* and accurate knowledge of the same parameter. Unsupervised means instead that the NN identifies clusters or patterns using only some statistical properties of the input data without any need for *a priori* knowledge. To be more explicit, let us focus on a specific application, namely the well known star galaxy classification problem, a task which can be approached using both supervised and unsupervised methods. In this case the input space will be a table containing the astrometric, morphological and photometric parameters for all objects in a given field.

Supervised methods require that for a rather large subset of data in the input space there must be an *a priori* and accurate knowledge of the desired property (in this case the membership into either the Star or the Galaxy classes). This subset defines the "training set" and, in order for the NN to learn properly, it needs to sample the whole parameter space. The "a priori knowledge" needed for the objects in the training sets needs therefore to be acquired by means of

[1] Financed through a MURST-COFIN grant and an ASI (Italian Space Agency) grant.

either visual or automatic inspection of higher S/N and better angular resolution data, and cannot be imported from other data sets, unless they overlap and are homogeneous to the one which is under study. Supervised methods are usually fast and very accurate, but the construction of a proper training set may be a rather troublesome task [7].

Unsupervised methods do not require a training set and can be used to cluster the input data in classes on the basis of their statistical properties only. Whether these clusters are or not significant to a specific problem and which meaning has to be attributed to a given class is not obvious and it requires an additional phase, the so called "labeling". The labeling can be carried out even if the desired information (label) is available only for a small number of objects representative of the desired classes (in this case, for a few stars and a few galaxies).

It has to be stressed, that - does not matter whether they are supervised or unsupervised - all these techniques, in order to be effective, require a lengthy procedure to be optimized and extensive testing is needed to evaluate their robustness against noise and inaccuracy of the input data.

The AstroMining codes are written in MatLab and in what follows we shall shortly summarise the most relevant adopted methodologies, namely the Self Organizing maps, the Generative Topographic Mapping and the Fuzzy C Sets.

2.1 Self-Organizing Maps or SOM

The SOM algorithm [8] is based on unsupervised competitive learning, which means that the training is entirely data-driven and the neurons of the map compete with each other [9]. A SOM allows the approximation of the probability density function of the data in the training set (id est prototype vectors best describing the data), and a highly visualized approach to the understanding of the statistical characteristics of the data. In a crude approximation, a SOM is composed by neurons located on a regular, usually 1 or 2-dimensional, grid. Each neuron i of the SOM may be represented as an n-dimensional weight:

$$\mathbf{m}_i = [m_{i1}, \ldots, m_{in}]^T \tag{1}$$

where n is the dimension of the input vectors. Higher dimensional grids are not commonly used since in this case the visualization of the outputs becomes problematic.

In most implementations, SOM's neurons are connected to the adjacent ones by a neighborhood relation which dictates the structure of the map. In the 2-dimensional case, the neurons of the map can be arranged either on a rectangular or a hexagonal lattice, and the total number of neurons determines the granularity of the resulting mapping thus affecting the accuracy and the generalization capability of the SOM.

The use of SOMs as data mining tools requires several logical steps: the construction and the normalization of the data set (usually to 0 mean and unit variance), the initialization and the training of the map, the visualization and the analysis of the results. In the SOMs, the topological relations and the number of neurons are fixed from the beginning via a trial and error procedure, with the neighborhood size controlling the smoothness and generalization of the mapping. The initialization consists in providing the initial weights to the neurons and, even though the SOM are robust with respect to the initial choice, a proper initialization allows faster convergence.

AstroMining allows three different types of initialization procedures: *random initialization*, where the weight vectors are initialized with small random values; *sample initialization*, where the weight vectors are initialized with random samples drawn from the input data set; *linear initialization*, where the weight vectors are initialized in an orderly fashion along the linear subspace spanned by the two principal eigenvectors of the input data set. The corresponding eigenvectors are then calculated using the Gram-Schmidt procedure detailed in [9].

The initialization is followed by the training phase. In each training step, one sample vector $\mathbf{x}$ from the input data set is randomly chosen and a similarity measure is calculated between it and all the weight vectors of the map. The Best-Matching Unit (BMU), denoted as c, is the unit whose weight vector has the greatest similarity with the input sample $\mathbf{x}$. This similarity is usually defined via a distance (usually Euclidean). Formally speaking, the BMU can be defined as the neuron for which:

$$\|\mathbf{x} - \mathbf{m}_c\| = \min_i \|\mathbf{x} - \mathbf{m}_i\| \tag{2}$$

where $\|\cdot\|$ is the adopted distance measure. After finding the BMU, the weight vectors of the SOM are updated and the weight vectors of the BMU and of its topological neighbors are moved in the direction of the input vector, in the input space. The SOM updating rule for the weight vector of the unit i can be written as:

$$\mathbf{m}_i(t+1) = \mathbf{m}_i(t) + h_{ci}(t)[\mathbf{x}(t) - \mathbf{m}_i(t)] \tag{3}$$

where t denotes the time, $\mathbf{x}(t)$, the input vector and $h_{ci}(t)$ the neighborhood kernel around the winner unit, defined as a non-increasing function of the time and of the distance of unit i from the winner unit c which defines the region of influence that the input sample has on the SOM. This kernel is composed by two parts: the neighborhood function $h(d,t)$ and the learning rate function $\alpha(t)$:

$$h_{ci}(t) = h(\|\mathbf{r}_c - \mathbf{r}_i\|, t)\alpha(t) \tag{4}$$

where r_i is the location of unit i on the map grid. The AstroMining package allows the use of several neighborhood functions, among which the most commonly used is the so called Gaussian neighborhood function:

$$\exp(-\|\mathbf{r}_c - \mathbf{r}_i\|^2/2\sigma^2(t)) \tag{5}$$

The learning rate $\alpha(t)$ is a decreasing function of time which, in the AstroMining package is:

$$\alpha(t) = (A/t + B) \tag{6}$$

where A and B are some suitably selected positive constants. Since also the neighbors radius is decreasing in time, then the training of the SOM can be seen as performed in two phases. In the first one relatively large initial α value and neighborhood radius are used, and decrease in time. In the second phase both α value and neighborhood radius are small constants right from the beginning.

2.2 Generative Topographic Mapping or GTM

Latent variable models [10] aim to find a representation for the distribution $p(x)$ of data in a D-dimensional space $x = [x_1, ..., x_D]$ in terms of a number L of latent variables $z = [z_1, ..., z_L]$ (where, in order for the model to be useful, L must be much smaller than D). This is usually achieved by means of a non linear function $y(z, \mathbf{W})$, governed by a set of parameters $\mathbf{W}$, which maps the points $\mathbf{W}$ in the latent space into corresponding points $y(z, \mathbf{W})$ of the input space. In other words, $y(z, \mathbf{W})$ maps the hidden variable space into an L-dimensional non euclidean manifold embedded within the input space.

Therefore, a probability distribution $p(z)$ (also known as "a prior" distribution of z) defined in the latent variable space will induce a corresponding distribution $p(y|z)$ in the input data space. The AstroMining GTM routines are largely based on the Matlab GTM Toolbox [10] and provide the user with a complete environment for GTM analysis and visualization. In order to render more "user friendly" the interpretation of the resulting maps, the GTM package defines a probability distribution in the data space conditioned on the latent variables and, using the Bayes Theorem, the posterior distribution in the latent space for any given point x in the input data space, is:

$$p(z_k|x) = \frac{p(x|z_k, \mathbf{W}, \beta)p(z_k)}{\sum_{k'} p(x|z_{k'}, \mathbf{W}, \beta)p(z_{k'})} \tag{7}$$

and, provided that the latent space has no more than three dimensions ($L = 3$), its visualization becomes trivial.

2.3 Fuzzy Similarity

Fuzzy sets are usually defined as mappings, or "generalized characteristic functions", from a universal set U into the real interval $[0, 1]$ [11] which plays the role of the set of the truth degrees. However, if further algebraic manipulation has to be performed, the set of truth values must be structured with an algebraic structure natural from the logical point of view. A general structure satisfying these requirements has been proposed in [12]: the complete residuated lattice which, by definition, is an algebra $\mathbf{L} = \langle L, \wedge, \vee, \otimes, \rightarrow, 0, 1\rangle$ where:

- $\mathbf{L} = \langle L, \wedge, \vee, 0, 1\rangle$ is a complete lattice with smallest and greatest elements equal to 0 and 1, respectively.
- $\mathbf{L} = \langle L, \otimes, 1\rangle$ is a commutative monoid, i.e $\otimes$ is associative and commutative, and the identity $x \otimes 1 = x$ holds

- $\otimes$ and $\rightarrow$ satisfy the adjointness property, i.e. $x \leq y \rightarrow z$ iff $x \otimes y \leq z$ holds.

Now, let p be a fixed natural number and let us define on the real unit interval I the binary operations $\otimes$ and $\rightarrow$ as:

$$x \otimes y = \sqrt[p]{max\{0, x^p + y^p - 1\}} \tag{8}$$
$$x \rightarrow y = min\{1, \sqrt[p]{1 - x^p + y^p}\}. \tag{9}$$

Then $\mathbf{I} = \langle L, \wedge, min, max, \rightarrow, 0, 1\rangle$ is a residuated lattice called *generalized Łukasiewicz structure*, in particular, *Łukasiewicz structure* if $p = 1$ [13]. This formalism needs to be completed by the introduction of the *bi-residuum*, id est an operator which offers an elegant way to interpret fuzzy logic *equivalence* and *fuzzy similar relation.* The bi-residuum is defined as follows:

$$x \leftrightarrow y = (x \rightarrow y) \wedge (y \rightarrow x) \tag{10}$$

In Łukasiewicz algebra, the bi-residuum is calculated via:

$$x \leftrightarrow y = 1 - max(x, y) + min(x, y) \tag{11}$$

Let A be a non-void set and $\otimes$ a continuous t-norm. Then, a *Fuzzy Similarity* (FS) S on A is a binary fuzzy relation such that, for each $x, y, z \in A$:

$$\begin{aligned} S\langle x, x\rangle &= 1 \\ S\langle x, y\rangle &= S\langle y, x\rangle \\ S\langle x, y\rangle \otimes S\langle y, z\rangle &\leq S\langle x, z\rangle \end{aligned}$$

Trivially, fuzzy similarity is a generalization of the classical equivalence relation also called *many-valued equivalence.* Now, let us recall that a fuzzy set X is an ordered couple (A, μ_X), where the reference set A is a non-void set and the *membership function* $\mu_X : A \rightarrow [0,1]$ tells the degree to which an element $a \in A$ belongs to a fuzzy set X, then we have that any fuzzy set (A, μ_X) on a reference set A generates a fuzzy similarity S on A, defined by

$$S(x, y) = \mu_X(x) \leftrightarrow \mu_X(y) \tag{12}$$

where x, y are elements of A. If we consider n Łukasiewicz valued fuzzy similarity S_i, $i = 1, \ldots, n$ on a set X, then:

$$S\langle x, y\rangle = \frac{1}{n} \sum_{i=1}^{n} S_i\langle x, y\rangle \tag{13}$$

is the *Total Fuzzy Similarity* (TFS) [14].
So far in the AstroMining tool, fuzzy similarity has been implemented to perform only object classification (such as, for instance, star-galaxy separation). The idea

behind this part of the tool is that a few prototypes for each object class can be used as reference points for the catalog objects.

The prototypes selection is accomplished by means of a Self Organizing Maps (SOM) or Fuzzy C-means. In this way, using a method, using the fuzzy similarities, it is possible to compute for each object of the catalog, its degree of similarity with respect to the prototypes.

3 The AstroMining Package

In Fig. 1 we depict the overall structure of the currently implemented package which, for the sake of simplicity, may be split into three sections.

The first section allows to import and manipulate both headers and data (in the form of Tables where each record corresponds to an object and each column to a feature) and to perform simple statistical tests to look for correlations in the data. In every day use not all features are relevant for a given application, and therefore AstroMining allows to run some statistical or neural tests to evaluate which features have to be kept (on the basis of significance) for the subsequent processing.

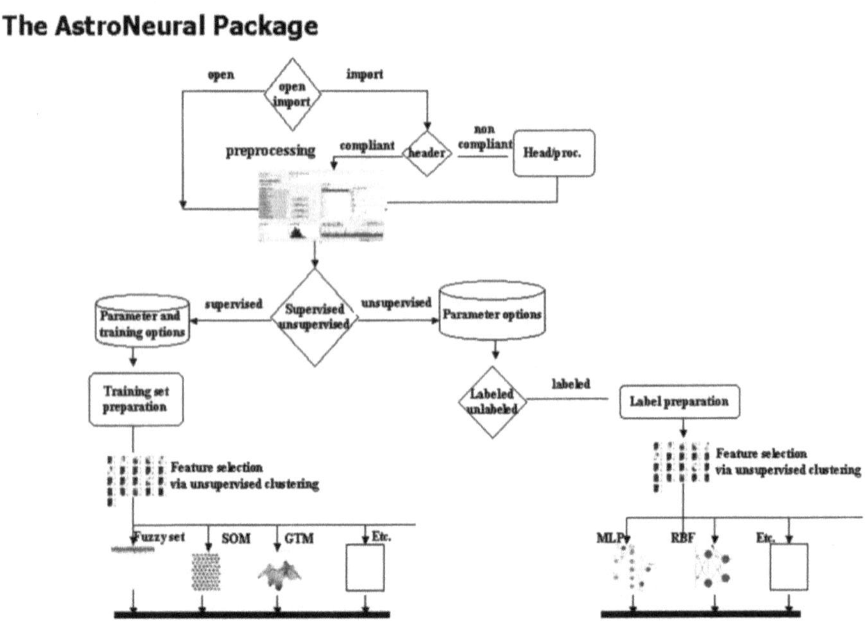

Fig. 1. Structure of the AstroNeural package

The second section allows to choose the modality (supervised or unsupervised) and select among a variety of possible options (SOM, GTM, Fuzzy similarity, etc.) accordingly to the specific task to be performed.

Finally, the third section allows to select the operating parameters of each option and the visualization modalities for the results.

4 Two Examples of Applications

4.1 A Supervised Application to the Derivation of Photometric Redshifts for the SDSS Galaxies

Early in 2000, the Sloan Digital Sky Survey project released a preliminary catalogue (Early Data release - EDR; [15]) containing astrometric, morphological and photometric (in 5 bands) data for some 16 million galaxies and additional spectroscopic redshift for a subset of more than 50.000 objects. This data set is an ideal test ground for the techniques outlined in the previous paragraphs. The problem which was addressed is the derivation of reliable photometric redshifts which, in spite of the advances in multi object spectroscopy, still are (and will remain for a long time) the only viable approach to obtaining distances for very large samples of galaxies. The availability of spectroscopic redshifts for a rather large subset of the objects in the catalogue allows, in fact, to turn the problem into a classification one. In other words, the spectroscopic redshifts may be

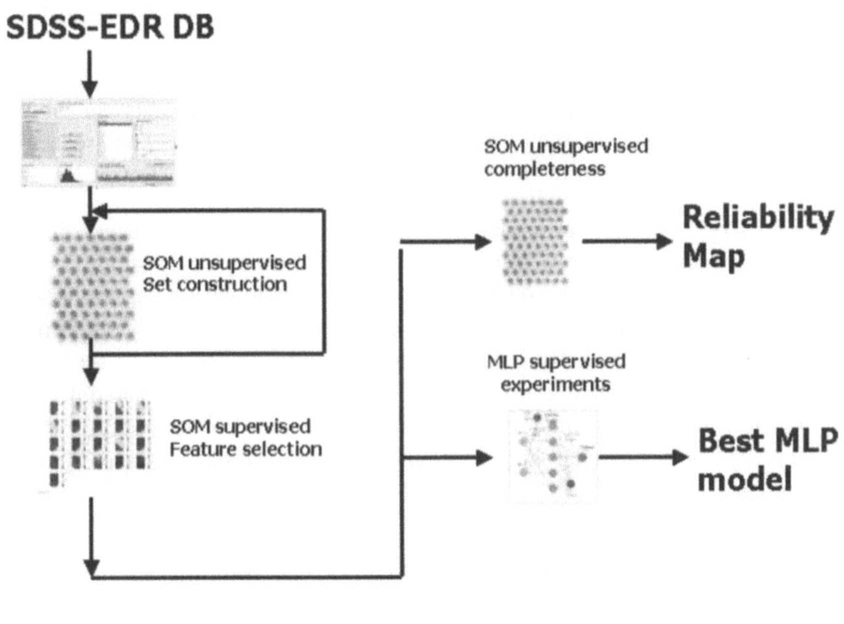

Fig. 2. Schematic diagram of the procedure followed to derive the photometric redshifts. As it can be seen the various modules of the AstroMining package can be combined accordingly to the needs of a specific task

used as the training set where the NN learn how to correlate the photometric information with the spectroscopic one. From a conceptual point of view this method is equivalent to the well known method of using the spectroscopic data to constrain the fit of a polynomial function mapping the photometric data [16]. The implemented procedure may be summarised as follows [17]:

- The spectroscopic subset is divided into three disjoined data sets (namely training, validation and test data sets) populated by objects with similar distributions in the subspace defined by the photometric and morphological parameters. This latter condition is needed in order to avoid losses in the generalization capability of the NN induced by a poor on non uniform sampling of the parameter space. For this purpose, we first run an unsupervised SOM clustering on the parameter space which looks for the statistical similarities in the data. Then objects in the three auxiliary data sets are extracted in order to have a uniform sampling of all significant regions in the parameter space.
- The dimensionality of the parameter space is then reduced by applying a feature elimination strategy which iteratively eliminates the less significant features leaving only the most significant ones [7] which, in our case turned out to be the photometric magnitudes in the five bands, the petrosian fluxes at the 50 and 90% levels, and the surface brightness (cf. [15]).
- The training set is then fed to a MLP NN which operates in a Bayesian framework [18] using the validation data set to avoid overfitting. When the training stops, the resulting configuration is applied to the photometric data set and the photometric redshifts (z_{phot}) are derived. Errors are computed by comparing the z_{phot} to the spectroscopic ones for the objects in the test set. An application to the above described data sets leads to an average robust error of $\Delta z \simeq 0.021$. In Fig. 3 we show the (z_{phot}) versus the spectroscopic redshifts for the objects in the test set.

We wish to stress that this approach offers some advantages and several disadvantages with respect to more traditional ones. The main disadvantage is the need for a training set which, coupled to the poor extrapolation capabilities of NN's, implies that photometric redshifts cannot be derived for objects fainter than the spectroscopic magnitude limit. The main advantages are, instead, the fact that the method offers a good control of the biases existing in the data set, and a very low level of contamination. In other words, the NN does non produce results for objects having characteristics different from those encountered in the training phase [17].

4.2 An Unsupervised Application to the TNG Telemetry

The Long Term Archive of the Telescopio Nazionale Galileo (TNG-LTA) contains both the raw data and the telemetry data collecting a wide set of monitored parameters such as, for instance, the atmospheric and dome temperatures, the operating conditions of the telescope and of the focal plane instruments, etc. Our

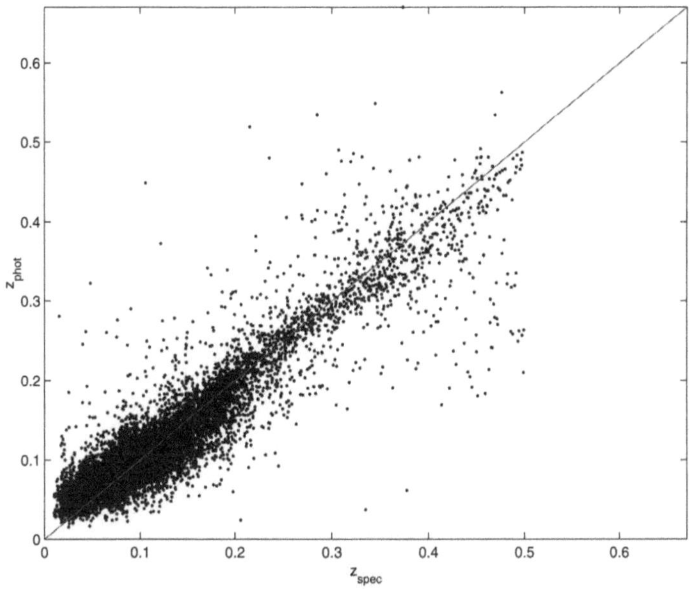

Fig. 3. The spectroscopic versus the photometric redshifts for the galaxies in the test set

experiment was devoted whether there is any correlation among the telemetry data and the quality (in terms of tracking, seeing, etc.) of the data. The existence of such a correlation would allow both to put a quality flag on the scientific exposures, and (if real time monitoring is implemented) to interrupt potentially bad exposure in order to avoid waste of precious observing time.

We extracted from the TNG-LTA a set of 278 telemetric data monitored (for a total of 35.000 epochs) during the acquisition of almost 50 images. The images were then randomly examined in order to assess their quality and build a set of labels (we shall limit ourselves to the case of images with bad tracking (elongated PSF) and good tracking (round PSF). The telemetry data were first passed to the feature analysis routine which identified the 5 most significant parameters (containing more than 95% of the total information) and then to SOM and GTM unsupervised clustering routines. The results of such clustering are shown in the lower panel of Fig. 4 and clusters of data can be easily identified.

In order to understand whether these clusters correspond or not to images of different quality we labeled the visualized matrix, id est we identified which neurons were activated by the telemetry data corresponding to the acquisition of the few labeled images. The result is that "good" images activate the neurons in the lower cluster, while "bad" ones activate neurons in the upper left and upper right corners.

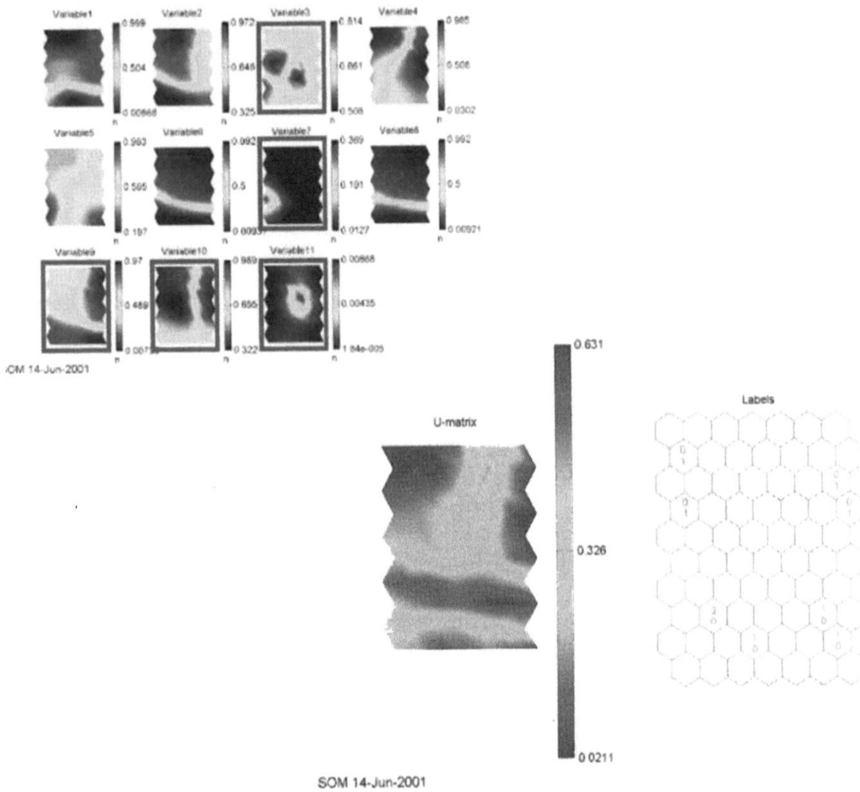

Fig. 4. Upper panel: U-matrix visualization of a set of individual input variables showing (black contours) the 5 most significant input features. The upper left U-matrix is the final one for the whole set of 278 parameters. In the lower panel we show the U-Matrix derived using only the most significant features and (on the right) the neurons activated in correspondence of the labeled images. Each hexagon is a neuron and the upper and lower numbers give the numbers of "good" and "bad" images activating that neuron

5 Conclusions

The AstroMining package offers a wide variety of neural tools to perform simple unsupervised and supervised data mining tasks such as: feature selection, clustering, classification and pattern recognition. The various routines, which are written in MatLab, can be combined in any (meaningful) way to a variety of tasks and are currently under test on a variety of problems. Even though the package is still in its testing phase, its running version can be requested at any of the following addresses: longo@na.infn.it, rtagliaferri@unisa.it).

Acknowledgements

This work was co-financed by the Italian Bureau for University and Scientific and Technological research trough a COFIN grant, and by the Italian Space Agency - ASI.

References

1. Djorgovski, S.G., Brunner R.J., Mahabal A.A., et al., *Exploration of large digital sky surveys*, in Mining the Sky, Banday Zaroubi & Bartelmann, eds., Springer, p.305, 2001
2. Djorgovski, S.G., in Proceed. of *Toward an International Virtual Observatory*, Garching June 10-14, this volume, 2002
3. Brunner, R.J., *The new paradigm: Novel Virtual Observatory enabled science*, in ASP Conf. Series, 225, p. 34, 2001
4. Bishop C.M, *Neural Networks for pattern recognition*, UK:Oxford University Press, 1995
5. Lin C.T., Lee C.S.G., *Neural fuzzy systems: a neurofuzzy synergism to intelligent systems*, Prentis Hall, 1996
6. Tagliaferri R., Longo G., D'Argenio B. Tarling D. eds., *Neural Networks - Special Issue on the Applications of Neural Networks to Astronomy and Environmental Sciences*, 2003, in press.
7. Andreon S., Gargiulo G., Longo G., Tagliaferri R., Capuano N., MNRAS, 319, 700, 2000
8. T. Kohonen, *Self-Organizing Maps*, Springer, Berlin, Heidelberg, 1995.
9. J. Vesanto, *Data Mining Techniques Based on the Self-Organizing Map*, Ph.D. Thesis, Helsinki University of Technology, 1997.
10. Svensen M., *GTM: the Generative Topographic Mapping*, Ph. D. Thesis, Aston Univ., 1998
11. Zadeh L. A., Information Control, 8, 338,1965
12. Goguen J. A., J. Math. Anal. Appl., 18, 145-174, 1967
13. Turunen E., *Mathematics Behind Fuzzy Logic, Advance in Soft Computing*, Physica-Verlag,1999
14. Turunen E., Niittymaki J., *Traffic Signal Control on Total Fuzzy Similarity based Reasoning*, submitted to Fuzzy Sets and Systems
15. Stoughton C., et al., AJ, 123, 485, 2001
16. Brunner R.J., Szalay A.S., Koo D.C., Kron R.G., Munn J.A., AJ, 110, 2655, 1995
17. Tagliaferri R., Longo G., Andreon S., Capozziello S., Donalek C., Giordano G., A. & Ap. submitted (astro-ph/0203445), 2002
18. Longo G., Donalek C., Raiconi G., Staiano A., Tagliaferri R., Sessa S., Pasian F., Smareglia R., Volpicelli A., *Data mining of large astronomical databases with neural tools*, in SPIE Proc. N.4647, "Data Analysis in Astronomy", F. Murtagh and J.L. Stark eds., 2003

Data Mining Facility

Sébastien Derriere[1], François Ochsenbein[1], Patricio F. Ortiz[2], Mark Allen[1], Thomas Boch[1], and Françoise Genova[1]

[1] CDS, Observatoire Astronomique de Strasbourg, 11 rue de l'Université, F-67000 Strasbourg, France
[2] Osservatorio Astronomico di Capodimonte, via Moiariello 16, I-80131 Naples, Italy

Abstract. A major goal of the Virtual Observatory is to build tools for exploring and extracting new information from heterogeneous (and possibly very large) data sets. There are two different facilities that a proper data mining facility tool should provide: first, a selection tool to locate relevant datasets; second, a tool to combine information from the selected datasets. We present data mining tools that have been developed at CDS, as practical implementations of a V.O. application (the first development was begun in the frame of the ESO/CDS data mining project). These tools rely extensively on long-term standards and emerging metadata (e.g. UCDs). They include: a selection tool to perform catalogue selection in the VizieR data federation on various criteria (keywords, UCDs); a positional cross-correlator to perform matching between different datasets.

1 Introduction

One of the Virtual Observatory (VO) objectives is to provide tools for exploring and extracting new informations from heterogeneous and possibly very large datasets. This requires at least two facilities:

- one to locate and select the relevant datasets. This task will use the metadata associated to the datasets;
- one to combine the informations contained in the relevant datasets. The typical tool will be a cross-identification engine.

This paper describes some tools and prototypes related to these topics that have been developed at the CDS. These tools are mainly dedicated to catalogue and survey data, and rely on the VizieR catalogue service.

2 Selection of Datasets

2.1 The VizieR Database

The VizieR database of astronomical catalogues [2] is a collection of over 3000 astronomical tables and catalogues coming from various origins. Owing to the wide diversity of sources for the catalogues, and to all the astronomical fields that are involved, the original contents of the data are highly heterogeneous.

An homogeneous description has thus been adopted for all the catalogues, and is stored in a ReadMe file. This file contains a byte-by-byte description of the data structure, and also global metadata describing the catalogue.

The VizieR interface[1] allows to select relevant catalogues by searching on various parameters:

- catalogue identifier;
- catalogue title, description, author;
- mission name, wavelength/frequency/energy domain, astronomical keywords.

These are in fact the main informations stored in the ReadMe file, and are used as metadata for catalogue selection.

2.2 UCDs for Catalogue Selection

The metadata present in the ReadMe do not always allow, however, a precise selection of tables on the basis of their contents. To circumvent that problem, and for other data mining operations, a new kind of metadata has been created during the ESO/CDS Data Mining Project [1]: the Unified Content Descriptors (UCD).

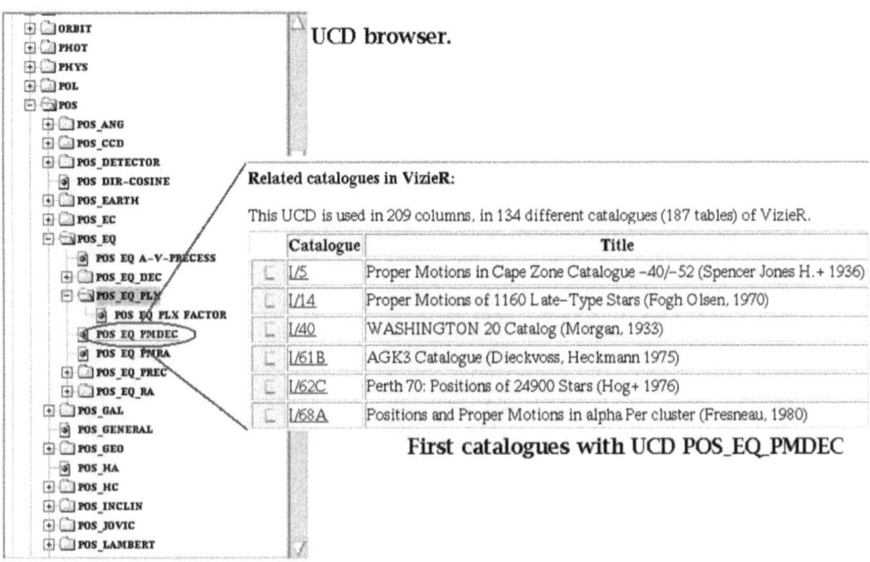

Fig. 1. Example of UCD usage for selecting catalogues in VizieR. To find catalogues containing a measure of proper motions, the relevant UCD has been selected in the browser on the left. The list of catalogues containing the UCD POS_EQ_PMDEC is presented on the right

[1] http://vizier.u-strasbg.fr/

A UCD is attached to every column in VizieR, and provides an homogeneous semantic description of its content. A set of tools have been developed around UCDs[2] [3]. With UCDs, it is possible to find which are the catalogues containing a given quantity.

Consider for example the case where one is interested in measures of K-band magnitudes in the Johnson system. It is possible to search for all the catalogues in the "infrared" domain, or to search for catalogue descriptions where the K band is explicitly mentioned. But these selections will likely give a lot of irrelevant catalogues, or incomplete results. On the contrary, a selection on the UCD "PHOT_JHN_K" will give the exact list of catalogues matching the query.

Another example of UCD usage is presented on Fig. 1. The proper UCD can be selected by exploring the full list of UCDs, with a browser, or can be derived from plain text.

3 Cross-Identifications

3.1 Positional Cross-Correlation

The most frequent operation involving combination of astronomical datasets is the cross-identification of sources, that is to make associations between different observations of the same sources seen in various spectral domains or at several epochs.

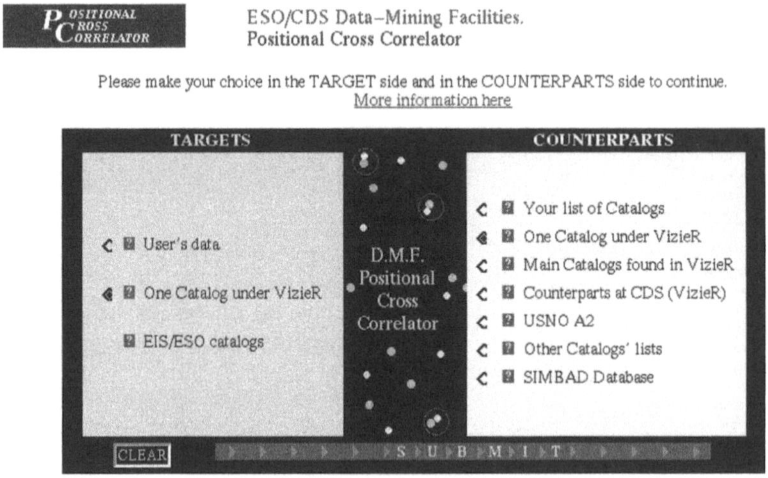

Fig. 2. Main interface of the positional cross-correlator. The "target" contains the reference sources for which counterparts will be searched in other datasets

[2] `http://vizier.u-strasbg.fr/UCD/`

For this, the celestial position of sources is a key criteria: except for high proper motion objects, the different observations of a source will be made on very close positions on the sky.

A positional cross-correlator was developed in the ESO-CDS data mining project (Fig. 2).

The "target" is the reference dataset: for each source of the target, counterparts will be searched in other datasets. The search of counterparts around target sources is done in a radius which depends in practice on the astrometric accuracy and on the source density in the catalogue. In general, matches that are found at small distances will be the most reliable.

Figure 3 shows the first result of the cross-matching, where only statistics on the number and distance of counterparts matches are printed. The full list of matches can be retrieved by selecting the corresponding catalogues.

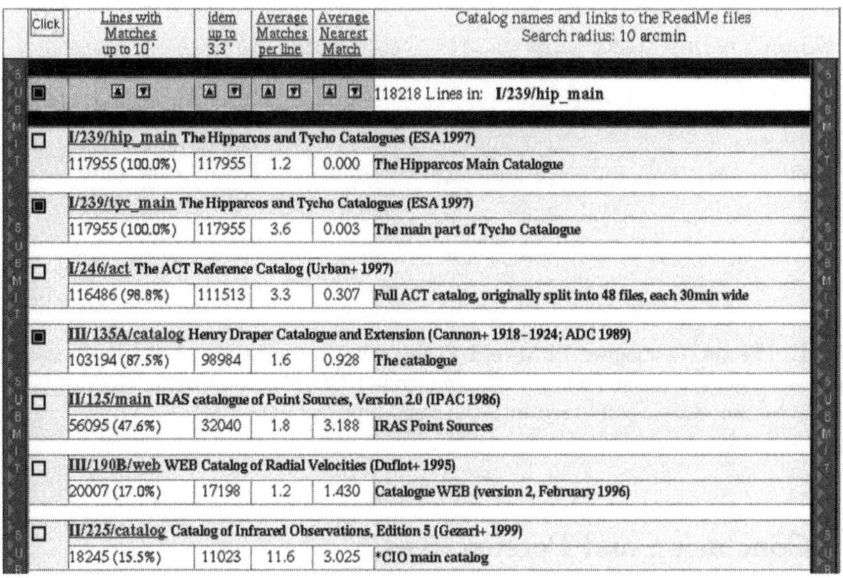

Click	Lines with Matches up to 10'	idem up to 3.3'	Average Matches per line	Average Nearest Match	Catalog names and links to the ReadMe files Search radius: 10 arcmin
					118218 Lines in: **I/239/hip_main**
	I/239/hip_main The Hipparcos and Tycho Catalogues (ESA 1997)				
	117955 (100.0%)	117955	1.2	0.000	The Hipparcos Main Catalogue
	I/239/tyc_main The Hipparcos and Tycho Catalogues (ESA 1997)				
	117955 (100.0%)	117955	3.6	0.003	The main part of Tycho Catalogue
	I/246/act The ACT Reference Catalog (Urban+ 1997)				
	116486 (98.8%)	111513	3.3	0.307	Full ACT catalog, originally split into 48 files, each 30min wide
	III/135A/catalog Henry Draper Catalogue and Extension (Cannon+ 1918-1924; ADC 1989)				
	103194 (87.5%)	98984	1.6	0.928	The catalogue
	II/125/main IRAS catalogue of Point Sources, Version 2.0 (IPAC 1986)				
	56095 (47.6%)	32040	1.8	3.188	IRAS Point Sources
	III/190B/web WEB Catalog of Radial Velocities (Duflot+ 1995)				
	20007 (17.0%)	17198	1.2	1.430	Catalogue WEB (version 2, February 1996)
	II/225/catalog Catalog of Infrared Observations, Edition 5 (Gezari+ 1999)				
	18245 (15.5%)	11023	11.6	3.025	*CIO main catalog

Fig. 3. Result of the cross-identification for the Hipparcos main catalogue with other catalogues from VizieR

3.2 Using Other Parameters

Other parameters than position can be used when cross-identifying catalogues. For example, if a magnitude is available in both catalogues, the two measures m_1 and m_2 can be combined to derive an extended distance

$$d = d_{\text{sky}} + c\left((m_1 - m_2)^2/\sigma_{m_1}\sigma_{m_2}\right)^{1/2},$$

that can be used as a criteria to compute a likelihood for the association to be valid.

Parameters like magnitudes or colors can also be combined to put constraints on the result of cross-identification (e.g. search for extremely red objects by combining V from an optical and K from an infrared catalogue, and selecting on $V - K$).

The UCDs can help in combining various parameters between catalogues. The knowledge of UCDs and units allows automated conversion of columns.

A prototype tool can identify common UCDs between two catalogues. This gives the fields which can be compared between the two catalogues. The comparison of columns having the same UCD is suggested. Unit conversion is performed when necessary, as can be seen on Fig. 4.

I/146/ppm1 Positions and Proper Motions - North (Roeser+, 1988) Catalogue PPM-North

RAJ2000	DEJ2000	pmRA	pmDE
"h:m:s"	"d:m:s"	s/yr	arcsec/yr
17 57 24.373	+04 36 09.20	-0.0014	0.032

I/239/tyc_main The Hipparcos and Tycho Catalogues (ESA 1997) The main part of Tycho Catalogue

RAhms	DEdms	RA(ICRS)	DE(ICRS)	pmRA	pmDE
		deg	deg	mas/yr	mas/yr
17 57 24.42	+04 36 09.0	269.35174824	4.60249678	41.60	37.50

I/239/tyc_main *converted columns :*

recno	pmDE
	arcsec/yr
35715	0.0375
35741	

Fig. 4. The two catalogues PPM and Tycho contain a measure of the declination of the proper motion (same label pmDE, and same UCD). For comparing the two values, the Tycho measure is converted into the same unit as in the PMM

4 Conclusion and Perspectives

The use of standard metadata such as UCDs for data-mining operations is very valuable. We have already applied them for:

- selecting relevant catalogues;
- finding comparable columns between catalogues;
- converting units between identical fields.

The definition of common metadata will be very important for all the data-mining operations in the VO, whether it be for selecting interesting datasets, or for performing operations on them. The definition of the VOTable standard [4], where metadata (via the "UCD" attribute) are associated to the tables, is very important in this context.

References

1. P. F. Ortiz, et al.: *ESO/CDS Data-mining Tool Development Project.* In: *ADASS VIII*, ASP Conf. Ser. **172**, p.379 (1999)
2. F. Ochsenbein, P. Bauer and J. Marcout: *The VizieR database of astronomical catalogues*, Astron. Astrophys. Suppl. Ser. **143**, p.23-32 (2000)
3. S. Derriere, et al.: *UCDs: Metadata for the VO*, this conference
4. F. Ochsenbein, et al.: *A Proposed XML Format for Astronomical Tables*, this conference

Automatic Detection–Extraction–Classification for a Large Number of Stellar Spectra from Slitless Wide Field Spectroscopy. A Project for AVO

E. Kontizas[1], I. Bellas–Velidis[1], E. Bratsolis[2], A. Dapergolas[1], M. Kontizas[3], D.H. Morgan[4], F. Pasian[5], and R. Smareglia[5]

[1] National Observatory of Athens, Institute of Astronomy and Astrophysics, P.O. Box 20048, GR-11810 Athens, Greece
[2] ENST, Dpt of TSI, Telecom Paris, France
[3] National and Kapodistrian University of Athens, Dept. of Physics, Section Astrophysics Astronomy and Mechanics, Zografos, GR-15784 Athens, Greece
[4] Institute of Astronomy, University of Edinburgh, U.K.
[5] Osservatorio Astronomico di Trieste, Via G.B. Tiepolo 11, I-34131 Trieste, Italy

Abstract. The distribution of spectral types in extended stellar systems in galaxies can be a powerful mine of information regarding:

1. not only the properties of the stellar content (such as age, composition) but also
2. the spatial distribution of certain types in order to trace specific structures (such as star forming regions and their geometry)

Objective prism plates are ideal data for this analysis, since thousands of spectra have been reported simultaneously. Even more so, when using low dispersion prism plates we are dealing with a pool of stellar spectra.

The full exploitation of digitized spectra in 1.2m UK Schmidt plates for a large number of stars involves three main automatic stages.

- Detection of the spectra
- Extraction of their images
- Classification

Automated methods for detection, extraction and classification published previously (Bratsolis et al. 1998, 2000a) have been implemented in a small SMC area very successfully.

1 Introduction

One of the most difficult tasks to be accomplished in astronomy is identifying and classifying objects. Several attempts have been made in various astronomical fields of interest to perform these tasks automatically.

Spectral classification is not only a tool for labeling individual stars but is also useful in studies of stellar population synthesis. Extracting physical quantities from digitized spectral plates involves tree main stages: Detection of spectra, extraction of their images and classification. The detection problem of stars for the digitized objective prism Schmidt plates was discussed by Bratsolis et al. (1998). The purpose of this paper is to present a fully automated method for the extraction and classification of spectra. High quality film copies of IIIa-J plates, taken with 1.2m UKST in Australia, have been used. The spectral plates have a dispersion of $244nm/mm$ at H_γ and spectral range from $320nm$ to $540nm$. The photographic material have been digitized at the Royal Observatory of Edinburgh using the SuperCOSMOS machine.

2 Automated Detection and Extraction

The automated detection of objects is a common problem in image processing for astronomy and astrophysics. Previous work for automated classification of stellar spectra (von Hippel et al., 1994), for spectral analysis of quasars (Hewett et al. 1995), for galaxy classification (Lahav et al., 1996) or for galaxy redshift measurements (Tucholke & Schuecker, 1996), applied spectra detection based on already known coordinates of the corresponding stars (either determined for the purpose with the automated plate measuring system or taken from catalogues). Our method is based on processing only the digitized prism plates without the need of the corresponding direct plates and utilizes signal processing techniques (Bratsolis 1997).

The most feature on which it is possible to rely for spectra detection on an objective prism plate is the beginning of each spectrum, characterized by the presence of intense gradients. Let us consider part of a digitized objective prism plate (Fig. 1a). This subimage consists of $N \times M$ pixels, and let $I(x, y)$ denote the value of the pixel (x, y) where the x–axis is down the dispersion. The subimage itself can be considered as a set of M discrete one-dimensional signals $I_y(x) = I(x, y)$. The energy E_y of a particular signal and its variation in decibels SNR_y are defined as:

$$E_y = \sum_{x=1}^{N} I_y^2(x), \qquad SNR_y = 20log_{10}\left\{\frac{E_y}{min(E_y)}\right\} \tag{1}$$

where the row with the $\min(E_y)$ is considered as background. Analysing this variation with an appropriate peak-picking algorithm we can identify energy peaks corresponding to stellar spectrum rows and separate them from spurious peaks (Fig. 1b). The detection of such rows is in the first step of the algorithm.

The second step is to fix in each detected row the beginning of the spectrum image. Let us know consider the one–dimensional signal of row 306 (Fig. 1d). In advance we applied to the digitized image appropriate filters resulting in an optimally smoothed spectrum scan (Fig. 1e). Its derivative (Fig. 1f) is used to determine the very beginning of the spectrum, the great change from the background. We find it by fixing the position of the maximal derivative which is leftward of the position of the row's maximum value. This helps us to overcome false detections (absorption and emission features, defects) and to detect only the spectrum beginning, as the crosses on Fig. 1c show.

3 Classification

3.1 Classification Using Linear Correlation

The field studied contains the associations NGC456, NGC460a,b and NGC465 of SMC. These associations occupy only 10% of the studied region. The majority of the stellar spectrum owing to the associations, superimposed to HII regions, has been saturated and excluded from our sample.

The most widely used method to measure the dependence of two variables, is the linear correlation r. The spectra are normalized and they are considered as vectors in R^N with N =128.

Let $D_{ij} = D_i(\lambda_j); j$ =1,...,128 be the normalized density value for the ith stellar spectrum and $S_{kj} = S_k(\lambda_j); j$ =1,...,128 be the normalized density value of the kth class standard stellar spectrum with k =1,...,6. For k =1,...,6 the standard stellar spectra are OB, A, F, G, K, M. The correlation coefficient for the ith stellar spectrum for the kth class is:

$$r_{ik} = \frac{\sum_{j=1}^{128}(D_{ij} - \overline{D}_i)(S_{kj} - \overline{S}_k)}{\sqrt{\sum_{j=1}^{128}(D_{ij} - \overline{D}_i)^2}\sqrt{\sum_{j=1}^{128}(S_{kj} - \overline{S}_k)^2}} \tag{2}$$

with $\overline{D}_i$ being the mean value (over the j variable) of the ith spectrum, i =1,...,426 and $\overline{S}_k$ the mean value of the kth class standard spectrum. The

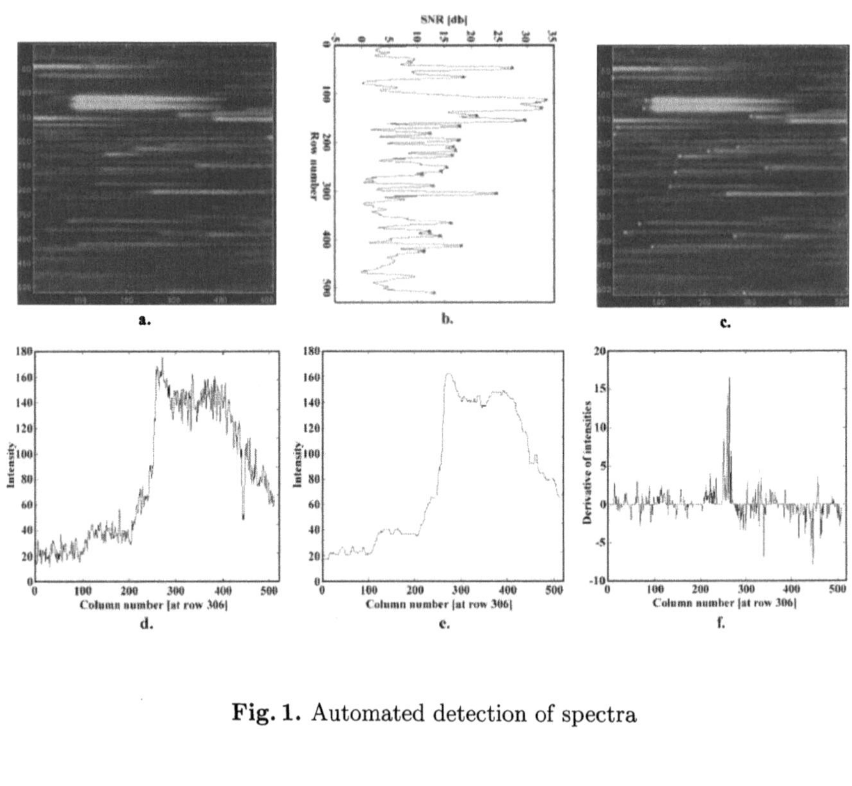

Fig. 1. Automated detection of spectra

correlation coefficient r_{ik} for the ith spectrum for the class k was calculated with displacement from the detection algorithm caused by the local background. For these seven correlation coefficients for every class k, the maximum value was chosen. The final classification was given by the maximum value of the coefficient r_i of all the r_{ik} coefficients as

$$r_i = arg(max \;\; r_{ik}), \qquad k = 1, ...6 \tag{3}$$

3.2 Classification by Use of Minimum Distance

The normalized spectra were standardized and a metric was introduced in vector space (Vieira & Ponz 1995). After standardization, we obtain

$$D_{ij}^{new} = \frac{(D_{ij}^{old} - \overline{D}_i)}{\sigma_i} \tag{4}$$

with $\overline{D}_i$ being the mean value (over j variable) and σ_i the standard deviation of the ith spectrum; i =1,...,426. The spectra have zero mean and unit standard deviation. The metric is the standard unweighted Euclidean distance between two real–valued vectors d_{ik} given by

$$d_{ij}^2 = \sum_{j=1}^{128} (D_{ij} - S_{kj})^2 \tag{5}$$

The minimum distance d_{ik} for the ith spectrum for the class k was calculated with displacement of ± 3 pixels to predict a possible displacement from the detection algorithm caused by the local background. The final result was given by

$$d_i = arg(min \;\; d_{ik}), \qquad k = 1, ..., 6 \tag{6}$$

3.3 Artificial Neural Networks

The method has been tested on spectra for which visual classification was available. The accuracy of this classification is ± 1 spectral class (Dapergolas et al. 1991). The down limit for the faint spectra was the same as in visual classification. The method has been developed on photographic objective prism plates but it can be equally well applied to CCD objective prism images.

Artificial Neural Networks (ANN) are connectionist systems consisting of many primitive units which are working in parallel and are connected via directed links. The main processing principle of these units is the distribution of activation patterns across the links similarly to the basic mechanism of the human brain. The knowledge is stored in the structure of the links, their topology and weights which are organized by training procedures. The link connecting two units is directed, fixing the source and the target unit. The weight attributed to the link transforms the output of the source unit to an input of the target. Depending on the weight the transmitted signal can take a value ranging from highly activating to highly forbidding.

The basic function of a unit is to accept inputs from other units acting as sources, to activate itself, and to produce one output that is directed to units–targets. Based on their topology and functionality, the units are arranged in layers. The layers can be generally divided into three types: input, hidden and output. The input layer consists of units that are directly activated by the input pattern. The output one is made by the units that produce the output pattern of the network. All the other layers are hidden and directly inaccessible.

A supervised ANN has to be used for the problem of spectral classification. It must be trained by applying a learning procedure during which the weights are properly adjusted, and there are many different learning algorithms. Choosing the proper one is one of the first steps in developing an ANN. On the other hand, in order to train, verify and to test such system, sets the input patterns with the corresponding outputs need to be prepared in advance. The most efficient implementation of ANN has been achieved by developing special hardware systems (Hecht-Nielsen, 1990). But the increase in processing power speed of computer systems gave rise to very efficient software simulators of ANNs running on general purpose computers. One is the freely distributed Stuttgart Neural Network Simulator – SNNS (Zell et al. 1996). It has powerful and flexible tools for developing, training, and testing different ANN architectures and provides a user friendly graphical interface

4 Results

The classification was tested by four different methods. The results from human expert (HE), linear correlation (LC), minimum distance (MD) and Artificial Neural Network (ANN) are shown in Table 1.

Table 1. Details for the different classification methods

Spectral Type	OB	A	F	G	K	M
Class	1	2	3	4	5	6
HE	144	58	27	52	68	77
LC	148	48	32	61	62	75
MD	145	41	38	64	65	73
ANN	137	58	35	62	66	68

To estimate the agreement between different classification methods, we have calculated the mean error me_{hehe} between two human experts a and b, me_{hemc} between human expert and maximum correlation classification, me_{hemd} between human expert and minimum distance classification and me_{henn} human expert and artificial neural network. The corresponding dispersions σ_{hehe}, σ_{hemc}, σ_{hemd} and σ_{henn} (Bratsolis et al., 2000b) were calculated. The results are summarized in Table 2.

Table 2. Statistical properties of different classification methods

Test	*me*	σ
hehe	0.23	0.47
hemc	0.15	0.43
hemd	0.16	0.45
henn	0.13	0.37

References

1. M. Bratsolis: In: *Wide-Field Spectroscopy*, ed. E. Kontizas et al., (Dordrecht, Kluwer) p. 119 (1997)
2. E. Bratsolis, I. Bellas–Velidis, E. Kontizas, F. Pasian, A. Dapergolas and R. Smareglia: A&AS **133**, 293 (1998)
3. E. Bratsolis, I. Bellas–Velidis, A. Dapergolas, E. Kontizas and M. Kontizas: A&AS **142**, 339 (2000a)
4. E. Bratsolis, F. Maragoudaki, I. Bellas–Velidis, A. Dapergolas, E. Kontizas and M. Kontizas: Baltic Astronomy **9**, 541 (2000b)
5. A. Dapergolas, E. Kontizas, M. Kontizas F. Pasian, M. Pucillo, and M. Santin: A&AS **87**, 97 (1991)
6. R. Heht–Nielsen: In: *Neurocomputing*, Reading, Addison–Wesley Publishing Company, (1990)
7. P.C. Hewett, C.B. Foltz, and F.H. Chaffee: AJ **109**, 1498 (1995)
8. O. Lahav, A. Naim, L. Sordé, and M.C. Storrie–Lombardi: MNRAS **283**, 207 (1996)
9. P. Schuecker: MNRAS **279**, 1057 (1996)
10. H.J. Tucholke, P. Schuecker: PASP **104**, 704 (1992)
11. E.F. Viera, J.D. Ponz: A&AS **111**, 393 (1995)
12. T. von Hippel, L.J. Storrie–Lombardi, M.C. Storrie–Lombardi, M.J. Irwin: MNRAS **269**, 97 (1994)
13. A. Zell et al. In: *SNNS Stuttgart Neural Network Simulator, User Manual*, Institute of parallel and Distributed High–Performance systems, University of Stuttgart (1996)

Star-Forming Regions at High Resolution: Interferometry for Virtual Observatories

Anita M.S. Richards and Simon T. Garrington

AVO and MERLIN/VLBI National Facility, Jodrell Bank Observatory,
University of Manchester, SK11 9DL, U.K.

Abstract. Star forming regions must be studied in 3-D with resolutions from milliarcseconds to degrees, to understand the birth of stars and planets and the evolutionary implications of the interactions between jets, discs and the energy output of massive young stars. No one image or even one telescope or array can give the full picture. Virtual observatories are needed to correlate the mass of disparate data which is available and to provide easy access to the relevant parts of huge interferometric data sets which are becoming available.

1 Introduction

Two of the most intriguing questions for astronomers and the public alike are "Are there other planets like ours?" and "How did the Universe begin?". Studying young stellar objects (YSO) and their environment is a vital part of understanding the occurrence and composition of planets. At the other extreme, the chemical and physical changes through successive cycles of star formation are a fundamental aspect of cosmology. Multi-scale observations are needed to investigate proto-planetary discs and to classify star-forming regions (SFR) by age, mass or metallicity and piece together a jigsaw of multi-frequency observations.

The examples in this paper show combinations of observations from many instruments. The data required a range of time-consuming procedures including astrometric and photometric alignment and allowances for different spatial and spectral scales and epochs. They illustrate controversies over how to distinguish between high- and low-mass proto-stars, or between bipolar outflows and disc winds. Some authors assert that masers are signposts of outflows from massive YSO, others suggest they also trace protoplanetary discs around younger versions of the Sun. Most current models take a few million years to form planets, but many stars form with close neighbours which could gravitationally disrupt or evaporate nascent systems more rapidly.

Such apparent contradictions arise because current investigations are limited to the study of a few SFR or isolated phenomena at high resolution, or to surveys of many regions at low resolution. The correlation of all known properties of YSOs and their surroundings is too large a project to be contemplated with present resources and models predicting properties such as the initial mass function (IMF) of an SFR contain too many free parameters in relation to currently manageable observational constraints. If a VO query engine could locate and align the relevant data sets, it might be possible to find reliable diagnostics for YSO mass and even accurate predictions of the IMF from the physical and chemical conditions in the parent cloud.

2 Protostellar and Protoplanetary Discs

The early stages of star formation are concealed deep inside dusty clouds. Figure 1 *Left* shows a helical protostellar jet imaged by MERLIN at 1.6 GHz [1] despite 150 magnitudes of extinction. Higher frequency interferometry at 5 – 100 GHz using MERLIN, the VLA and the BIMA array show the core contains a circumbinary envelope surrounding two separate discs ([2] and references therein). (Sub-)millimetre and IR imaging interferometry is technically challenging but the visibility amplitudes alone can be used to measure disc diameters and show the temperature structure on scales comparable to the size of the Solar System.

IR speckle interferometry produces stunning images such as that of S140 IRS1 Fig. 1 *Right* [3]. They interpret this as emission from the cavities cleared by outflows seen in CO, orthogonal to NE-SW radio emission which traces the wind from an evaporating disc. They propose that this is characteristic of high-mass YSOs, in contrast to more collimated radio jets from low-mass objects. However [4] suggest similar IR observations show multiple jets. We find that H_2O masers near IRS1 are closely associated with the radio emission, suggesting it is of a different nature to the IR regions. A systematic comparison of the properties of other objects with apparent multiple outflows would confirm (or disprove) the presence of disc winds from massive YSOs.

The masers were imaged as a follow-up to decades of single-dish 22-GHz monitoring [5] (Fig. 2). They modelled the variability of the triple-peaked spec-

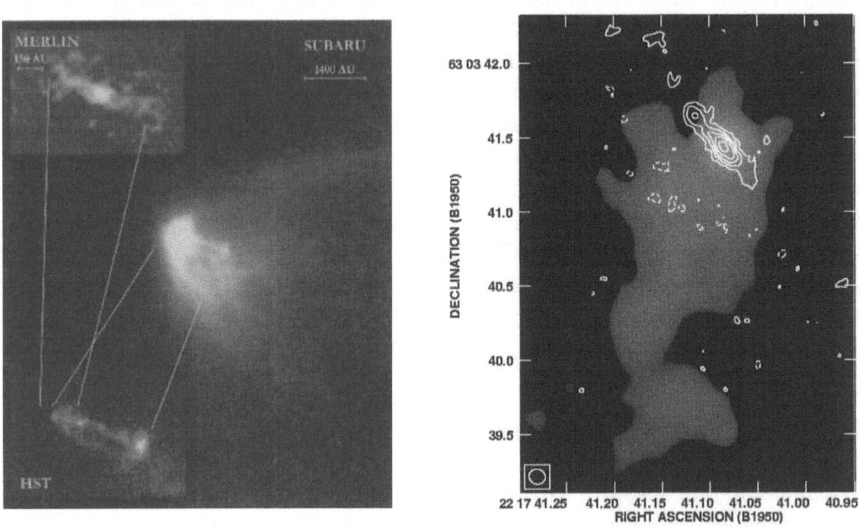

Fig. 1. *Left* Subaru, HST and MERLIN images of continuum emission from L1551 [1]; *Right* 5-GHz radio continuum contours superimposed on a IR K-band speckle interferometry image of S140 IRS1 [3] (corresponding to region (1) in Fig. 2)

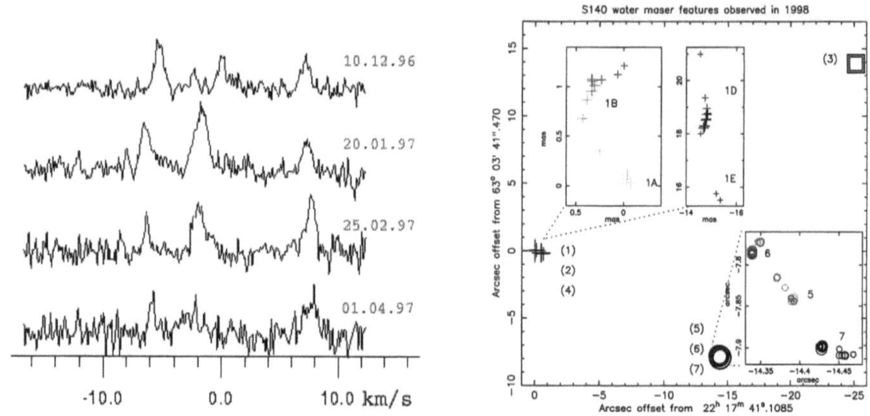

Fig. 2. *Left* Four epochs of 22-GHz spectra of S140 [5]; *Right* 22-GHz H_2O maser components fitted to high-resolution images, darker shading implies more blue-shifted masers. Region (1) (IRS1) has velocities between 4 – 12 km s^{-1}; regions (5) and (6) have velocities between –7 – 4 km s^{-1} and the other regions have velocities < -10 km s^{-1}

trum *Left* as a Keplerian disc with perturbations due to planetary condensation. However high-resolution maps show the masers come from several regions too far apart to be a single disc, although regions (5) (6) and (7) do have angular and velocity distributions consistent with the protoplanetary disc model. Small changes in physical conditions can cause rapid exponential maser brightness changes, so a single-epoch image can be misleading. Easier access to data is required for full interpretation of the complementary long-term single-dish monitoring of maser emission and few-epoch high-resolution interferometer maps which exist for many similar regions.

Methanol masers are also associated with both discs and outflows. They were thought to be signposts of high-mass star formation, but recent VLBI observations (e.g. [6]) show some have position-velocity gradients consistent with Keplerian discs on the 100-mas scale around low-mass cores. One such maser series is seen towards IRS1 in NGC7538, generally regarded as a high-mass SFR and at first sight it lies almost parallel to a bipolar CO outflow [7]. However, [8] find no evidence for such an outflow from IRS1 in IR line and continuum images and suggest the CO jets are associated with IRS2 instead.

3 The Proplyds and the Trapezium

Low-mass, long-lived stars with modest energy outputs seem best suited to supporting planet formation and possibly evolution into earth-like worlds. HST images of the M42 nebula in Orion show numerous tadpole-like objects with their tails pointing away from Θ^1 Ori C ([9] and references therein). These protoplanetary discs or proplyds are associated with low mass stars and would be ideal candidates for planet formation – if they can survive evaporation by the Θ^1 Ori C and the other Trapezium stars. Long-exposure MERLIN observations at 5 GHz resolved a number of proplyds. Figure 3 *Left* shows a cool proplyd in silhouette, with the light from the embedded star reflected at the poles. The *Centre* image shows the cusp and microjet of 167-317, and comparison with optical images shows some evidence for extinction implying the survival of neutral material [10]. Their multi-wavelength studies also suggest such low-mass discs can survive if they overlap the lifetime of nearby hot stars overlap by $< 10^5$ yr; once pebble-sized planetesimals have formed they are far less vulnerable.

The MERLIN data also revealed Θ^1 Ori A is a triple system [11], as the radio peak, Fig. 3 *Right*, did not coincide with either of the previously resolved optical stars. However recent IR speckle interferometry [12] show the third component to be a 4–5 $M_\odot$ pre-main-sequence star with an age of ~ 1 Myr – puzzling, since stars this old are not thought to have the surface convective zones required to produce radio emission via dynamo-generated magnetic fields.

These results were only possible because the radio and optical images could be aligned with an accuracy of ~ 30 mas. The position of phase-referenced interferometry data is tied to the International Celestial Reference Frame (ICRF, `http://maia.usno.navy.mil/ICRF/`), and the positions of the stars in Orion are determined using Hipparcos data.

The 3D structure of the Orion nebula [13] is one of the triumphs of interpretation of SFR, involving observations on angular scales from tens of milli-arcsec (mas) up to arcmin. The velocity structure and depth scale were bootstrapped from spectral imaging at cm to nm wavelengths of many neutral and species tracing a range of environments. But is it typical – do young OB stars inevitably appear and disrupt their surroundings for many cubic pc with ionising radiation and shocking jets, or can quieter low-mass stellar nurseries exist long enough for Solar-type systems to be commonplace?

The VO offers the opportunity to extend this understanding to a more complete sample of SFR as similar observations over a range of angular and frequency resolutions are becoming available for regions such as NGC7538 and S140.

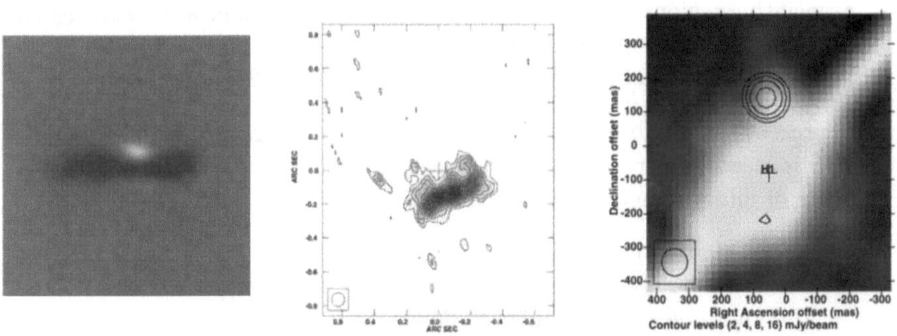

Fig. 3. *Left* HST image of proplyd 114-426 [14]. *Centre* MERLIN 5 GHz image of the the proplyd 167-317 [10]. *Right* MERLIN contours of 5 GHz emission from Θ^1 Ori A overlaid on the HST image [11]. The + marks the Hipparcos position

4 VO Contributions to Understanding SFR

Access to catalogues, observatory archives and tables from publications via CDS is expanding, but the integration of these services in a VO will lead to far more definitive models of star- and planet formation by facilitating complex operations on these data. These are some diagnostics for conditions conducive to planet formation which could be tested by access to large data-bases at multiple frequencies including high-resolution interferometry data:

- The rôles of magnetic fields and chemistry in determining the rate of cloud collapse and the size and dynamics of star-forming fragments e.g. [15], [16];
- A high D:H ratio as a diagnostic of low-mass star formation [17]
- Distinguishing characteristics of high- and low-mass YSOs [3], [18];
- Radio/IR/optical comparisons showing survival of neutral material in protoplanetary discs [14], [19] and references therein;
- Correlations in discs between a high sub-mm:IR ratio with grain growth and a high dust:gas ratio with evaporation of volatiles [16],[19].
- The evolutionary rate of multiple high-mass YSOs and their effects on low-mass neighbours [11];
- Use of IR colours to distinguish free-floating planets from brown dwarfs [20].

The next generation of radio – sub-mm arrays such as eMERLIN, ALMA, VERA and the eVLA will image thermal line and continuum emission with mas resolution, but produce Tb of data. The visibility data contains information which is usually too complex to be shown in a single map, needing $> 10^8$ pixels to show the whole field of view (f.o.v). Potential VO services include:

- Access to interferometer data via a user-friendly interface, obviating the need to transfer massive data sets or specialised software (under development for the MERLIN archive);
- Astrometric alignment using information provided in data sets, cross-references to catalogues of ICRF sources, proper motions, instrumental properties and ultimately algorithms for matching sources at different frequencies;
- Associating regions and objects in 3D, using models of ionisation and velocity fields and shock propagation to interpret multi-scale observations;
- Links to databases of chemical reactions e.g. UMIST [21] and library spectra.

At present the number of interferometer observing logs and images available via CDS is steadily increasing and the necessary metadata are being developed. Providers of interferometry data should include:

- Array/observation properties:
  - Duration, bandwidth (sensitivity, velocity coverage);
  - *Maximum baseline* (angular resolution);
  - Minimum baseline (most extended structure detectable)
  - Primary beam (f.o.v.)
  - Phase-reference and flux calibration source details (astrometric and photometric accountability).

- Correlation parameters:
  - Integration time per complex visibility (f.o.v.);
  - Spectral resolution (f.o.v., velocity resolution);
  - Polarization products (magnetic field measurements).

This will allow imaging anywhere in the f.o.v. and also extraction of visibility data products such as radio light curves for variable sources and amplitude/baseline length plots for compact source modelling.

Acknowledgements

We thank AstroGrid and the AVO for supporting our attendance at this conference. We thank the astronomers whose work we have drawn on, including Cohen, Diamond, Holloway, MacDonald & Meaburn (JBO); Hutawarakorn & Gasiprong (Bangkok), Gray (UMIST), Hoare (Leeds), Lekht (INOEP), Phillips (JIVE), Minier & Pestalozzi (Onsala) and Samodourov et al. (Moscow).

References

1. S. Curiel et al. in prep.
2. L.W. Looney, L.G. Mundy, W.J, Welch: ApJ **529**, 477 (2000)
3. M.G. Hoare, A. Glindemann, A. Richichi: In: *The Role of Dust in the Formation of Stars, Garching, 1995*, ed. by H.U. Käufl, R, Siebenmorgen (Springer-Verlag, Berlin 1996) p. 35
4. G. Weigelt: A&A **381**, 905 (2002)
5. E.E. Lekht, R.L. Sorochenko: ARep **45**, 113 (2001)
6. V. Minier, R.S. Booth, J.E. Conway: NAR **43**, 569 (1999)
7. C.J. Davis, G. Moriaty-Schieven, J. Eislöffel, M.G. Hoare, T.P Ray: AJ **115**, 1118 (1998)
8. J.D. Bloomer et al.: ApJ **506**, 727 (1998)
9. J. Bally, C.R. O'Dell, M.J. McCaughrean: AJ **119**, 2919 (2000)
10. W.J Henney, C.R. O'Dell, J. Meaburn, S.T. Garrington, J.A. Lopez: ApJ **566**, 315 (2002)
11. S.T. Garrington, H.J. van Langevelde, R.M. Campbell, A.G. Gunn: In: *Proceedings of the 6th EVN Symposium, Bonn, 2002*, ed. by E. Ros, R.W. Porcs, A.P. Lobanov, J.A. Zensus (MPIfR, Bonn) p. 259
12. G. Weigelt, Y. Balega, T. Preibisch, D. Schertl, M. Schöller, H. Zinnecker: A&A **347**, L15 (1999)
13. C.R. O'Dell: PASP **113**, 29 (2001)
14. M. McCaughrean, C.R. O'Dell: ST Sci OPO PRC95-45c (1995)
15. V. Mignenes, R.J. Cohen, G.C. Brebner: MNRAS **254**, 501 (1992)** A.P. Boss: ApJ **568**, 743 (2002)
16. S.B. Charnley, S.D. Rodgers, P. Ehrenfreund: A&A **378**, 1042 (2001)
17. H. Roberts, G.A. Fuller, T.J. Millar, J. Hatchell, J.V. Buckle: A&A **381**, 1026 (2002)
18. W.R.F. Dent, H.E. Matthews, D. Ward-Thompson: MNRAS **301**, 1049 (1998)
19. Zuckerman: ARA&A **39**, 549 (2001)
20. P.W. Lucas, P.F. Roche: MNRAS **326**, 695 (2001)
21. T.J. Millar, P.R.A. Farquhar, K. Willacy: A&AS **121**, 139 (1997)

The Astronomical Data Warehouse

Clive G. Page

Department of Physics & Astronomy, University of Leicester, Leicester, LE1 7RH, UK

Abstract. The idea of the astronomical data warehouse has arisen as the natural extension of current trends in astronomical data archives, and from an analysis of the types of query likely to be sent to a virtual observatory. Data warehouses will be centres providing both data and computational facilities.

1 Querying the Virtual Observatory

Members of the AstroGrid project have some background in the provision of data archive services and of using them, but in designing the virtual observatory we have been driven principally by some recent collections of typical astronomical problems and use-cases:

- The set of some 47 science problems set out on the Astrogrid Wiki pages [1].
- The collection of (so far) 29 use-cases assembled by the US-NVO Project [2].
- A set of 35 queries for the Sloan Digital Sky Survey [3].

Most of these involve a series of interactions with the virtual observatory, which can be further broken down into more basic data access operations. These can be roughly classified as follows: firstly positional queries, second observational queries, and thirdly all other types (which will be called 'non-positional' for convenience).

1.1 Positional Queries

Positional queries are ones seeking information about a specific small patch of sky, or on a named object. An object name can usually be translated to a position using a name resolver such as Simbad or NED. Given the errors in coordinates, in either case the query is actually one about a small area of sky, typically a circular region about a given (RA, DEC): this has become known as the cone search. Typical examples might be:

- Are there any X-ray sources near HD123456?
- Can I have all infra-red images centred on PSR0123-456?

Because of their importance in astronomy, positional queries make up a high proportion of all queries to on-line resources, but there may be some bias caused by the fact that many existing archives specialise in the cone search, and some support little else.

It now seems generally accepted that one of the prime requirements of the VO is some sort of Resource Registry, which will direct user's queries to the sites most likely to provide them with relevant information. Pilot projects to implement the cone search using modern XML-based protocols such as SOAP and WSDL, and to standardise the results using the VOTable format, are under way (as noted by other papers in these proceedings).

1.2 Observational Queries

Much useful on-line data arises not from surveys but from specific observations of the sky. Those with sufficient expertise can locate observations which may have covered their objects of interest and download the raw or semi-processed data to their own computer for local re-analysis. Here are two examples of such queries:

- Can I download the HST data from all observations of NGC9876?
- Are the XMM-Newton data from observing HD123456 made public yet?

Such queries are currently limited to those who have expertise in the relevant branch of astronomy, the appropriate software, and enough computing power necessary to make good use of the resulting data. With the advent of the VO, and with more observatories running standard data reduction pipelines, such downloads may decline in importance. Even experts, however, are likely to benefit from the resource discovery mechanisms being proposed for the VO, and from the security architecture, which should provide a uniform mechanism for access to both public and private data.

1.3 Non-Positional Queries

This category is extremely diverse, since it includes everything not covered in the other two. Examples include:

- Cross-matching objects in catalogues from different wavebands: the basis of many types of more detailed investigation.
- Simple selections on ranges of parameters, such as selecting on (B-V) or X-ray/Optical flux ratio.
- Statistical investigations such as finding rare objects as outliers, and regression analysis to find trends and correlations.
- Clustering algorithms, classification algorithms, and measurements of spatial correlation functions.
- Visualisations and other methods of using the human eye to aid pattern discovery.
- Time-series operations: finding and classifying variability, using power-spectra to find periodicities, etc.
- Discovering moving objects by comparing images or positional measurements taken over a period of time.

2 Handling Non-Positional Queries

Many of the operations identified above will be I/O-intensive, or cpu-intensive, or both. Some of them can be carried out using a relational database management system (DBMS), but many others (time-series analysis, clustering algorithms, image processing) will require specialised software. This may be available from existing packages (AIPS, CIAO, FTOOLS, IRAF, MIDAS, Starlink, XMM-SAS, etc.) or may require the development of new algorithms and software.

To get VO up and running quickly we should concentrate initially on operations that can be done well using existing software, such as those supported by a relational DBMS, which is likely to be installed at almost every data archive site. A notable feature of the DBMS is the ability to create and use indices.

As tabular datasets increase in size the efficiency gain from indexing also grows, almost linearly. Suppose we try to find a given row in a table such as the USNO-A2 catalogue, which has 626 million rows: if a suitable index has been created this should require only two or three disc seeks, taking only 20 or 30 ms. But even the simplest operation which requires trawling through the entire table will take several hundred seconds. So the ratio in this case is of the order of 10,000:1.

Unfortunately not all operations can be speeded up with indices. Many tables have simply too many columns for this to be practical: several tables in the Vizier collection [4], for example, have over 100 columns, most of them fields which a user might want to search. In addition some users will want to select on say the difference between two magnitude columns (or ratio of two fluxes): here indices created on individual columns do not help. There are also operations which cannot avoid a complete sequential scan, for example some statistical queries, and of course index creation itself. Most join operations (and all outer joins) need to access every row in one of the two tables involved.

Many existing data archive sites limit external users to operations which make no more than modest use of resources (for example the cone-search with an index on celestial position); a few have a more liberal policy, but of course external users take their place in the queue with everyone else, and there is the ever-growing risk of abuse of such open facilities.

As noted earlier, scientifically valuable results often arise from combining data from two (or more) different sites: this usually involves making substantial data transfers across the wide area network. Although network capacities are continually rising, with many international links running at 155 Mb/s or even higher, the effective bandwidth for file transfers seems considerably slower in practice. The particle physics community [5], whose networking needs are even greater than ours, have found in hard to get file transfers running internationally at more than 100 Mb/s, and that typical rates on most routes are around 20 Mb/s, i.e. around one terabyte/week.

For many DBMS operations, low latency is also vital, so the speed of light presents a serious obstacle to performing some database operations over a wide-area network.

All these considerations lead to the conclusion that while the simpler types of query, such as the cone-search, can be carried out on remote servers, many of the more complex ones will need the data to be located close to the processing power – but not necessarily close to the user. In cases where adequate processing power is not available attached to the discs, or not available to external users of a site, then the best alternative is to obtain a copy of the relevant dataset and install it on a more powerful computer facility.

3 The Data Warehouse Concept

The term "Data Warehouse" has been used in commerce for some time: in large organisations it is often impractical to carry out complex querying over a large and disparate collection of DBMS, especially if these are spread over a number of physical sites, or are continually being updated through new transactions. The idea of the data warehouse is to set up a facility with a copy of all the important datasets from around the organisation, and get them all in the same format, and indexed optimally.

Since many of the same considerations apply to the world's collection of astronomical data, the term Astronomical Data Warehouse seems reasonably appropriate. What is proposed is a facility which contains:

- A collection of the most popular standard datasets permanently installed;
- Ample spare disc space so that users can download and install such other datasets as they require;
- Adequate processor power, probably though the use of Beowulf or other cluster systems;
- A relational database management system and the usual astronomical packages pre-installed;
- Good network connections to the outside world.

Although there is no reason why data warehouses should not be set up as separate facilities, it seems to me that they will probably be set up as a natural extension of the facilities already provided by the major data centres.

4 Towards the Data Warehouse

A trend is already apparent towards moving the data to where processing power is located. Many important datasets and data collections are already present at two or more locations around the world. The HST archive is, of course, mirrored also in Garching; similarly the Chandra data archive has a mirror at LEDAS (Leicester), while the NASA Astrophysics Data System (ADS) also has a mirror in the UK at Nottingham. The popularity of the Vizier collection of catalogues and other tabular data, produced by the Centre de Donées Astronomiques de Strasbourg, can be seen from the fact that it now has eight mirror sites. Some important source catalogues (such as the GSC and USNO-A2 catalogues) and images collections (such as the Digital Sky Survey) are already distributed on CD-ROM and installed on disc at many sites. It will soon be exceptional for an important data resource to be located at only one site.

One caveat must be noted: astronomical datasets are produced by complex instruments which need to be fully understood by anyone carrying out the basic data reduction stages and making the results available to external users. The existence of mirror sites does not at all lessen the case for having as curators of the original datasets the astronomers who produced them.

5 Software for the Data Warehouse

The software to be installed at an astronomical data warehouse will probably include all of the following:

- A suitable database management system.
- Standard astronomical software packages.
- Statistical packages: there seem to be a many similar packages around, but almost all depend on being able to read the dataset of interest into physical memory. Since many astronomical datasets are already larger than feasible memory sizes, this presents a problem. A good review of available packages is given by by Wegman et al. [6].
- Web Services interfaces using SOAP and WSDL.
- Grid Services based on the Open Grid Services Architecture [7], so virtual observatory users can take advantage of computational and data grid facilities.

The AstroGrid project is conducting an evaluation of a number of commercial and freeware relational DBMS: these investigations of DBMS are still incomplete, but a summary of the findings so far is available on-line [8]. One of the prime application areas is likely to be access to catalogues of celestial objects, in which efficient indexing on celestial position (RA,DEC) is essential, especially when performing a fuzzy join between one catalogue and another. It has been assumed that to to this efficiently we needed a DBMS with spatial indexing capability, such as the R-tree indexing in Postgres. Recently a new algorithm has been developed and tested which makes use of an equal-area pixelisation of the sky such as those of HTM [9] or HEALPix [10], though currently the latter is used. This PCODE [11] method seems to be significantly faster than the use of an R-tree index, and it can use any DBMS since uses an equi-join of integers, and has no need for specialised spatial indexing.

6 Potential Problems

Many problems remain to be solved, of course, for example

- Will data curators be happy to allow remote data warehouses to take copies of complete datasets?

- How can large datasets be copied from one site to another? If current network links are inadequate, we need to find alternatives. The VISTA project [12] plans to ship data on physical discs from Chile to Europe when it goes on-line around 2006, and the SDSS project, having similar problems, has proposed terascale sneaker-net [13] as a solution.
- How can we interface DBMS with other scientific packages? This is especially difficult because few systems can import or export more than simple text files, and have practically no support for metadata. Perhaps the adoption of XML-based data formats such as VOTable will help.
- How can astronomers write their own software to interface to large databases?

Acknowledgements

I would like to acknowledge many valuable discussions with my colleagues in the AstroGrid, AVO, and US-NVO projects, without whom these ideas would have been even less than half-baked.

References

1. 'AstroGrid science problem list', `http://wiki.astrogrid.org/bin/view/VO/ScienceProblemList`
2. US-NVO Project, '100 science problems for the NVO', `http://monet.astro.uiuc.edu/~rplante/VO/metadata/100questions.html`
3. J. Gray, D. Slutz, A. Szalay, A. Thakar, J. Vandenberg, P. Kunszt, C. Stoughton, 'Data Mining the SDSS SkyServer Database', MSR-TR-2002-01 (2002), `http://research.microsoft.com/~grey`
4. 'VizieR Catalogue Service', `http://vizier.u-strasbg.fr/`
5. Personal communication from David Newbold of Bristol University.
6. E. J. Wegman, D. B. Carr, R. D. King, J. J. Miller, W. L. Poston, J. L. Solka, and J. Wallin, 'Statistical Software, Siftware and Astronomy', `http://www.galaxy.gmu.edu/papers/astr.html`
7. 'Open Grid Services Architecture', `http://www.globus.org/ogsa/`
8. 'Astrogrid DBMS Evaluations', `http://wiki.astrogrid.org/bin/view/Astrogrid/DbmsEvaluations`
9. 'Hierarchical Triangular Mesh (HTM)', `http://taltos.pha.jhu.edu/htm/`
10. 'Hierarchical Equal Area isoLatitude Pixelisation (HEALPix)', `http://www.eso.org/science/healpix/`
11. C. Page, 'Indexing the Sky', `http://wiki.astrogrid.org/bin/view/Astrogrid/SkyIndexing`
12. 'Visible & Infrared Survey Telescope for Astronomy', `http://www.vista.ac.uk/`
13. J. Gray, W. Chong, T. Barclay, A. Szalay, and J. Vandenburg, 'TeraScale SneakerNet', MS-TR-02-54 (2002), `http://research.microsoft.com/~grey`

The Real Astronomy Experience: Making the IVO Effective for International Planetaria and Science Centers

Carl Pennypacker

Space Sciences Laboratory and Lawrence Hall of Science,
University of California at Berkeley, USA
Grizzly Peak Blvd, Berkeley, CA 94720, USA

Abstract. The Real Astronomy Experience, or RAE, is a new exhibit just beginning development in the late fall of 2002,hopefully to be funded by the US National Science Foundation. RAE will allow public visitors in science centers, museums, and planetaria to operate and acquire images from a either a remote telescope in the southern hemisphere or from IVO data bases. RAE can operate from Europe and North America during daylight at the exhibit site. This project will be implemented also by international Global Hands-On Universe ("GHOU") collaborators. Greater international success will be facilitated by other GHOU nations joining the collaboration to build a viable network of small, inexpensive telescopes, and federating their data bases for the IVO.

1 Introduction

Science Center and Planetaria visitors are eager for real-science-based, hands-on activities. Over the past three years, pilots of remote telescope usage at the Tokyo Museum of Science the "Universe Live-how", usage of Japanese and Australian telescopes from Germany for teacher workshops, and also usage of telescopes from the Cite des Sciences et de l'Industrie at La Villette in Paris, have shown that people are indeed ready to use such a system. In addition, Yerkes Observatory staff in Wisconsin, US, regularly connect with visitors to the Tokyo Museum of Science and allow them to "steer" their telescope from Japan. Many people wait for the chance to control a remote telescope, watch it move, and then study the image they have just captured.The Telescopes in Education project ("TIE") has a long and meritorious record of enabling students to use their telescope on Mt. Wilson remotely. However, usage of these systems has been intermittent at best, as the robust weather, easy telescope control systems, and enthusiasms of the owners to share their telescopes have not facilitated usage on a regular basis. Hence, to meet the growing and projected demands, the first step in building such a system is underway. The projected visitor user "kiosk" is shown below in Fig. 1.

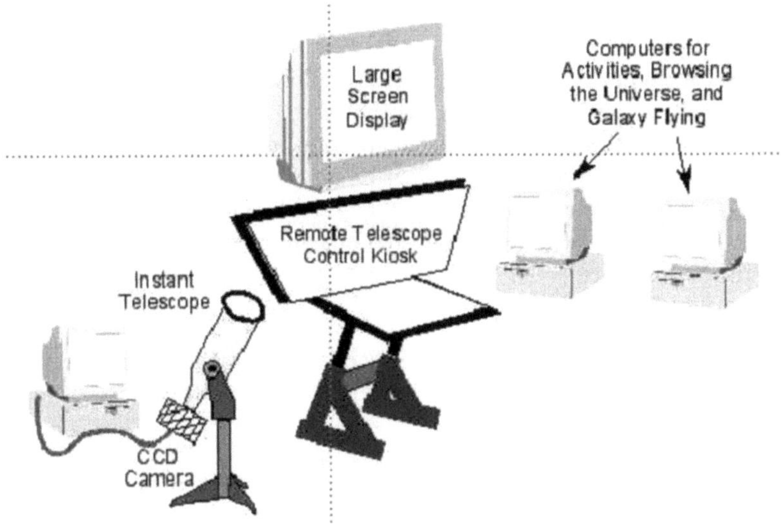

Fig. 1. The Real Astronomy Experience Exhibit

2 RAE Activities for the Visitor

Visitors will be able to undertake a number of activities with RAE, eventually with the aid of GHOU, give this system much more power. Such activities include, but are not limited to the following astronomy and remote telescope activities. Visitors can:

1) control and view world-wide remote robotic telescopes via the Internet,

2) use image processing software to understand the images they capture,

3) capture images with an exhibit-based hands-on functional telescope equipped with a Charge-Coupled Device (CCD) camera to learn how a modern telescope image capture system works,

4) learn key astronomical concepts while browsing through the RAE astronomical

3 Details of the System

Such a remote system is now only feasible based on many simultaneous developments in software and hardware. Up to now, it has been hard for telescopes to point and capture a target, track well enough, work over the Internet, be powered by Planetaria software. But things have changed. Such advances include

1) ACP Observatory Control Program by BobDenny/DC3 Dreams.

This software allows scripting and elaborate control and response to requests, scheduling, control of filters, focus, and other observatory system.

2) "Pinpoint" Registration/World Coordinate System Software by BobDenny/DC3 Dreams.

This software, using the USNO2 catalog, allows an image to be astrometrically registered to within a half arc second or so. Pinpoint is essential for this system, as the hardware in our implementations has not been capable of pointing the telescope to closer than a few arc minutes regularly. Pinpoint analyzes a short image to register coordinates near the target, then provides an update for the final positioning of the telescope. The image is always in the middle of the field with this software.

3) Graeme White and Australia

GHOU has an incredibly strong and good collaborator in Graeme White in Australia. Graeme is eager to site and service such telescopes.

4) Maxim DL

This software allows image control and manipulation at the observatory, and also a scaled down version for image processing at the Planetaria. Maxim

5) Paramount or Losmandy Precision Pointing Drives:

Several companies are now making very high quality automated drive systems – both Software Bisque and Losmandy Astronomical Products – among others – are selling off-the-shelf drives that can point well and track well, under automated control. These breakthroughs allow Internet operation of telescopes at an unprecedented level of throughput and accuracy, at a reasonable cost.

6) Starry Nights Planetarium Software

Fig. 2. Starry Night Software Interface for RAE

When the user approaches the exhibit, a friendly planetarium software system, by Starry Nights, greets the visitor, and allows the visitor a view of the visible sky. Visitors can then click on a star, galaxy, planet, or nebulae, and then watch the telescope slew to the target.

7) Internet Web Cameras

After the visitor starts to move the telescope, the user can view the telescope in Australia move over an Internet Web Camera. This is an exciting event! In addition, the web camera sends video of the stars moving across the sky, to give a sense of the southern sky.

4 Conclusion

It is clear such a system has the potential to reach many planetaria and science center visitors. The power of the IVO, coupled with the IVO community holds the genuine promise of bringing very powerful images, data sets, tools, and understanding to a broad segment of the world in a meaningful way.

Enabling Outreach with Virtual Observatories

G. Mark Voit

STScI, 3700 San Martin Drive, Baltimore, MD 21218, USA

Abstract. Virtual observatories are potentially excellent vehicles for public outreach and education because they can put anyone in direct contact with the data that professional astronomers use. However, our vast wealth of sophisticated information poses a daunting challenge. How are we to present astronomical data to the public in ways that will intrigue them and inspire exploration instead of driving them away, bewildered and confused? This article outlines some approaches to outreach being planned by the National Virtual Observatory collaboration.

1 Introduction

Virtual observatories offer an unprecedented opportunity for the general public to share in the excitement of astronomical discovery. In principle, an entire universe of observations could be at the fingertips of anyone with the imagination and curiosity to explore them. An extensive community of educators, museum professionals, and software developers is eager to begin working on educational activities and outreach products based on virtual observatories, but will these observatories be suited to their needs? The conceptual vocabulary that seems so natural to professional astronomers can seem quite arcane to nonprofessionals. Likewise, the questions that interest us often require an enormous amount of background information to appreciate. If we want our virtual observatories to be used and comprehended by the general public, then we will need to include public-friendly features in their design. Such features should minimize the amount of astronomical background needed by non-expert visitors and maximize their ability to explore the questions they find most interesting. To that end, the NVO framework project is partnering with a diverse community of outreach professionals to anticipate the needs of the general public and to incorporate features that address those needs into the design of NVO.

2 Importance of Partnerships

Public outreach is often spoken of as a single activity, but effective outreach involves a variety of approaches tailored to many individual types of audience. For example, one might be interested in reaching elementary school students, museum visitors, college students studying to be teachers, web surfers, and amateur astronomers, to name just a few possibilities. For outreach to be most effective, each of these audiences requires a specialized approach.

Scientists are experts in astronomical data gathering and interpretation, but we are not necessarily well versed in the interests of all the audiences who might want to interact with virtual observatories. That is why partnerships with educators are critically important. People who earn their living serving the types of audiences we want to reach are the most likely people to know the educational needs of those audiences and the appropriate level for presenting information. To that end, the NVO collaboration has cultivated partnerships with the following organizations:

- STScI Office of Public Outreach
- Space Sciences Laboratory, Berkeley
- NOAO Outreach Office
- Chandra Science Center Outreach Office
- Sloan Digital Sky Survey Outreach Office
- AAS Outreach Office
- American Museum of Natural History
- CLEA Project, Gettysburg College (college-level lab exercises)
- NCSA Outreach Office
- American Association of Variable Star Observers
- Starry Night (desktop planetarium software)
- Hands-on Universe Project (K-12 education)
- Hands-on Astrophysics Project (K-12 education)
- Sky & Telescope Magazine
- Spitz Planetariums
- Loch Ness Productions (planetarium shows)

We are actively seeking more partnerships and invite anyone with a serious interest in astronomy outreach to work with us.

3 Infrastructure for Outreach

The NVO collaboration envisions an outreach environment that reflects the architecture of the NVO itself. We anticipate that many astronomical outreach organizations, especially our partners, will develop educational activities involving NVO, just as many astronomy data archives will develop scientific services for NVO. We are therefore focusing our efforts toward building an outreach infrastructure that simplifies the process of producing outreach materials based on virtual observatories and that enables our audiences to locate the activities and data services most likely to be of interest to them.

In order to identify the most urgent and desirable tools and services for our outreach infrastructure, we held a workshop in Baltimore on July 11-12, 2002. Among the highest-priority items to be recommended at that workshop were the following:

- **Outreach-Related Metadata.** Astronomical datasets contain volumes of information describing attributes of interest to scientists, but non-expert audiences often have vastly different priorities. For example, our experience at STScI resoundingly affirms that most of the public interest in astronomy revolves around the spectacular pictures that astronomers produce. Thus, one of the most useful forms of outreach-related metadata would indicate where a certain dataset lies on the continuum between raw telemetry and color photography. Obviously, we would like our outreach-related search tools to find the color photography of a given object, if such pictures exist.

Outreach metadata will also be needed to identify the content and intended audience of educational activities incorporated into NVO. For example, suppose a group of teachers develops an exercise for high-school students based on observations of the Crab Nebula. Appropriate metadata would help speed the process of identifying this activity to anyone who is interested in teaching high-school students or in learning more about supernova remnants themselves.

- **Protocol for Outreach Queries.** There will likely be many entry points into the realm of NVO education and outreach resources. For example, users of a desktop planetarium software package like *Starry Night* might want to select a particular astronomical object and then access data on it through NVO. In order for the desktop planetarium to retrieve the appropriate data, it will need to know how to query NVO, specifying the outreach-related metadata most likely to identify items of interest to the user. A protocol for outreach-related queries should enable a broad spectrum of educational services to query NVO for appropriate data.

- **Popularity Index for Astronomical Objects.** The vast majority of non-professional users of NVO will be interested in only a minute fraction of the data accessible through NVO. In order to avoid burying their curiosity under a flood of information, we need to ensure that their queries quickly locate items that will pique their interest. One way to do that is to rank objects in order of popularity, so that general searches for astronomical information will place the most popular objects at the top of the list.

- **Directory of Educational Resources.** Volumes of educational activities based on astronomy data currently exist, and we hope that many more activities based explicitly on NVO will soon follow. However, because most of these activities have not been systematically indexed, it can be very difficult for a teacher who would like to teach astronomy to unearth much of this material. One such indexing effort can be found at

  `teachspacescience.stsci.edu`

  One of the most useful things NVO could do on behalf of astronomy outreach would be to create a broader directory of astronomy-related educational resources.

- **Amateur Photography Archive.** Today's amateur astronomers create many high-quality images that would be broadly appreciated by the public. Creating an amateur photography archive as part of NVO would enable these images to be indexed and searched in the same way as the professional archives. Such an archive would have multiple benefits: (1) it would greatly expand the number of beautiful images available to stimulate the public imagination, (2) it would provide spatial and temporal sky coverage within NVO over and above that provided by professional observatories, and (3) it would generate great enthusiasm for NVO in the amateur community, who are likely to be the most avid non-professional users of NVO.

- **Mechanisms for Accessing Fresh Data.** Educators consistently report that students get more excited about recently taken data than they do about archived data. Fresh data therefore serves as a useful motivator for further exploration of a data archive. Thus, we need to find ways to put students in contact with this kind of data. Several telescopes dedicated to on-line use by students currently exist. The NVO collaboration is currently investigating how to integrate them into virtual observatories.

4 Training and Dissemination

Any effort to develop educational resources for astronomy must include a strategy to disseminate those resources to teachers and to train them in how to use those resources. Again, well-chosen partners are an invaluable ally. Organizations who already have the attention of teachers, such as the National Science Teacher's Association (NCSA) or the National Council of Teachers of Mathematics (NCTM) in the United States, are excellent for recruiting enthusiastic teachers to your cause and training them to use your resources. Even better are teachers who are already part of an astronomy outreach program, like Hands-on Universe or Hands-on Astrophysics.

In order to maximize the number of people creating and using NVO outreach services, the NVO framework project is planning to organize workshops that will train teachers and other developers of educational resources in how to create educational activities based on NVO. Many of the infrastructure tools that the NVO collaboration is planning are intended as middleware that will ease the task of building education and outreach activities. Our hope is that by simplifying and streamlining the task of resource development, we will empower the broader astronomy outreach community to unleash its creativity on behalf of NVO.

Poster Papers

Building the MEGA Image-Subtraction Pipeline in the Era of Virtual Observatories

David R. Alves, A.P.S. Crotts, A. Bergier, P. Cseresnjes, and A. Gersch

Columbia Astrophysics Laboratory, 550 W. 120th St., New York, NY 10027, USA

Abstract. We are inventing techniques to increase variable source yields from image subtraction analyses. An example innovation that will improve detection efficiency by about 10% is presented. The MEGA image-subtraction pipeline is a software instrument that could evolve into a tool to do microlensing science with virtual observatories.

1 Introduction to the MEGA Pipeline

What is dark matter and where are the primordial baryons? The pioneering first generation of microlensing surveys have provided tantalizing clues that both the M31 and Milky Way dark halos harbor a significant number of MACHOs [1,2]. However, no lenses have yet been proven to actually reside in either halo. Our next generation survey, the Microlensing Exploration of the Galaxy and Andromeda or MEGA, will map the spatial distribution of a large sample of M31 microlensing events with unprecedented accuracy. This event distribution and a model of our experimental detection efficiency can together be used to prove whether or not the majority of lenses are located in M31's dark halo.

As developers of the MEGA image-subtraction pipeline, we have imagined interacting with Virtual Observatories (VOs) in several ways. It seems possible that VO technologies may one day federate our data, but also, that the software written for the MEGA pipeline could serve as a basis for powerful new experiments in the era of VOs. Indeed, huge amounts of microlensing events could be harvested from the data flow produced by all-sky surveys like Pan-STARRS, LSST, and GAIA [3] with a VO-enabled image-subtraction pipeline. Our innovations could be implemented, for example, in an instrument that calculates and monitors the microlensing detection power of VO-federated image data.

2 Optimal Image Subtractions

We are inventing techniques to increase the yield of microlensing events and variable stars [4] from subtractions of time-series image data. The following is an example of the kind of optimization implemented for the MEGA pipeline.

Recall that to detect microlensing, a high signal-to-noise reference image is created and from it relative flux variations are measured for a time-ordered sequence of images of the same field. The reference is made by registering and coadding some or all of the images in the dataset. The best choice for the origin of the reference image is the most typical pointing so that all of the other images overlap as much as possible. A first-order approximation of the percentage of non-overlapping image area relative to any particular image j in a dataset of N images with dimensions X and Y and origins (or "pointings") x_j and y_j is:

$$D(j) = \sum_{i=1}^{N} \frac{|x_i - x_j|}{X} + \frac{|y_i - y_j|}{Y} \tag{1}$$

Eqn. (1) is also known as the "Manhattan dissimilarity" for being the shortest distance between two points in city blocks. The image j with the smallest dissimilarity $D(j)$ is the best origin for a single reference image.

Subtracting from 2 reference images further reduces the efficiency loss due to telescope pointing errors. It is also possible to replace bad data present in a single reference image with good data from the second image. Understanding how to pick the 2 best reference images is really an introduction to partitioning problems. When partitioning a set of objects into K clusters the goal is to find clusters, the objects of which show a high degree of similarity, while objects belonging to different clusters are as dissimilar as possible [5]. Clusters can be constructed by assigning each object to the nearest of a set of K representative objects. A representative object, or in our case a reference image, is also called the "K-mediod" of a cluster. Thus we want K=2 clusters. In addition to pointing (x and y), we add a third dimension to the parameter space being partitioned that quantifies the on-sky signal-to-noise of each image, and we weight the images by exposure and seeing. Modeling results in Table 1 show that double-subtraction and use of K-mediod cluster finding will increase our source yield by about 10%. Note that the largest loss is due to saturation of M31's bulge.

Table 1. Examples of MEGA Pipeline Efficiency Loss

Source of Data Loss (mosaic-8k; N=10)	$K = 1$	$K = 2$
interstices (or gaps) in reference image(s)	2.0%	0.02%
saturated pixels in reference image(s)	6.8%	0.55%
pointing errors in time-series image data	1.8%	0.70%

References

1. C. Alcock et al.: ApJ **542**, 281 (2000)
2. R. Uglesich et al.: BAAS **201**, 109.04 (2002)
3. `www.ifa.hawaii.edu/pan-starrs; www.noao.edu/lsst; astro.estec.esa.nl/GAIA`
4. D.R. Alves et al.: in *New Horizons in Globular Cluster Astronomy*, ed. by G. Piotto, G. Meylan, G. Djorgowski & M. Riello (ASP, San Francisco 2003)
5. L. Kaufman, & P.J. Rousseeuw: *Finding Groups in Data*, (Wiley-Interscience, New York 1990)

Visualisation Tools for Very Large Amounts of Data

S.G. Ansari

European Space Agency, ESTEC, Noordwijk, The Netherlands

Abstract. Can we perceive a billion stars? How would you go about classifying them? How quantitatively good would your analysis be? How reliable would your results be? In the coming decade an enormous amount of data will be made available to the Astronomical Community. The access to data is being addressed through initiatives such as the AVO and the NVO. With it, the analytical tools to allow serious research to take place will also need to evolve in parallel. Concepts such as Volume Visualisation and other man-machine interfaces will need to be investigated and implemented in parallel to the effort of accessing data uniformly.

1 Introduction

As time goes by, the quality of information dramatically improves. Each new generation instrument adds details of unprecedented quality. The pixelsize decreases, the resolution increases and in general this leads to an almost exponential increase in data acquisition.

This trend is also true in the quantity of data acquired in the various surveys being carried out today and ones that are defined for the future. While Hipparcos collected positional data for about 118,000 objects and was launched in 1989, its successor: GAIA targeted for launch in 2010 will collect data for over 1 billion objects. And as the sample increases, our statistical results will improve. Such fundamental missions will have a remarkable impact on the whole of Astronomy.

2 The Search Paradigm

Up until today, most of our searches are carried out by interrogating individual objects and individual databases. Or we carry out our searches within single catalogues. No attempt has really been made in "identifying a structural pattern", "collecting data for a single filter band", "surveying the archives for a particular spectral feature". Most of our research tools are still very rudimentary in this respect. The main obstacles being of course the quality and *interpretation* of the gathered data.

Let us suppose, however, that such obstacles can be overcome with proper tools to distinguish the good data from the bad, how can we evolve the interface to a degree, where we create a *virtual universal laboratory*, a place where we begin to test our theories in a more efficient way by concentrating the mind on the problem at hand? What would typically be the search paradigms, which would allow us to more easily find what we really seek? How can we manage the data more efficiently?

3 Evolution of Archive Research

With the advent of Virtual Observatories and the enormous amount of data that will be made available in the coming ten years, it is evident that the man-machine interface will need to be reviewed in a very fundamental way. The Astronomer cannot continue to spend her/his time in writing and rewriting software to reduce a minimum amount of data. This leads to a clear waste of resources and time, which we cannot afford. It is worth stepping back and considering *how* to carry out science and interpret it, rather than try to resolve the "nitty-gritty" issues. Verification of results should become a straight-forward matter. The observer at the telescope should be able to immediately confirm or refute an observed phenomenon.

In an attempt to join computing forces across the globe and to establish a common "virtual" laboratory, the Universities of Washington, Chicago, Illinois (NCSA), Rutgers in the U.S. and the Max-Planck Institut für Gravitationsphysik in Germany created the first Astrophysics Simulation Collaboratory (`http://www.ascportal.org/`). Initially foreseen to investigate numerical problems in General Relativity, this ambitious programme, once completed can be applied to a variety of problems of astrophysical nature. This computational laboratory is a typical example of how *Grid* applications will take shape in the future as the technology becomes simplified and more readily useable.

4 A New Research Paradigm

It is clear that new ways must be invented to deal with the large databases of the future. Techniques discussed elsewhere in this book will need to become more accessible and more *scientist-friendly*.The sample and complexity of data analysis is about to take a huge step. We cannot afford to *tinker* with the microworld of computing. We must begin to investigate large-scale physical phenomena in the universe by combining multi-disciplinary astronomical data.

In an attempt to carry out a prototyping activity to envisage what this interface may look like in the future, and based on a popular astronomical software, Maris Technologies in collaboration with ESA is embarking on creating an Astrophysical Virtual Desktop. The interface will make use of AVO's VOTable standard, allowing the user to upload to a local application any data from any archive using VOTable as a valid interface. The objective of this activity is to investigate new ways to carry out bulk data research in multiwavelength/multidisciplinary domains, and finding new ways to visualise large amounts of objects and to *navigate* through a Virtual Astronomical Universe.

Beyond these typical features, a method is being devised to *interactively* be able to share data in an Astronomical context. Here, a user may collaborate with (an)other colleague(s) online within the same interface or "Zone", where multiple users share the same virtual sky to discuss various Astronomical phenomena. Each user would be able to either load a catalogue on the Internet or upload proprietary data visible to all other users, while carrying out a live conversation.

Most of these concepts are already very mature, it just requires putting them together into a concise and typically research context.

References

1. GAIA Science Advisory Group: 'GAIA Composition, Formation and Evolution of the Galaxy, Concept and Technology Study Report'.(European Space Agency ESA-SCI(2000), July 2000)
2. M.A.C. Perryman: American Astronomical Society, 191st AAS Meeting,68.01; Bulletin of the American Astronomical Society, Vol. 29, p.1314.
3. I. Foster, C. Kesselman, S. Tuecke, 2001: *The Anatomy of the Grid*, (`http://www.globus.org/research/papers/anatomy.pdf`)
4. O'Dell, C.R., Wen, Zheng, Genetti, Jon, Johnson, Greg, Nadeau, D.R., Napear, S., Snavely, A., 1999: 'Galactic MPIRE: Flying through the Digital Galaxy', ENVISION VOl. 15, No. 4.

Laplacean Ideology for Preliminary Orbit Determination and Moving Celestial Body Identification in Virtual Epoch

O.P. Bykov

Pulkovo Astronomical Observatory, Russia
oleg@OB3876.spb.edu

Abstract. Any CCD frames with stars or galaxies or clusters and other images must be studied for a searching of moving celestial objects, namely asteroids, comets, artificial Earth satellites inside them. At Pulkovo Astronomical Observatory, new methods and software were elaborated to solve this problem.

The introducing of CCD technique in astronomical practice allows us to elaborate a new approach to the problem of orbit determination and identification of any moving celestial body. The fundamental base for such approach is the Classic Laplacean Method of initial orbit calculations and its development – the Apparent Motion Parameters Method (the AMP-method) which was created by Dr. A.A. Kisselev and his colleagues at Pulkovo Observatory in the period 1970–2000.

Mass positional CCD observations of celestial bodies moving against the background of reference stars are now characterised by a high accuracy, a large efficiency and unlimited density of distribution of observed positions along a short arc of celestial sphere. Due to the accurate star catalogues and computers one can obtain not only several astrometric positions of any celestial object but its angular velocity by means of a simple statistical processing of the set of object's coordinates practically in a real-time of the CCD observations. These four parameters (namely α, δ, $\dot{\alpha}$, $\dot{\delta}$ for utilisation of the Laplacean method or topocentric angular velocity μ and its position angle ψ in the case of the AMP-method) are usually enough to determine its circular or parabolic preliminary orbit which may be useful for identification of the observed object or for ephemeris service during several close nights or weeks. One ought to remember that both methods require three or more observed positions during a single night.

In these circumstances the observer himself can identify all known celestial objects which could be seen in his telescope's field of view during a time of observations. It may be done by means of a preliminary calculation of orbital catalogue positions for all known objects at given UT moment and a comparison of observed and calculated asteroid positions and velocities. After this identification, initial circular, or parabolic, or sometimes elliptical orbits for the unknown asteroids having several positions per night or per three successive nights may be calculated. In this manner we can investigate each CCD frame for the search of small Solar system bodies.

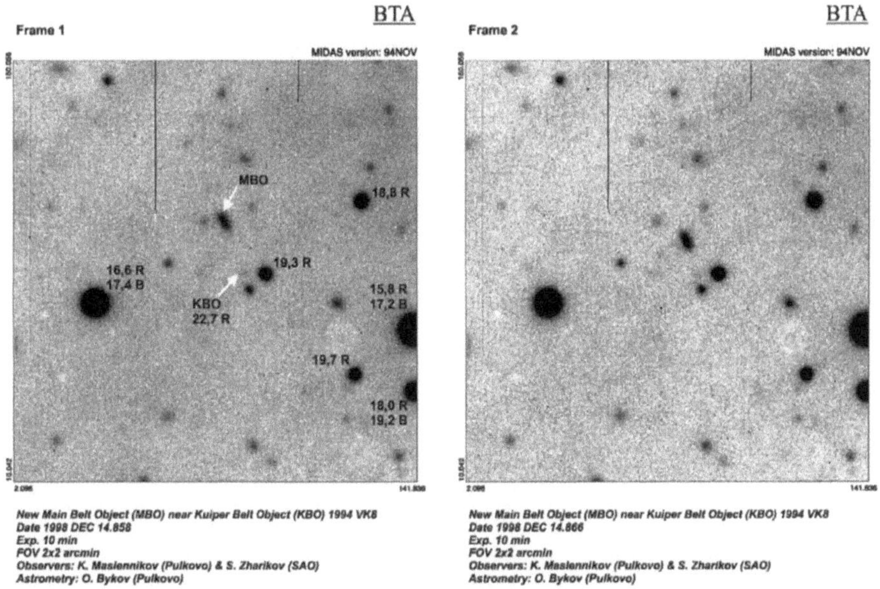

Fig. 1. Frames 1 and 2 are the pictures of a Field Of View of the Russian 6-m Telescope with the images of stars and moving Main Belt Object (MBO, between 21-22 mgn, long stroke) which was found by the observers 14 Dec. 1998 as accident asteroid near the Edgworth-Kuiper Belt Object named 1994 VK8 (KBO, practically invisible in the Frames). A comparison of the First and the Second Frames obtained in 10 minutes interval shows a motion of the MBO in direction to the Double Star with 19.3 mgn.
The magnitude of stars were taken from USNO A2.0 Star Catalogue. The Field Of View is 4 square arcmin, CCD matrix made by "Electron" and SAO (Russia), an exposition is 10 min.
This MBO is not identified with any known asteroid up to the present. A following of the new MBO immediately was not possible due to current Observational Program. Thus, the MBO was lost but its coordinates and topocentric velocities were calculated and the object awaits an identification in years to come

The AMP-method was successfully applied by the author for calculations of initial orbits of the Near Earth Asteroids, Edgworth-Kuiper Belt objects and Artificial Earth Satellites with the use of short arc positional observations when the Parameters of Apparent Motion (namely: values of angular velocity μ and its position angle ψ, angular acceleration $\dot{\mu}$ and curvature C of trajectory) or the first and the second derivatives of spherical coordinates of observed objects were accurately obtained.

The algorithms and EPOS (Ephemeris Program for Objects of Solar system) software were developed by Dr. V. L'vov and his colleagues at Pulkovo observatory for the fast analysis of any CCD frame where the moving celestial objects could be detected. Thanks to this approach we can get a new Celestial Mechanics information as "by-product" of dominating astrophysical CCD observations obtained, for example, with the Russian 6-m Telescope.

Our possibilities may be used in various Virtual Observatory projects for the immediate identification and fast orbit determination of any small celestial body moving across telescope's field of view. Any archives CCD frames can be also investigated.

Using XML-Schema to Model Data from Present and Future Astronomical Databases

Bernard Debray

CNRS UMR6091, Observatoire de Besançon, BP1615, F-25010 Besançon Cedex, France

Abstract. XML (eXtensible Markup Language) has now become an almost inescapable tool for the exchange of formatted data over the World Wide Web and is now becoming extensively used in astronomical data bases to interweave them, therefore making it an important tool for the building of the Virtual Observatory. Document Type Definitions (DTD) have been initially used to describe XML documents formatting; they offer very limited possibility about data typing, constraints or bounds for the handling of data-oriented documents. These limitations can now be overcome thanks to the XML Schema recommendation. The potential benefit of XML Schema for astronomical databases is illustrated through an ongoing implementation of XML Schema for formatted data output in the Besancon Double and Multiple Star database.

1 XML and DTDs

XML, the eXtensible Markup Language, was designed to permit the exchange of formatted informations over the World Wide Web in providing a standardized framework for the description of both data and metadata. Since its start in 1998, XML has been used among the scientific community for both the management of information documents such as publications, and the modelling and exchange of scientific data (see e.g. [5,1]). Originally, XML documents make use of a Document Type Definition (DTD) to define their structure and syntactic organization. DTDs make use of a non-XML syntax (SGML), which does not allow to process them using XML tools such as parsers. They offer limited possibilities to insert documentation, limited data types and limited possibilities to specify data formats and constraints ; such informations have to be included in dedicated comments to be possibly processed by user programs run downstream from the XML parsing process.

2 Moving from DTDs to XML Schema

Released in May 2001 as a full recommendation of the World Wide Web Consortium, XML Schema 1.0 allow now to reflect completely the data types, constraints and structure of a data set using the XML syntax itself (see e.g. [6]). The data model can now be handled itself through standard XML parsers and softwares. XML parsers can be used to handle the data before handing over the processing of the data to user programs up to a higher level than with DTDs. XML Schema allow the definition of complex data types from numerous basic data types (string, byte, integer, float, time, uri, ...) and other derived data types as well as the inheritance of data structures. As an XML structure, the data model can also be included within the XML document itself. As a first step to move to XML Schemas, one can use DTD to XML Schema translators such as the *dtd2xs* software [4] which was designed to preserve as much as possible the information contained in DTDs, especially in comments.

3 XML Schema in Astronomical Databases

At the Besançon Observatory, an XML output to data base queries following the Astrores DTD [1] was implemented in the BDB binary stars data base [3]. To concretely assess the use of XML Schema, we are now testing the production of XML Schema validated XML output. As a first step, we converted the Astrores DTD into XML Schema. The resulting XML Schema file can be found at `http://bdb.obs-besancon.fr/xml/astrores.xsd`. The VOTable format for representing a data table in XML was recently released in the framework of the Virtual Observatory initiatives [2] ; a draft version of an XML Schema description of the format was included in the document. Our developments will therefore be turned into this format from now on.

4 XML Schema and the VO

Although XML Schema are "heavier" than DTDs in terms of overhead, they offer a standard way within the XML framework to formulate the models of scientific data, their types and constraints of values. On the other hand, software tools are now rapidly emerging. All this works in favour of XML Schema to emerge as a helpful tool in the process of data validation when interweaving scientific databases and, consequently, for the development of the Virtual Observatory.

References

1. F. Ochsenbein, M. Albrecht, A. Brighton, P. Fernique, D. Guillaume, R.J. Hanisch, E. Shaya, A. Wicenec; In: *ASP Conf. Ser., Vol. 216, Astronomical Data Analysis Software and Systems IX*, ed. by N. Manset, C. Veillet, D. Crabtree (San Francisco: ASP, 2000) p. 83, 2000
2. F. Ochsenbein, R. Williams, C. Davenhall, D. Durand, P. Fernique, D. Giaretta, R. Hanisch, T. McGlynn, A. Szalay, A. Wicenec; this proceedings
3. E. Oblak, B. Debray, E. Lastennet, T. Kundera. In: *Proceedings of SF2A annual meeting*, (EDP-Science, 2002), to be published
4. R. Schweiger, S. Hoelzer, K.U. Heitmann, D. Dudeck: Medical Informatics and the Internet in Medicine, **26 (4)**, 297 (2001)
5. E. Shaya, J. Gass, J. Blackwell, B. Thomas, B. Holmes, C.Y. Cheung. In: *ASP Conf. Ser., Vol. 216, Astronomical Data Analysis Software and Systems IX*. ed. N. Manset, C. Veillet, D. Crabtree (San Francisco: ASP, 2000), p. 87
6. K. Williams, M. Brundage, P. Dengler, J. Gabriel, A. Hoskinson, M. Kay, T. Maxwell, M. Ochoa, J. Papa, M. Vanmane: *Professional XML Databases* (Wrox Press, 2000)

The Virtual Observatory as a Tool to Study Star Cluster Populations in Starburst Galaxies

Richard de Grijs[1] and Uta Fritze–v. Alvensleben[2]

[1] Institute of Astronomy, University of Cambridge, Madingley Road, Cambridge CB3 0HA, U.K.
[2] University of Göttingen, Universitätssternwarte, Geismarlandstraße 11, 37083 Göttingen, Germany

1 The Cluster Luminosity Function

The cluster luminosity function (CLF) is one of the most important diagnostics in the study of old globular and young compact star cluster populations. While the CLFs of young star cluster systems are closely approximated by power laws, as predicted by standard cluster formation models (e.g., [2,4]), for old globular cluster systems the shape of the CLF, and also of the *initial* mass function, is well-established to be roughly Gaussian, with an apparently universal turnover luminosity (mass).

Processes responsible for the transformation of the CLF of young star cluster systems into the lognormal distributions characteristic of older CLFs include differential fading due to aging of the cluster population, the preferential depletion of low-mass clusters by evaporation and tidal interaction with the gravitational field of the parent galaxy, and the preferential destruction of high-mass clusters by dynamical friction. The relative importance of these destruction processes is still controversial (see, e.g., [4–6]).

One should be cautious, however, to generalize any of the "standard" models, since they only apply to quiescent, non-interacting galaxies. In the high-pressure environments and the time-varying potentials of interacting and merging galaxies we expect significant differences for the cluster formation and destruction processes, mass spectra, and time-scales for the dynamical evolution of star clusters due to, e.g., external pressure and shocks.

2 Scientific Questions and the Virtual Observatory

Using the wealth of archival data from the *Hubble Space Telescope* and large ground-based facilities, we are currently using the ST-ECF ASTROVIRTEL tools[1] to obtain CLFs and colour distributions in several optical and/or near-infrared passbands, in order to address these issues statistically.

We caution strongly that it is obviously very important to obtain accurate ages for the individual clusters and to correct the observational CLF to a common age before interpreting the "intrinsic" CLF, see Fig. 1 [1,3].

[1] E.g., *querator* (http://archive.eso.org/querator) for imaging observations and *listator* (`http://archive.eso.org/listator`) for spectroscopic observations.

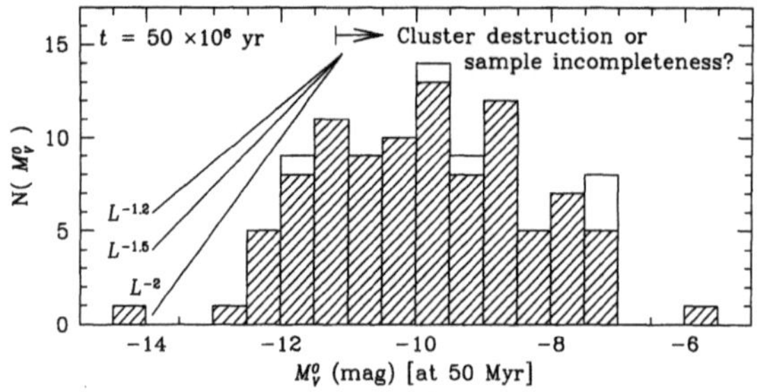

Fig. 1. CLF for the cluster sample in M82's fossil starburst region "B", corrected individually to a fiducial age of 50 Myr [1]. The CLF may be broadened due to dynamical destruction of lower mass clusters, although the degree to which our cluster sample is incomplete for these brightnesses is uncertain. Questions like these can now be addressed with the large amount of data available in the archives

The ASTROVIRTEL initiative allows us, for the first time, to address questions such as:

1. Is the *intrinsic* CLF (or, alternatively, the cluster initial mass function) a power law in all cases?
2. What are the time-scales for the evolution from the power-law young compact CLFs into old globular CLFs?
3. Is there a clear difference between OB associations/open clusters and compact globular clusters at formation or do they all form a continuum mass spectrum?
4. Can we "observe" dynamical cluster destruction by comparison of young star cluster systems in ongoing mergers and merger remnants of various ages?
5. Is there an environmental dependence on the efficiency and therefore on the time-scale of low-mass star cluster depletion?
6. Is the (old) globular CLF *universal*? To what extent and how reliably can it be used as a (secondary) distance indicator and to estimate H_0?
7. Do stronger starbursts favour the formation of more massive clusters?
8. Does the ratio between field star formation and star formation in clusters depend on the galaxy or interaction properties?
9. Can we determine the violent star and cluster formation histories of galaxies based on star cluster colour distributions and deep luminosity functions?

References

1. de Grijs R., O'Connell R.W., Gallagher J.S., 2001, AJ, 121, 768
2. Elmegreen B.G., Efremov Y.N., 1997, ApJ, 480, 235
3. Fritze–v. Alvensleben U., 1999, A&A, 342, L25
4. Gnedin O.Y., Ostriker J.P., 1997, ApJ, 474, 223
5. Vesperini E., 2000, MNRAS, 318, 841
6. Vesperini E., 2001, MNRAS, 322, 247

The Master Catalogue of Stars Towards the Magellanic Clouds

Nausicaa Delmotte[1,3], Cécile Loup[2], Daniel Egret[1], Maria–Rosa Cioni[3], and Francesco Pierfederici[3]

[1] CDS, 11 rue de l'Université, 67000 Strasbourg, France
[2] Institut d'Astrophysique de Paris, CNRS, 98 bis Bd Arago, 75014 Paris, France
[3] ESO, K.-Schwarzschild-Str.-2, D-85748 Garching, Germany

Abstract. The Master Catalogue of stars towards the Magellanic Clouds (MC2) is a multi–wavelength reference catalogue. We performed a systematic cross-matching of the DENIS [3] Catalogue towards the Magellanic Clouds (DCMC) [1] with 2MASS [4] and the optical (GSC-II[1], UCAC1 [6]) catalogues. It provides an unprecedented wealth of data on the stellar populations of the Magellanic Clouds (MCs). New cross–matching procedures for very large catalogues have been developed and detailed results on the astrometric and photometric accuracy of the cross–identified catalogues were derived. The cross–matching of large surveys is an essential tool to improve our understanding of their specific contents in the Virtual Observatory context. This study has been partly supported by ASTROVIRTEL[2], a project that aims at improving access to astronomical archives as virtual telescopes, and funded by the European Commission under FP5 Contract No. HPRI-CT-1999-00081.

1 Why Cross–Matching Multispectral Catalogues?

Cross–matching catalogues is highly relevant for: (i) completing the spectral or spatial coverage when there are missing or unpublished data, (ii) searching for astrometric or magnitude shifts between surveys and check the internal accuracy of each catalogue, and (iii) multiwavelength studies of astronomical objects.

Each catalogue has its own characteristics. The observational strategy influences the homogeneity of the final data. The passbands, as well as the characterized stellar populations and the number of sources are different. These factors have a strong impact on the results of the cross–matching.

2 Accuracy of the Cross–Identified Catalogues

Field distortions in the DCMC affect the quality of the astrometry. The relative shifts in R.A. and Dec. are a function of the pixel coordinates of the camera.

Restricting to the range [10, 14] in J and [8, 12] in K_S, the derived mean relations between DCMC and 2MASS magnitudes are as follow:

$$J_{\mathrm{DCMC}} = J_{\mathrm{2MASS}} - (0.11 \pm 0.06) \qquad K_{\mathrm{sDCMC}} = K_{\mathrm{s2MASS}} - (0.14 \pm 0.05)$$

[1] `http://www-gsss.stsci.edu/gsc/gsc2/GSC2home.htm`
[2] `http://www.stecf.org/astrovirtel/`

These relations have been computed in the case of the DCMC catalogue and may not be valid for the whole DENIS survey.

A *strip by strip* cross–matching strategy has been implemented to correct from the residual discrepancies in the near–IR catalogues, using both position and magnitude criteria. Adding the GSC2.2 and UCAC1 catalogues is done with a cross–matching in position only. More details can be found in Delmotte et al. (2002) [2].

3 Multispectral Views of the LMC

Combining IR with optical wavelength (see Fig. 1) enables us to discriminate between dwarf and giant stars. Red and blue supergiants are also easy to trace.

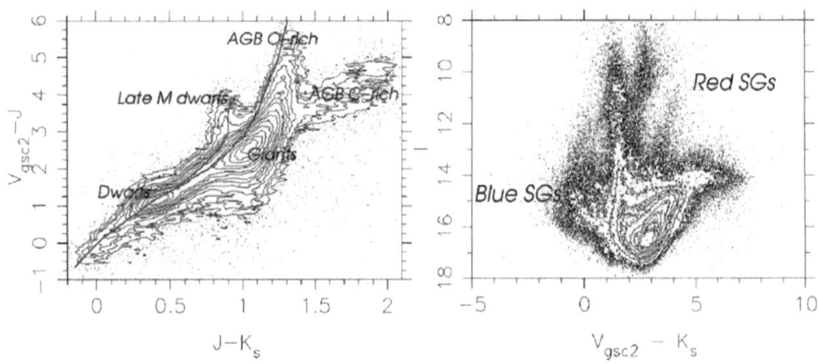

Fig. 1. The J and K_s bands are from 2MASS. The colour/colour dwarf and giant tracks superimposed on the left diagram are from Wainscoat et al. (1992) [5]

4 Conclusions

The MC2 is available on the web at CDS: `http://vizier.u-strasbg.fr/MC2/` Links for each source allow to access the complete data from the original catalogues through the VizieR search engine. Cross–identification of large multispectral surveys is an essential step towards the Virtual Observatory of the future.

References

1. M.–R. Cioni, et al.: A&AS **144**, 235 (2000)
2. N. Delmotte, et al.: accepted for publication in A&A (astro-ph/0209207)
3. N. Epchtein, et al.: The Messenger **87**, 27 (1997)
4. M.F. Skrutskie, et al.: ASSL **210**, 25 (1997)
5. R.J. Wainscoat, et al.: ApJS **83**, 111 (1992)
6. N. Zacharias, et al.: AJ **120**, 2131 (2000)

UCDs: Metadata for the VO

Sébastien Derriere[1], Thomas Boch[1], François Ochsenbein[1], and Patricio F. Ortiz[2]

[1] CDS, Observatoire Astronomique de Strasbourg, 11 rue de l'Université, F-67000 Strasbourg, France
[2] Osservatorio Astronomico di Capodimonte, via Moiariello 16, I-80131 Naples, Italy

1 Introduction

The original descriptions of astronomical tables come from many different sources and are therefore highly heterogeneous. Retrieving astronomical data stored in catalogues, like many other Virtual Observatory (VO) applications, is much easier if a uniform description of the contents is available (using standardized metadata).

The UCDs (Unified Content Descriptors), first developed in the ESO/CDS data mining project [1], are metadata describing precisely the contents of the individual fields (columns) of tables available from a data center. They have been applied to describe the content of the 10^5 columns available in the different VizieR tables [2].

We briefly describe the use of UCDs for various purposes[1].

2 UCD Exploration

A *browser* has been developed to explore the hierarchical structure of UCDs (4–levels tree with approximately 1500 elements). The javascript and the applet versions of the browser present the UCD tree in a familiar form, resembling a graphical view of a filesystem, where folders are nodes in the UCD tree, while documents are the leaves.

Clicking on a leaf gives access to:

- a short definition for this UCD;
- statistics on column labels and units associated to this UCD;
- usage statistics for this UCD in VizieR.

3 Catalogue Selection

The UCDs allow to select catalogues which exactly contain a given measurement. As opposed to selecting 'optical' catalogues, hoping to have a measure of V magnitude, one can select directly all catalogues which contain a column with the UCD PHOT_JHN_V.

[1] The available tools are on `http://vizier.u-strasbg.fr/UCD/`

The catalogue selection can be done via the browser: there is a direct link to VizieR for querying the catalogues having the selected UCD in at least one column.

In addition, it is possible to translate plain text into relevant UCDs. One provides one or several terms to describe in natural language the desired quantity (e.g. 'proper motion'). The answer is a list of corresponding UCDs, tentatively ordered by relevance. It is then easy to select the proper UCD and query the relevant catalogues.

4 Automated Column Comparisons

One very promising use of UCDs is to allow automated conversion and comparison of columns during data mining or cross-identification operations. Two tables can be merged on the basis of their common UCDs: if a UCD is used in both tables, it means that the same quantity is present, and then it is possible for the user to ask for a comparison of the values in those columns, with possible on-the-fly unit conversion when necessary.

5 Other Applications

For each photometric UCD, a range in the electromagnetic spectrum (corresponding for example to the filter transmission) can be assigned. It is then possible to select catalogues '*with measure between wavelengths (or frequencies) x and y*', for example.

Given a catalogue, it is possible using UCDs to find other catalogues similar *in content*, because they will have many UCDs in common.

6 Conclusions

UCDs design to encode the semantics of astronomical content, beyond column labels, to enable cross catalogue/archive interoperability. Their use should be extended in the future to other archives. Owing to the wide diversity of table contents, UCDs constitute an excellent starting point for a hierarchical description of astronomy, for general data mining purposes.

References

1. P. F. Ortiz, et al.: *ESO/CDS Data-mining Tool Development Project.* In: *ADASS VIII*, ASP Conf. Ser. **172**, p.379 (1999)
2. F. Ochsenbein, P. Bauer and J. Marcout: Astron. Astrophys. Suppl. Ser. **143**, pp.23-32 (2000)

Applicability of Emerging Resource Discovery Standards to the VO

Markus Dolensky

European Southern Observatory, Karl-Schwarzschild-Straße 2,
85748 Garching bei München, Germany

Abstract. The Virtual Observatory (VO) will be one of the first applications belonging to a new generation of network services which are commonly referred to as the Semantic Web. As the name indicates the Semantic Web is supposed to represent its contents and usage in a systematic and decentralized way. In order to achieve this goal a number of new protocols and standards are currently defined. The Semantic Web is still in its early stages and therefore it is not surprising that there is a moderate number of evolving standards and systems out there with more or less comparable scope. It is the purpose of this paper to evaluate the applicability of a number of those evolving technologies to the VO.

1 Overview

From the viewpoint of a system architect the major obstacle towards the Semantic Web is a lacking standard on how to express meaning, i.e., how to model data and how to exchange meta data [1]. An initiative of the World Wide Web Consortium (W3C, `www.w3.org`) called the Resource Description Framework (RDF) aims at supporting knowledge exchange on the Web. All technologies introduced below are solving part of the problem and are more or less tightly coupled to RDF (`www.w3.org/RDF`) and the eXtensible Markup Language XML (`www.w3.org/XML`).

In the framework of an ESO-CDS data mining project a scheme of Unified Content Descriptors (UCD) [2] has been developed. The original goal was to map database columns with a distinct set of well described parameter identifiers. It turns out that these UCDs can be extended into a taxonomy for Astronomy. This UCD vocabulary may serve as meta data for annotating data and building a resource registry for Astronomy.

Various design decisions have to be taken when building the VO with its heterogeneous data stores and Grid based computing facilities. One of them is to choose a meta data architecture for registering services and automated resource discovery like UDDI plus WSDL or RSS. To express meaning in a system independent way it takes data modeling. With data models described in UML or DAML+OIL ontology systems can be build to translate queries between the meaning domain and the various system specific domains. Various query specifications and languages are popping up lately like XQuery or MDL to facilitate data mining on information with XML structure. To orchestrate queries across various archives and distributed processing there are new protocols (SOAP, VOTable) and various systems supporting remote procedure calls (XML-RPC, SOAP, J2EE).

2 Conclusion

Obviously there is an awful lot of terms and acronyms. Any of them, however, denotes a technology which may contribute part of the functionality required by a VO. It is a matter of separating the hype from visionary thinking, though. By latching onto evolving standards of the Semantic Web and carefully selecting the right technology a large fraction of the resources will remain available to support features specific to Astronomy.

In historical terms, with the Semantic Web we're moving from the Stone Age to the Bronze Age. With a VO we may soon learn how to combine copper with zinc to produce a gold appearing metal called brass.

3 Glossary

DAML+OIL	DARPA Agent Markup Language + Ontology Interchange Language
J2EE	Java 2 Enterprise Edition
MDL	Meaning Definition Language
RDF	Resource Description Framework
RSS	RDF Site Summary
SOAP	Simple Object Access Protocol
UCD	Unified Content Descriptor
UDDI	Universal Description, Discovery, and Integration
UML	Unified Modeling Language
VO	Virtual Observatory
VOTable	Proposed XML Format for Astronomical Tables
W3C	World Wide Web Consortium
WSDL	Web Service Description Language
XML	eXtensible Markup Language
XML-RPC	XML-Remote Procedure Call
XQuery	XML Query Language

References

1. K. Ahmed et. al., 2001, XML Meta Data, Wrox Press Ltd., Birmingham, UK
2. P.F. Ortiz, F. Ochsenbein, A. Wicenec, M. Albrecht: 'ESO/CDS Data-mining Tool Development Project'. In: *Astronomical Data Analysis Software and Systems VIII at Urbana, USA, 1998*, ed. by D.M. Mehringer, R.L. Plante, D.A. Roberts (ASP Conference Series, Vol. 172, 1999) pp. 379–382

The Astronomy Digital Library and the VO

Guenther Eichhorn[1], Alberto Accomazzi[1], Carolyn S. Grant[1], Michael J. Kurtz[1], and Stephen S. Murray[1]

Harvard-Smithsonian Center for Astrophysics, Cambridge, MA 02138, USA

Abstract. The NASA Astrophysics Data System (ADS) is the central part of the Digital Library in Astronomy. It provides the users with a search system for the astronomical literature and an extensive system of links to on-line resources like data tables, catalogs, on-line journals, etc. The extensive system of linking between on-line resources in Astronomy is the foundation on which the Virtual Observatory will be built.

The ADS is funded by NASA and is accessible world-wide free without restrictions at:

`http://ads.harvard.edu`

1 Current ADS

The NASA Astrophysics Data System Abstract Service is by now a central facility in astronomy. It is a key element in the emerging digital information resource for astronomy, which has been dubbed Urania [2]. The ADS is tightly interconnected with the major journals of astronomy, and the major data centers. A detailed description of the ADS has been published in a special issue of Astronomy & Astrophysics Supplement in April, 2000 (Overview: [5], Search Engine and User Interface: [3], System Architecture: [1], Data: [4]).

1.1 Abstracts

The first major part of the ADS is the Abstract Service. Abstracts come from many different sources (see [4]). As of July, 2002 there are 807,000 astronomy references indexed in the ADS, the database is nearly complete for the major journals articles beginning in 1975. In the Physics database there are 1.25 million references, and in the Instrumentation database there are 628,000 references. Approximately half of all references have abstracts, the other half only have titles, authors, and journal information.

1.2 Scanned Articles

The second major part of the ADS is the Article Service. It contains scanned full journal articles for most of the astronomical journal literature going back to volume 1. The ADS has obtained permission to scan, and make freely available on-line, page images of the back issues of all the major journals and most smaller journals in astronomy. This will provide (along with the indexing and the more recent archives held by the journals) a complete electronic digital library of the major literature in astronomy.

1.3 Links

The ADS responds to a query with a list of references and a set of hyperlinks showing what data are available for each reference (see [3]). There are over 6 million hyperlinks in the ADS, of which almost 50% are to sources external to the ADS project.

The largest number of external links are to SIMBAD, the NASA Extragalactic Database (NED), the Space Telescope Science Institute (STScI), and the electronic journals. A rapidly growing number, although still small in comparison to the others, are to data tables created by the journals and maintained by the Centre de Données Astronomiques de Strasbourg (CDS). These links are an extremely important aspect of the ADS.

2 ADS and VO

The astronomical community in the last decade has built a system of interlinked data centers that is unique among the sciences. This system has been called Urania. It is composed of the closely integrated operations of the ADS, CDS, NED, the Data and Archive Centers, and the Journals. It began functioning in 1993 with the ability to make joint queries to the ADS and the SIMBAD database at the CDS. It now provides the user with a tightly integrated web of information links. We believe that the continuous expansion and strengthening of the indexing centers in Urania (ADS, CDS, NED) along with similar indexing efforts at the data and archive centers will be a key requirement for the success of the transition from the current system to a fully integrated AVO.

3 Acknowledgement

Funding for this project has been provided by NASA under NASA Grant NCC5-189.

References

1. Accomazzi, A., Eichhorn, G., Kurtz, M. J., Grant, C. S., & Murray, S. S. 2000, A&AS, 143, 85
2. Boyce, P. B. 1995, American Astronomical Society Meeting, 27, 1333
3. Eichhorn, G., Kurtz, M. J., Accomazzi, A., Grant, C. S., & Murray, S. S. 2000, A&AS, 143, 61
4. Grant, C. S., Accomazzi, A., Eichhorn, G., Kurtz, M. J., & Murray, S. S. 2000, A&AS, 143, 111
5. Kurtz, M. J., Eichhorn, G., Accomazzi, A., Grant, C. S., Murray, S. S., & Watson, J. M. 2000, A&AS, 143, 41

The Digitized Hamburg Objective Prism Surveys

D. Engels[1], H.-J. Hagen[1], N. Christlieb[1], D. Groote[1], D. Reimers[1], L. Wisotzki[2], and F.-J. Zickgraf[1]

[1] Hamburger Sternwarte, Gojenbergsweg 112, D-21029 Hamburg, Germany
[2] Astrophysikalisches Institut Potsdam, An der Sternwarte 16, D-14482 Potsdam, Germany

Abstract. The digitized objective prism plates of the Hamburg Sky Surveys will be made available via the World-Wide Web, providing the raw data in FITS format for small areas around positions defined by the user. The digitized database will be part of the German Virtual Observatory (GAVO). Currently a test server is available.

1 Introduction

The Hamburg objective prisms surveys consist of two complementary parts: The northern Hamburg Quasar Survey (HQS) ($|b| > 20°, \delta > 0°$) and the southern Hamburg/ESO Survey (HES) ($|b| > 30°, \delta < +2.5°$). The plates of the HQS were taken 1985–1997 with the Calar Alto Schmidt telescope, and contain usable spectra with a seeing limited resolution of ~ 45 Å/mm at Hγ in the magnitude range $14 < B < 19$ [1]. The southern plates were obtained with the ESO Schmidt telescope 1990–1996, have a spectral resolution of ~ 15 Å/mm at Hγ and contain usable spectra in the magnitude range $13 < B_J < 18$ [2]. Both surveys used the KODAK IIIa-J emulsion providing sensitivity over a wavelength range between 3200 Å and 5400 Å. The HQS archive contains 1288 plates for 567 fields, while for the HES one spectral plate was taken for each of 428 fields. Plates for both surveys are scanned in full mapping mode by the Hamburg PDS microdensitometer and the digitized database is used for a variety of astronomical applications, in addition to the search for (mostly bright) quasars. Further information of HQS and HES is provided by `http://www.hs.uni-hamburg.de/surveys.html`.

2 Web Access

We are developing tools to allow an access to the digitized objective prism data using the World Wide Web. It is planned to provide online the raw data of small fields around a given position in FITS format. The spectra have to be extracted by the user e.g. with standard CCD reduction software. Astrometric information, brightness calibration and a catalog with sample spectra will be provided. Currently a test server is installed, allowing data retrieval for one field. The digitized plates will be stored on a 1 Terabyte RAID system in order to make the data accessible to the community. Initially data of the northern sky and selected areas of the southern sky will be provided. Incremental releases of the remaining southern areas will follow later.

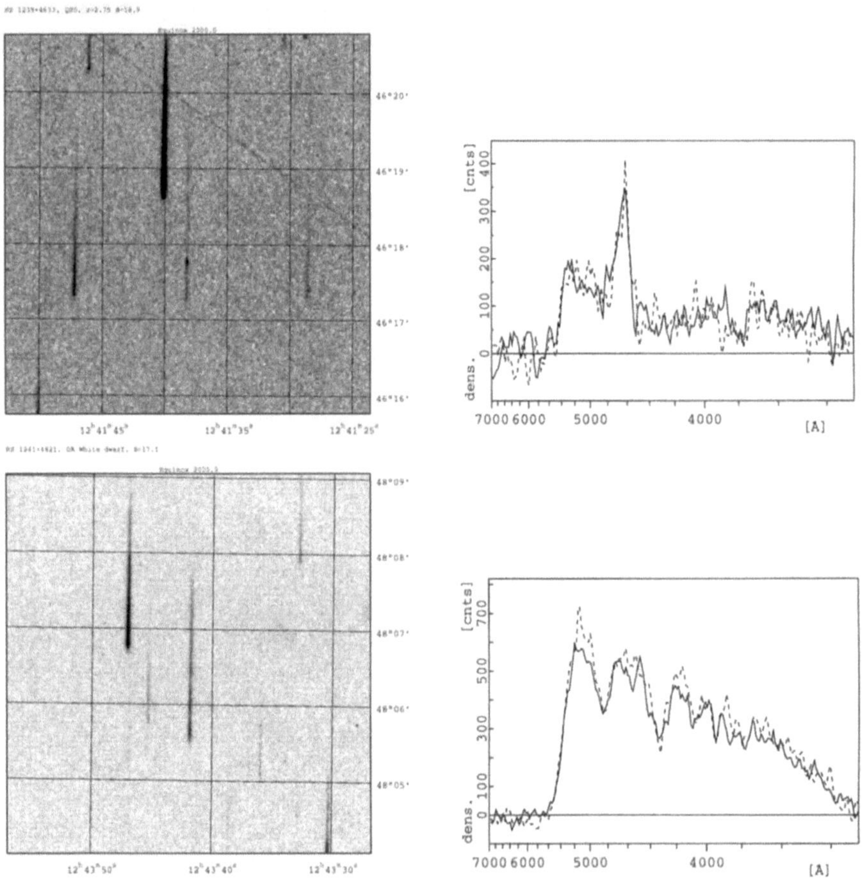

Fig. 1. Data samples from the HQS. Upper panel: HS 1239+4633 is a z=2.75 quasar with B=18.9. The strong Ly-α line can easily be seen on the image at left as a black dot superposed on the spectrum. On the righthand side the extracted spectra from two plates are shown. Lower panel: A white dwarf of type DA, displaying strong Balmer absorption lines

The online Hamburg surveys will be a unique database of digital spectra which can be used e.g. for optical identification of sources found by surveys in other wavelength ranges (Zickgraf et al., these proceedings).

References

1. H.J. Hagen, D. Groote, D. Engels, D. Reimers: Astron & Astrophys. Suppl. **111**, 195 (1995)
2. L. Wisotzki, N. Christlieb, N. Bade, V. Beckmann, T. Köhler, C. Vanelle, D. Reimers: Astron & Astrophys. **358**, 77 (2000)

Aladin in the VO Context

Pierre Fernique[1], François Bonnarel[1], Mireille Louys[2], Anaïs Oberto[1], Thomas Boch[1], Mark Allen[1], Olivier Bienaymé[1], and Daniel Egret[1]

[1] CDS, Observatoire de Strasbourg, UMR CNRS 7550, 11, Rue de l'Université, F-67000 Strasbourg, France
[2] Université Strasbourg I, LSIIT, UMR CNRS 7005, Bd S. Brant, F-67400 Illkirch, France

Aladin, the CDS interactive sky atlas is one of the important interoperability tools of the astronomy Web. Several recent improvements prepare it for use as a portal to the Phase A of the Astrophysical Virtual Observatory (AVO), as shown by its use as a testbed for the AVO-January 2003 demonstration (Quinn, P., 2002).

As part of the tasks of the AVO Work Area 2, we have included in Vizier and Aladin several logs of ground-based observatories and space missions, sometimes with links to images stored in the archives. The more recent inclusions are ISO (logs), MERLIN (log + images), INT(logs), ESO/VLT (logs)and XMM (logs). This adds to the previously existing access to IUE logs, VLA FIRST and HST images and logs, SUPERCOSMOS sky survey images and catalogues, as well as the recently installed links to Chandra logs and images, and to the basic, MAMA, DSS1, DSS2 and 2MASS reference image content of our image server. This means that data from a large panel of archives, as well as reference survey data, can really be merged in the same interface.

New functionalities have been developed in the Aladin interface, allowing comparison of images at various wavelengths, various epochs, various resolutions or sensitivities. The first feature is RGB composition of three images, after registration and resampling of two of the images with respect to the third one. The lookup table of each color can be adapted independently from the two others allowing various feature enhancements. In addition, the "images subtraction" facility allows to display directly "Photometric" color images... We are currently including registration and resampling facilities in the Aladin server, allowing both faster processing and larger set of interpolation algorithms, based on the Swarp Software (Bertin, 2002). Image mosaicking facility is also provided. All these server upgrades will be made public by fall 2002.

Isoluminosity contours can be computed on the fly and overlayed on any image at the same position, currently available in the interface.

To complete this set of image comparison tools an "Astrometric solution fixing" facility allows the user to assign appropriate WCS coefficients to the currently displayed image, either by editing a WCS frame or by entering manually projection type, center, orientation and scale, or through estimation of the parameters from a set of user chosen standards selected by clicking on the image and on the appropriate overlay.

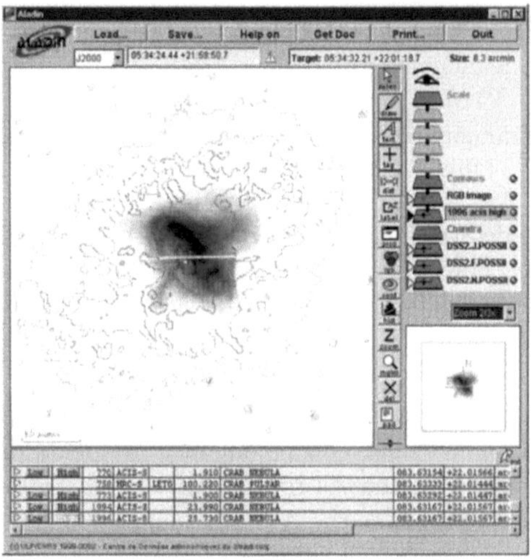

Fig. 1. R band contours overlayed on the Chandra image of the M1 nebula

Dealing with very large images (of typical size of 10K × 10K or more) is a challenge for interactive client tools like Aladin. By fall 2002 Aladin will provide three strategies:

- image cutouts of reasonable size will be provided by the server.
- a progressive vision functionality, based on multiresolution compression/decompression algorithms developed at CEA (Starck, J-L, 1998). Starting from the very low resolution displays of the images, it's possible to enlarge the scale by factor of 2 steps, adding the scale specific details at each step.
- Aladin standalone can manage locally in good conditions very large images, provided they are retrieved previously, and the computer get sufficient memory.

We are confident that the delivery of these new functionalities developed all over year 2002, has prepared Aladin to face the challenges of the interoperability objectives of the phase A AVO.

Acknowledgements: These developments have been made in the context of the European Commission funded AVO project and the French Ministère de la Recherche ACI-GRID IDHA project.

References

1. P.J Quinn, this conference
2. Bertin, E., 2002, `http://terapix.iap.fr/soft/swarp/`
3. Starck J.L., 1998, MR/1: The Multiresolution Analysis Software, User Manual, Version 1.0, (`http://www.multiresolution.com`)

Power Spectrum for the Distribution of Galaxies on the Sphere

Włodzimierz Godłowski[1], Magdalena Pietka[1], Jerzy Stelmach[2], and Marek Szydłowski[1]

[1] Astronomical Observatory of the Jagiellonian University, 30-244 Krakow, ul. Orla 171, Poland
[2] Institute of Physics University of Szczecin, Wielkopolska 15 70-451 Szczecin, Poland

Abstract. A method of Gorski of calculation of the power spectrum for CMB for incomplete COBE data set is applied to the distribution of galaxies on the sphere. The power spectrum is found for the distribution of radiogalaxies in the Condon-Machalski sample [1] containing radiogalaxies with the flux greater than 2.5 J and the brightness less than 18 mag selected from the LCRS and NVSS catalogues. The calculated power spectrum is then compared with an appropriate power spectrum expected from the theory.

1 Introduction

Power spectra of the Cosmic Microwave Background (CMB) and of the Large Scale Structure (LSS) of the Universe are the notions which play a crucial role in the theoretical and observational cosmology. This is because their form depend strongly on different observable astronomical parameters and and can be therefore used to test cosmological scenarios.

However, important problem arise in CMB as well as in the LSS case because observational data is not complete - some regions of the sky are not observable. Mathematically it means that we cannot find expression coefficients of the temperature contrast of CMB and of density contrast of LSS in appropriate basis since the space on which these densities are defined is not complete. The standard way to solve this problem is using the window function (see [2,3]) which is the Fourier transform of the survey window $W(r)$ defined to be unity inside the survey region, and zero outside. The window function method although straightforward has different disadvantages (for example is not suitable in the case of the case of the thin strips as in the LCRS catalogue [3]). More appropriate and mathematically rigorous method was proposed by Gorski [4] in the case of CMB, where the data are collected from only an part of a sky sphere. According to his method in the considered fragment of a sphere, an orthogonal set of functions on the cut sphere can be constructed from linear combination of the spherical harmonics. Some modification of this method was also use by Mortlock, Chalinor and Hobson [5] because effectiveness of the original Gorski method is limited to maximum Legendre multipoles of $l_{max} < 100$ due to numerical problems occurring during computations. In the case of LSS it is natural to use Gorski method to find two dimensional angular power spectrum in order to examine a galaxy distribution from the data collected from incomplete sphere.

In the three-dimensional case the classical method of determining the power spectrum bases on the orthogonality of functions $\exp(i\boldsymbol{k}\cdot\boldsymbol{r})$ in the cube. These functions form an orthonormal set of eigenfunctions in such a space. However the region from which astronomical data are collected is hardly approximated by a cube. Therefore it is necessary to employ the window function. We propose the alternative method called 'cut ball', which is the generalization of Gorski's method for the three-dimensional power spectrum. Let us note that if we were given a sample of radiogalaxies from catalogues covering the full sky it would be natural to consider a three-dimensional ball with a radius R. For this ball we know the orthonormal basis of eigenfunctions. The definition and the discussion of the power spectrum in such an approach ($3D$ case) will be given in our next paper, however in the present short communication we want to show how the method work in 2D case.

2 Description of the Method

We follow the method of Gorski. The only differences is that instead of temperature contrast $\Delta T/T$ expected in the case of CMB now we deal with the galaxies density contrast $\Delta n/<n>$ distributed on the sphere. It can be expanded into series:

$$\delta(\theta,\varphi) = \frac{\Delta n}{\langle n\rangle} = \sum_{l=1}^{\infty}\sum_{m=-l}^{l} a_{lm}Y_{lm}(\theta,\varphi). \tag{1}$$

In general this expansion is infinity. However, in practice the function $\delta(\theta,\varphi)$ is known with limited accuracy. Hence infinite basis is not necessary:

$$\delta(\theta,\varphi) = \sum_{l=1}^{l_{max}}\sum_{m=-l}^{l} a_{lm}Y_{lm}(\theta,\varphi). \tag{2}$$

The coefficients a_{lm} uniquely determine the power spectrum, which is defined:

$$P(l) = \frac{1}{2l+1}\sum_{m=-l}^{l} |a_{lm}|^2. \tag{3}$$

Let us assume now that the function $\delta(\theta,\varphi)$ is given only in some part V' of the sphere. Of course on this part our spherical function $Y_{lm}(\theta,\varphi)$ are not orthogonal. Hence we define a matrix:

$$W_{ij} = \int_{V'} Y_i(\theta,\varphi)Y_j^{\dagger}(\theta,\varphi)d\Omega. \tag{4}$$

Where the integration is carried out over the solid angle where the galaxies exist according to the considered catalogue and instead of two index l, m we used the single index i defined:

$$i = l^2 + l + 1 + m \qquad (i = 1, ..., l_{max} = (l_max + 1)^2). \tag{5}$$

The inverse mapping is:

$$l = integer(i-1)^{1/2}; \tag{6a}$$
$$m = i - (l^2 + l + 1);. \tag{6b}$$

In general $W_{i,j}$ is non-diagonal, symmetric, nonsingular and positively defined matrix. If so, there exists the so-called Choleski decomposition into the product of a lower triangular matrix $\boldsymbol{L}$ and its hermitian conjugate:

$$\boldsymbol{W} = \boldsymbol{L}\boldsymbol{L}^{\dagger}. \tag{7}$$

Let us define an inverse matrix of $\boldsymbol{L}$

$$\boldsymbol{\Gamma} \equiv \boldsymbol{L}^{-1} \tag{8}$$

and a new set of functions defined on a subset V'

$$\Psi_m(\theta,\varphi) = \sum_{n=1}^{i_{max}} \Gamma_{mn} Y_n(\theta,\varphi) \tag{9}$$

Let us notice that

$$\langle \Psi_m | \Psi_n^{\dagger} \rangle = \delta_{mn}. \tag{10}$$

Hence the functions $\Psi_n(\theta,\varphi)$ are orthonormal on bounded region V'. Hence we have

$$\delta(\theta,\varphi) = \sum_{m=1}^{i_{max}} c_m \Psi_m(\theta,\varphi) = \sum_{m=1}^{i_{max}} c_m \sum_{n=1}^{i_{max}} \Gamma_{mn} Y_n(\theta,\varphi) = \tag{11}$$
$$= \sum_{n=1}^{i_{max}} \left(\sum_{m=1}^{i_{max}} c_m \Gamma_{mn} \right) Y_n(\theta,\varphi),$$

and

$$a_n = \sum_{m=1}^{i_{max}} c_m \Gamma_{mn}, \tag{12}$$

since

$$c_m = \int_{V'} \delta(\theta,\varphi) \Psi_m^{\dagger}(\theta,\varphi) d\Omega = \sum_{n=1}^{i_{max}} \left(\int_{V'} \delta(\theta,\varphi) Y_n^{\dagger}(\theta,\varphi) d\Omega \right) \Gamma_{nm}^{\star} \tag{13}$$

we finally get

$$a_n = \sum_{m=1}^{i_{max}} \left(\int_{V'} \delta(\theta,\varphi) Y_m^{\dagger}(\theta,\varphi) d\Omega \right) \langle (W_{mn}^{-1}) \rangle, \tag{14}$$

what can be explain as vector:

$$\boldsymbol{a} = (\mathbf{W}^{\star})^{-1} \int_{V'} \delta(\theta,\varphi) \mathbf{Y}^{\star}(\theta,\varphi) d\Omega. \tag{15}$$

3 Results

Our paper is to test Gorski's method (applied for the first time to CMB) in the case of distribution of galaxies on the incomplete sphere. We use Condon-Machalski sample of radiogalaxies. To do that we calculate the coefficients a_m given by (14), next using the mapping (6) to find appropriate coefficients $a_l m$ and finally power spectrum $P(l)$. On should noted that in practice integration is replaced by the sum over these regions of the sky where the galaxies exist according to the considered catalogue. Also we represented the distribution of galaxies by the sum of delta functions. Our computation was done in Mathematica. Due technical limitations of that program, our preliminary results is obtained for values of $l \leq 11$.

Relations for the correlation function and the power spectrum are:

$$w(\theta) = \frac{1}{4\pi} \sum_{l} \sum_{m=-l}^{l} |a_{lm}|^2 P_l(\cos\theta), \tag{16}$$

$$|a_{lm}|^2 = 2\pi \int_{-1}^{1} w(\theta) P_l(\cos\theta) d\cos\theta, \tag{17}$$

where $P_l(\cos\theta)$ is a Legendre polynomial.

Substituting obtained power spectrum into the equation (16) we find angular correlation function $w(\theta)$ for our sample of radiogalaxies (Fig. 1). From the model $w(\theta) = \left(\frac{\theta}{\theta_0}\right)^{-\gamma}$ we obtain $\gamma \sim 1.4$, $\theta_0 \sim 0.015 rad$ On the Fig. 2. we present obtained power spectrum together with theoretical one obtained for above $w(\theta)$ from eq(17). For small l both power spectrum is in a good agreement. For $l \geq 7$ observed power spectrum is rapidly going down because our limit in $l = 11$. In

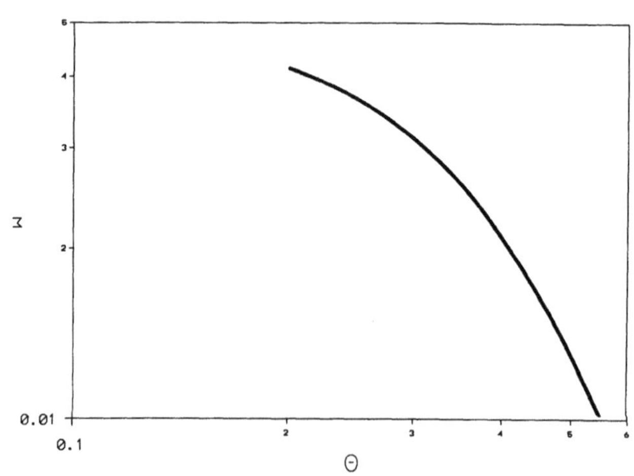

Fig. 1. Correlation function $w(\theta)$

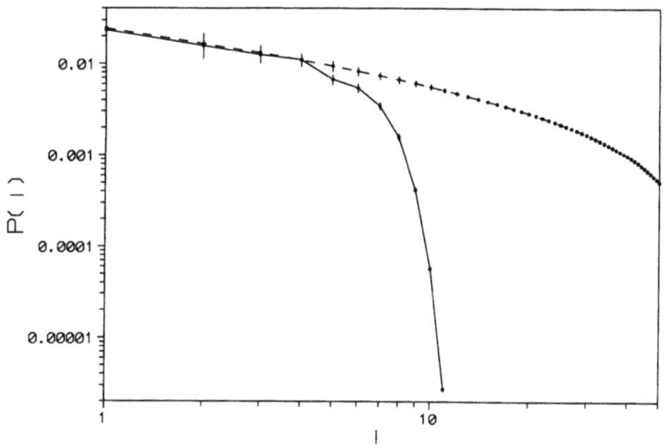

Fig. 2. Observed power spectrum (normalized by $4\pi a_0$ i.e.N^2 in classical method) together with theoretical on (dashed line)

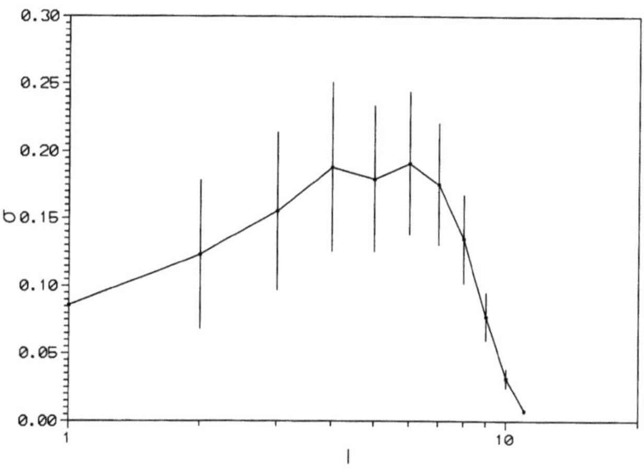

Fig. 3. Parametrized σ_l power spectrum

the Fig. 3 we presented power spectrum parametrize:

$$\sigma_l \equiv \left(\frac{l(l+1)|a_{lm}|^2}{2\pi} \right)^{1/2} \tag{18}$$

(as usual in the case of the of CMB). Our result is in agreement with that obtained for SDSS by Tegmark and Dodelson [6] (however power spectrum is again broken for $l \geq 7$ because our limit in numerical calculations). We realize that the Gorski method of finding the power spectrum is very effective. Even for small value of number l_{max} the expected behavior of the angular correlation function $w(\theta)$ and angular power spectrum is recovered. Our next step will be application of the described method to the three dimensional case.

Acknowledgements

The authors wish to thank Krzysztof Gorski for discussions during development of this project. Also we thanks Scott Dodelson and Alex Szalay for helpful comments. The paper was partially supported by KBN grants $2PO3D01417$ and $266/PO3/99/17$.

References

1. Machalski J. Condon J. APjS 123 p 41 1999.
2. Peebles APJ 185 p413 1973
3. Hauser M. Peebles P. APJ 185 p757 1973
4. Gorski K. ApJ 430 l85
5. Mortlock D. Challinor A. aph 0008083 MNRAS submitted
6. Tegmark M. Dodelson S. aph 0107418

The World Space Observatory (WSO/UV) Archive: Requirements for Interoperation

A.I. Gómez de Castro[1], D. Ponz[2], W. Wamsteker[2], and the WIC[3]

[1] IAG (CSIC-UCM), Univ. Complutense de Madrid, Spain
[2] VILSPA-ESA, P.O. Box 28080, Madrid, Spain
[3] The World Space Observatory Implementation Committee (WIC): Barstow, M., Binette, L., Brosch, N., Cheng, F-Z., Dennefeld, M., Gómez de Castro, A.I., Hakala, P., Haubold, H., van der Hucht, K.A., Kappelmann, N., Martínez, P., Pagano, I., Sahade, J., Shustov, B., Steshenko, N., Wamsteker, W.

Abstract. The World Space Observatory (WSO/UV) is a space project for an ultraviolet telescope which has to fulfill the needs and scientific requirements of a worldwide community. For this very reason, the project development is distributed among many countries, scientific institutes and companies. The technical requirements relevant for the AVO efficient performance have to be included and improved for the WSO/UV project. These requirements are briefly outlined in this contribution.

1 The World Space Observatory/UV Mission

The World Space Observatory (WSO/UV) is a 1.7 telescope with UV and optical imaging and UV spectroscopic capabilities (URL: `http://wso.vilspa.esa.es`). The spectroscopic instrumentation consist of a low resolution (R$\simeq$ 1000) long slit spectrograph and a high resolution (R$\simeq$ 50000) echelle spectrograph both working in the 103-310 nm range. The spectral capabilities are comparable to the HST/STIS echelle modes but the reduction of optical elements compared with the HST produces a considerable enhancement of the effective area. As a consequence, the WSO/UV is ideally suited for high resolution spectroscopy of weak astronomical sources and will have a major impact on our understanding of the intergalactic and interstellar medium.

The WSO/UV is an astronomical observatory rather than a targeted mission for a single science goal. The satellite will be operated at the second Lagrangian point (L2) allowing long uninterrupted observations, ideally suited for monitorings and surveys. WSO/UV will be continuously controlled from the ground also allowing rapid follow-up observations of targets of opportunity.

2 The WSO/UV Operations Model and Archive: Basic Requirements

The major requirements can be summarized as follows:

- Distributed operations to promote the international collaboration and reduce the operational costs. There will be a small number of mission operations

centres that will be responsible for spacecraft control and mission safety. They will be co-located with the ground antennas.
- An implementation model based on a layered structure with four basic layers: TT&C, mission operations, science operations and end users. Three active archives are defined (operations, data and science archive) that will work as active tools in communication between layers.
- A number of science operations centres that will share responsibilities on the mission science output and an unlimited number of end user nodes that will be integrated within the mission data flow cycle.
- World-wide distribution of the acquired data without restriction to enhance the scientific production of the mission.

A prototype for this collaborative model has been run for the NOT telescope at La Palma Observatory (Spain) (URL: `http://www.tcpi.es/dynacore/`).

3 The Science Archive Within the AVO Concept

The WSO project intends to define a scientific network that can work efficiently on planetary scales and to which countries with very different incomes can incorporate; therefore cost efficient computer architectures and open source software have to be adopted.

In the light of the discussions of the Virtual Observatory science working group, a common framework for the description of the astrometric, photometric, spectroscopic and temporal accuracy is required to guarantee interoperation between the astronomical science archives. The astronomical coordinates of the WSO/UV targets will be given in the International Celestial Reference System (IAU99) in a natural manner since the basic catalogue for the WSO/UV pointing will be USNO-A2. In the images the astrometry will be based in the USNO based coordinates of the plate center plus the plate constants. For further information, the qualification of the image distortions will be stored in the usual orthonormal functional space of Zernicky polynomials.

The photometric characteristics of the images will be coded though the normalized throughput of the optical system (telescope+filter+camera) as a function of the wavelength for each filter. UV (TD1) and optical photometric standards will be used for the calibration. For UV spectroscopy, the spectrophotometric standards (hot dwarfs and G191B2B) will be used [1,2].

Finally, mathematical tools are being developed for massive cross-correlation between catalogues and automated classification. We are currently working on algorithms based on a fuzzy approach and developing polynomial algorithms for pattern recognition.

References

1. Turnshek, D.A., Bohlin, R.C., Lupie, O.I., Koorneef, J., 1990; ApJ, 99, 1243
2. Gonzalez-Riestra et al., 2001, A&A, 373,730

A Common Data Analysis Environment: Do We Need One?

Preben Grosbøl, Pascal Ballester, and Klaus Banse

European Southern Observatory,
Karl-Schwarzschild-Str. 2, D-85748 Garching, Germany

1 Basic Needs

We present a short discussion of the needs for a common data analysis environment in the European astronomical community especially in view of the Virtual Observatory concept. Currently, several such environments are available but they were generally designed more than 10 years ago and do not offer adequate support for many important concepts (e.g. access to remote databases, distributed computing).

Whereas the creation of a single system may not be desirable, the definition of a common environment would improve efficiency by setting standards for interfaces between components used for analyzing scientific data. This would also enable different research groups to share their work and experience.

As the volume of data used by typical research projects increases, it becomes impractical to transfer and process the complete data at a single location. This suggests that distributed access and processing of data will become increasingly important. Some of these issues are addressed by the Virtual Observatory and GRID studies but a common environment for data analysis which integrates these concepts will be an essential tool to increase scientific productivity of the individual researchers.

Many major astronomical facilities are starting to provide standard data reduction pipelines for their instruments. Even with access to reduced data, the astronomer is still left with the major part of the work, namely to extract useful quantities from the data and analyze them in an astronomical context. In a typical case, one would need to consider the following issues: availability of web services, parallel and distributed computing, extraction of quantities, propagation of errors, statistical analysis, modeling, access to databases, interface to commercial packages, and use of composite data types.

A general issue is that, whereas technology evolves very fast, other parts of a system like algorithms have a much slower rate of change. It is therefore important to have a clean separation of these two areas so that one can improve either part independently. It is essential that a wide community of users can use the environment for research of new algorithms as this is the best way to ensure a growing set of applications and increase the sharing of developments also between different branches of science. The close interaction with computer science and applied mathematics is crucial for the development of efficient implementation of many numerical solutions to physical problems. The main objectives for a common Data Analysis Environment are summarized below:

- create open data analysis environment
- take full advantage of global communication (e.g. Internet)
- make simple usage simple
- encourage cross discipline development of algorithms
- be open for simple addition of new packages
- make access and sharing of raw and processed data simple
- design modular environment
- separate science algorithms and technology
- take advantage of new technologies (e.g. data mining and object orientation)
- support easy exchange of information with commercial packages
- support parallel and distributed computing

2 New Environment or Upgrading an Existing One

Most of the existing systems were designed over 10 years ago and therefore have not fully taken into account recent developments such as data mining, object orientation and fast Internet connections. They are, to a large extent, closed systems with a monolithic structure. Thus, it is difficult for non-expert users to add simple applications and integrate them with other packages.

A common European environment for data processing and analysis can only be established if a broad consensus is reached in the community. Beside the technical definition, a critical issue to achieve is that the environment will be used by a sufficiently large fraction of the community as only this will yield the benefits of sharing and re-usage. The most important factors for getting a wide acceptance are: 1) to satisfy the main need for data analysis as outlined above, 2) to make it attractive and efficient to develop new applications, 3) to ensure stability and backwards compatibility, and 4) to provide adequate maintenance and support including documentation.

In order to achieve a consensus, one would need first to identify institutes interested in supporting such an effort after which requirements and a high level design need to be established. The implementation could re-used many existing components (e.g. Python) and be done in an Open Source model.

3 Conclusion

It is clear that both individual astronomers and institutes would benefit from having a common data analysis environment as it would be easier to use and maintain than the current mixture of systems. However, the most important point would be an increase of collaboration, sharing of software and competition in development of new algorithms.

The main issue is less the definition and design than finding institutes which are willing to guarantee long term support for such an environment. Without that it would be very difficult to convince potential users that it would be worthwhile to make the personal effort in adopting a new environment.

Multiwavelength Studies of AGN

Dirk Grupe

Max-Planck-Institut für extraterrestrische Physik,
Giessenbachstr., D-85748 Garching, Germany

1 Eigenvector 1

Active Galactic Nuclei (AGN) are characterized by a whole set of observational properties such as emission line widths, line ratios, and spectral slopes. Many of these properties show correlations among each other. One of the final goals is to find an AGN analogon of the HR-Digram. One method to achieve this goal is by the **Principal Component Analysis** (PCA, e.g. Francis & Wills 1999 as a reference). PCA is a tool to simplify the large number of observable properties by a smaller set of parameters. The way it does it is to put n observed properties of p objects into an n-dimensional space and search for the longest axis of this point cloud. This major axis is called the First Principle Component or **Eigenvector 1** (EV1, e.g. Boroson & Green 1992). The next longest axis is then Eigenvector 2 and so on. These Eigenvectors can be representatives of physical parameters such as the mass of the central black hole, the accretion rate to mass ratio, the inclination angle, or the age of the AGN. The larger the number of objects involved in this study the more statistically relevant are the results. For this reason large datasets have to be analyzed.

2 Example of a PCA

I have studied a sample of 110 steep X-ray spectrum AGN (Grupe et al. 2001) selected from the ROSAT All-Sky Survey (RASS, Voges et al. 1999). Half of them belong to the class of Narrow-line Seyfert 1 galaxies (NLS1) which are at the extreme end of the Eigenvector 1 relation: Steep X-ray spectra correspond to narrow Hβ width, strong FeII emission and weak [OIII] lines. A PCA was performed with this sample. Optical spectroscopy have been obtained at La Silla and McDonald Observatory (Grupe et al. 1999). X-ray, UV, IR, and radio data have been retrieved from public archives and from the RASS (Grupe et al. 1998). Figure 1 shows one of the results of the PCA: a) NLS1, Seyfert 1s and Seyfert 1.5s are at separate locations on the EV1 - EV2 Diagram, b) NLS1 are at the extreme end of EV1, and objects with higher luminosity are located in the lower right part of the diagram. A possible interpretation of this result can be that EV1 represents the accretion rate to mass ratio and EV2 the mass of the central black hole, where the mass increases with decreasing values of EV2.

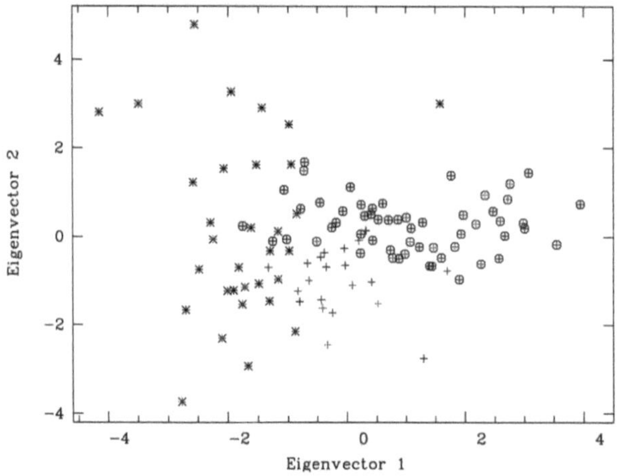

Fig. 1. Result of the Principal Component Analysis of 110 soft X-ray selected AGN. A cross inside a circle displays NLS1, a simple cross a Seyfert 1 type and the asterix displays a Seyfert 1.5 type AGN. Dark grey crosses are sources with log L_X >37.5 [W], light grey crosses 37.0<log L_X <37.5 [W] and black are log L_X < 37.0 [W]

3 Discussion

For our study of the soft X-ray selected AGN sample the following steps had to be performed: 1. Define criteria to selected sources for a sample, 2. Get X-ray data from RASS and pointed observations, 3. Obtain optical spectra for 110 AGN at optical telescopes, and 4. Time needed to get all these data: 10 years. The task of a Virtual Observatory to perform a similar PCA as described above will be to: a) provide a search engine to select whole samples of sources and b) search through public archives to retrieve public reduced data and data products, such as spectra and lightcurves. Good examples at least for searching for public data of individual objects are currently the ESO/ST-ECF Science Archive Facility and the NASA Extragalactic Databank (NED). The later has been extensively used to retrieve public data of the soft X-ray AGN sample described above. However, with the help of a Virtual Observatory we will be able to a) access data much faster and b) do new science with defining and accessing new samples, e.g. for comparison purposes.

References

1. Boroson, T.A., & Green, R.F., ApJS 80, 109, (1992)
2. Francis, P., & Wills, B.J., ASP Conf. Series Vol. 162, p363, (1999)
3. Grupe, D., Beuermann, K., Mannheim, K., et al., A&A 330, 25, (1998)
4. Grupe, D., Beuermann, K., Mannheim, K., Thomas, H.-C., A&A 350, 805, (1999)
5. Grupe, D., Thomas, H.-C., Beuermann, K., A&A367, 470, (2001)
6. Voges, W., Aschenbach, B., Boller, T., et al., A&A 349, 389, (1999)

The CCD Sky Patrols and Plate Archives

René Hudec[1], Jan Štrobl[1], and Peter Kroll[2]

[1] Astronomical Institute, Academy of Sciences of the Czech Republic, CZ–251 65 Ondřejov, Czech Republic
[2] Sonneberg Observatory, Sonneberg, Germany

Abstract. We report on the recent efforts to extract the data from the CCD and photographic sky patrol archives as well as to evaluate them and use for scientific analyses.

1 Photographic Sky Monitoring

The photographic sky monitoring is available for more than 100 years. However, only the recent development of photographic scanners and powerful computers allows an efficient extraction of scientific data. There are very numerous sky plate archives across the world, counting more than 2 million plates (Hudec, 1999). Some of the archives have very high quality plates achieving limiting magnitudes of up to 20 ... 23 (direct imaging) and/or 17 ... 19 (spectral with objective prism).

Some of the plate archives already have devices for digitisation of plates and few of them have started projects to develop high quality scanners to convert all plates into files/CD ROMs (e.g. The Royal Observatory Brussels). There are efforts to use these data for automated evaluation of objects on the plates and creating their light curves (de Cuyper et al., 2001).

Two of the photographic sky patrols are still in operation. The Ondrejov all sky patrol operated for monitoring of bright meteors has a sensitivity of mag 12 in the best case and a very large sky coverage (full visible sky hemisphere). The Sonneberg sky patrol is operated for variable stars simultaneously in two colours with limits 14 ... 15 but with a less extended sky coverage. Both programs exhibit suitable scanners to digitize and to evaluate the patrol images.

2 The CCD Sky Monitoring

The introduction and recent wide use of astronomical CCD cameras allows the development and the operation of CCD based sky monitors (e.g. Akerlof et al., 2001, Castro-Ceron et al., 2001). In the following we discuss few examples of CCD based sky monitors developed and/or operated recently.

ROTSE. One of the best examples is the ROTSE experiment operated by the University of Michigan. The ROTSE database represents almost complete sky monitoring down to declination -30 degrees. The ROTSE I experiment is based on an array of four electronic cameras mounted in a common equatorial platform. The optics used is Canon f=200 mm lens f/1.8. The CCD cameras are Thomson 2048×2048 14-micron imagers (Akerlof et al., 2000). Although the main goal of ROTSE and analogous experiments is the rapid follow-up of Gamma Ray Bursts, they are perfectly suited to create extended databases for long term analyses of various objects.

The Sonneberg All Sky CCD Monitor. The all-sky CCD monitor is based on the Zodiak 3.5/30 mm Fish Eye Optics and 7k × 4k CCD camera OES MM7k4k (Phillips chip 7168 pixels, 12 micron each, 16 bit, binning 4×4, readout time 60 sec). During he tests, the camera achieved 9 mag sensitivity limit for a 5 min exposure image. The system is operated routinely at the Sonneberg Observatory since the beginning of 2002.

The digitised data archives from both digitised plates as well as from CCD sky monitors require fully innovative algorithms and software for data analyses, photometry and astrometry. The evaluation of wide field images is related to specific problems absent for narrow field CCD images. We have developed and tested procedures for star identifications and data evaluation both for digitised plates as well as for CCD WF images (Štrobl, 2001). The goal of the study was to develop and to test method for generating of light curves of objects inside the ROTSE monitoring fields. The test field 54b has FOV approximately 8 deg. Our processing detected about 20000 stars on each picture. The whole process of all data (about 300 pictures) has taken tens of hours (on computer with processor Intel Celeron 566MHz). The ROTSE WF CCD monitoring data represent a unique database for study of variable astrophysical objects. We have tested procedures dedicated to create light curves for all objects into the FOV of particular fields. The accuracy of photometric measurements represent few hundreds of magnitude, hence real light changes of order of 0.1 mag can be revealed and analysed.

Acknowledgements. We acknowledge the participation of ROTSE team in the studies and analyses described here. We acknowledge the support by the ESA Prodex Project 14527, by the Grant Agency of the Academy of Sciences of the Czech Republic, grant A3003206, as well as by the Ministry of Education of the Czech Republic, Project ES002.

References

1. de Cuyper J. P. et al. In Astronomical Data Analysis Software and Systems X. ASP Conference Proc. 238, ASP 2001, 125 (2001).
2. Castro Ceron J. M. et al. In Proc. 4. INTEGRAL Workshop, ESA SP–459, 407 (2001).
3. Štrobl J. CCD sky surveys - statistics and correlations. Thesis, Charles University Prague(2001).
4. Akerlof C. et al. AJ 119, 1901 (2001).
5. Hudec R. and Wenzel W. Bull. Astron. Inst. Czech 27, 325 (1976).
6. Hudec R et al. in Blazar Continuum Variability, ASP Conf. Series 110, 129 (1986).
7. Hudec R. Acta Historica Astronomiae 6, 28 (1999).

Wide-Field X-Ray Monitoring as a Data Source for the Virtual Observatory

René Hudec[1], Adolf Inneman[2], and Ladislav Pina[3]

[1] Astronomical Institute, Academy of Sciences of the Czech Republic, CZ–251 65 Ondřejov, Czech Republic
[2] Czech Technical University, Faculty of Nuclear Science, Prague, Czech Republic
[3] Center for Advanced X–ray Technologies, Reflex s.r.o., Prague, Czech Republic

Abstract. We refer on novel X–ray telescopes with high sensitivity as well as large field of view of order of 1 000 square degrees or even more. The results are very promising, allowing the proposals for space projects with very wide–field Lobster–eye X–ray optics to be considered. The novel telescopes will monitor the sky with unprecedented sensitivity and angular resolution better than 1 arcmin.

1 Introduction

Wide field X-ray telescopes with imaging optics are expected to represent an important tool in future space astronomy projects, especially those for deep monitoring and surveys in X-rays over a wide energy range. The Lobster–Eye wide field X-ray optics has been suggested in 70ies by Schmidt (Schmidt, 1975, orthogonal stacks of reflectors) and by Angel (Angel, 1979, array of square cells) but has not been constructed until recently. Up to 180 deg FOV may be achieved. This novel X-ray optics offers an excellent opportunity to achieve very wide fields of view (FOV, 1 000 square degrees and more) while the widely used classical Wolter grazing incidence mirrors are limited to roughly 1 deg FOV.

2 The Recent Status

The construction of Lobster Eye Telescopes is feasible with recent innovative technologies. Various Lobster-eye X-ray telescope prototypes have been finished. As an example, the medium-sized prototype of the Schmidt geometry represents one module and consists of two perpendicular arrays of double-sided X-ray reflecting flats (36 and 42 double-sided flats 100 × 80 mm each). The flats are 0.3 mm thick and gold-coated. The microroughness is below 1 nm. The focal distance is 400 mm from the midplane. The FOV of one module is about 6.5 degrees. More such modules may create an array with substantially larger FOV. The optical and X-ray tests have confirmed performance close to those calculated and expected (e.g. by ray tracing).

The Mini–Schmidt modules are based on 0.1 mm thick glass plates 23 × 23 mm, gold coated, spaced at 0.3 mm. 60 such plates are used for one module, the double focusing device is created by two such modules. The aperture/length ratio is 80, the reflecting surface microroughness amounts to 0.2 - 0.5 nm. The FOV of the module is 2.5 deg. Recently, both large Lobster Schmidt modules (30 × 30 cm) as well as Micro Lobster Schmidt modules (3 × 3 mm) have been designed and developed.

For the alternative Angel geometry, numerous square cells of very small size (about 1×1 mm or less at lengths of order of tens of mm, i.e. with the size/length ratio of 30 and more) are to be produced. This demand can be also solved by modified innovative replication technology. The test modules with LE Angel cells have been successfully produced. First linear test module has 47 cells 2.5 × 2.5 mm, 120 mm long (i.e. size/length ratio of almost 50), surface microroughness 0.8 nm, f = 1.3 m. Second test module is represented by a L-shaped array of 2×18 = 36 cells of analogous dimension. The surface microroughness of the replicated reflecting surfaces is better than 1 nm. The various prototypes of lobster eye X-ray lenses of both the Schmidt as well as Angel geometries demonstrate the feasibility that the wide field telescopes may be constructed in the future based on this type of reflective X-ray optics.

3 Discussion

The use of very wide field X-ray imaging systems could be without doubts very valuable for many areas of X-ray and gamma-ray astrophysics. Results of simulations of lobster-eye telescopes have indicated that they will be able to monitor the X-ray sky at an unprecedented level of sensitivity, an order of magnitude better than any previous X-ray all-sky monitor. Limits as faint as $10^{-12} ergcm^{-2}s^{-1}$ for daily observation in soft X-ray range are expected to be achieved, allowing monitoring of all classes of X-ray sources, not only X-ray binaries, but also fainter classes such as AGNs, coronal sources, cataclysmic variables, as well as fast X-ray transients including gamma-ray bursts and the nearby type II supernovae. For pointed observations, limits better than $10^{-14} ergsec^{-1}cm^{-2}$ (0.5 to 3 keV) could be obtained, sufficient enough to detect X-ray afterglows of GRBs. The various prototypes of both Schmidt as well as Angel arrangements have been produced and tested successfully for the first time, demonstrating the possibility to construct these lenses by innovative but feasible technologies. This makes the proposals for space projects with very wide field lobster eye optics possible. These experiments will provide important data for the X–ray virtual observatory.

Acknowledgements. We appreciate the support provided by the Grant Agency of the Czech Republic, grant 102/99/1546 and by the Ministry of Industry and Trade of the Czech Republic, project FB–C3/29 "Centre of Advanced X–ray Technologies".

References

1. Angel J. R. P. Astroph. J. 364, 233 (1979).
2. Schmidt W. H. K. Nucl. Instr. and Methods 127, 285 (1975).

The UDAPAC Project – Uccle Direct Astronomical Plate Archive Centre: A New International Facility for Inherited Observations

Jean-Pierre De Cuyper[1], Elizabeth Griffin[2], and René Hudec[3]

[1] Royal Observatory of Belgium, Ringlaan 3, B-1180 Brussels-Uccle, Belgium
[2] Visitor to Department of Nuclear and Astrophysics, University of Oxford, 1 Keble Road, Oxford, OX1 3RH, UK
[3] Astronomical Institute, Academy of Sciences of the Czech Republic, CZ-251 65 Ondrejov, Czech Republic

Abstract. We discuss a project to create a centralized astronomical direct-plate archive and digitizing facility, with a view to making photographic data fully available for scientific research.

1 Scientific Rationale

All astronomical objects are changing irreversibly (e.g. proper motions, stellar evolution); many also change periodically or repeatedly (e.g. binary orbits, novae outbursts). A sequence of astronomical observations constitutes a unique record of such changes. The preservation of all past records is therefore vital to this science. Many changes take place on time-scales that are too long to be covered at all adequately by modern observing, however sophisticated the equipment. Innumerable questions – concerning (for instance) long-term trends in dynamical and astrophysical evolution, pre-outburst conditions of cataclysmic events, apsidal motion, or the causes of newly-discovered stellar variability of a longer period and larger amplitude than previously suspected – can only be solved or even tackled by recourse to archival material that spans many decades. Well-organized digital archives are thus an essential research tool to complement modern observations.

The world's collection of photographic images (more than 2 million, Hudec 1999) represents the costly output from over a century of devotion and skill. They are already nominally in the public domain, but as a universal resource they are seriously under-exploited. The main reasons are (a) lack of information in digital form about the plates, and (b) lack of digital versions of the observations.

The astronomical community itself must be responsible for creating a user-friendly on-line database of calibrated digital images from this resource. To create such a database requires a well-organized effort and substantial, but not exorbitantly high, resources. The gradual but steady loss of ageing expertise in the particular techniques implies a degree of urgency.

2 Goals and Objectives

One year after the international workshop on *'Treasure-Hunting in Astronomical Plate Archives'* held at Sonneberg Observatory, Germany (Kroll, La Dous and Bräuer (eds.) 1999, La Dous 1999), 16 astronomers from 9 different European countries met at the Royal Observatory of Belgium, Brussels (ROB) and agreed to attempt to create, within a time-frame of 5 years, a digital archive, in scientifically meaningful form, based on observations selected from some half a million direct photographic plates currently stored in Europe. CCD technology now makes the digitization of large volumes of plates feasible.

To that end, the group examined carefully the storage space being offered by the ROB for the creation of a central plate vault and an archiving and digitizing facility, and agreed to:

- plan the details of a project to store such plates in the ROB vault,
- identify research projects in which data of the age characteristic of such an archive can make new and unique contributions,
- advertise the scientific potential of the scheme via pilot projects,
- discuss the building of a public digital archive of calibrated observations,
- plan how to provide the astronomical community with a user-friendly interface to the new digital archive,
- identify sources to which applications for funding could be submitted.

A Core Team was established to coordinate the strategies and follow up the action commitments and schedule. The project was named UDAPAC, meaning Uccle Direct Astronomical Plate Archive Centre. A dedicated website has been set up at `http://udapac.oma.be`.

Acknowledgements. RH appreciates the support provided by the Grant Agency of the Academy of Sciences of the Czech Republic, grant A3003206 and by the Ministry of Education of the Czech Republic, project ES02.

References

1. Kroll P., La Dous C. and Bräuer H.-J. (eds.): 'Treasure-Hunting in Astronomical Plate Archives' Proceedings of the international workshop held at Sonneberg Observatory March 4-6, 1999, Historica Astronomiae 6, Verlag Harri Deutsch, Thun und Frankfurt am Main (1999).
2. La Dous C.:WGSS_Newsletter 11, 26, (`http://www.skynet.org/wgss_newsletter`) (1999).
3. Hudec R., Historica Astronomiae 6, 28 (1999).
4. Tsvetkov M. et al., Catalogue of Wide-Field Plate Archives, Sky Archive Data Center, Institute of Astronomy, Sofia, Bulgaria (`http://www.skyarchive.org`)(2000).

The Russian Virtual Observatory Portal. Convenience and Attractiveness

Nadejda Kalinina, Vladimir Vitkovskij, Olga Zhelenkova, Elena Kaisina, Vladimir Chernenkov, Aleksej Chepurnykh, and Vladimir Shergin

Special Astrophysical Observatory of RAS, Nizhnij Arkhyz, Russia

Abstract. Nowadays Web-site designers pay more attention on attractiveness in the presentation and convenient work with offered information for site users. Working on the project of creation of the Russian virtual observatory portal we decide to analyze the problem by an example a lot of sites in field of astronomy and other branches.

Nowadays Web-site designers pay more attention on attractiveness in the presentation of information, also convenient work with offered information for site users. Working on the project "Creation of the portal of the Russian Virtual Observatory" we decide to analyze the problem by an example of a sample of the European and American astronomical sites. The most part of astronomical sites are well organized, interconnected and balanced from the point of view of design. An example of the new approach to performance and data visualization can be served with the site of CDS. Though the essential discoveries in astrophysics occur not too frequently, the information placed into a portal should dynamically be updated. The most entertainment is the new images of astrophysical objects, and on sites there are the best images of a day. Besides the advertising of techniques of astronomical data processing means is important. The example of successful accommodation and support of astronomical software can be served by the ESO site.

Taking into account specificity of astronomy as philosophy science and the large interest of the people which are not being the professionals (amateurs of astronomy, schoolboys, students), is important, that the information placed on the virtual observatory portal is stated accessibly. There are should be a hierarchy of accessibility. An example of the well balanced data access for the decision of scientific tasks is the 'S' archive site. As the RVO project is wider, than simply access to astronomical information resources, and includes access to automatic telescopes both for educational, and for the scientific purposes, a specificity of the RVO site will be the advanced system of resource scheduling.

The work is carried out with support of the Federal Program "Scientific instrument development", project 25/01, and RFBR, the grant 00-07-90014.

IDHA Image Archive Model

Mireille Louys[1], François Bonnarel[2], François Ochsenbein[2], Pierre Fernique[2], Françoise Genova[2], Fionn Murtagh[3], Marc Wenger[2], and Pierre Didelon[4]

[1] Université Strasbourg I, LSIIT, UMR-CNRS 7005, Bd S. Brandt, F-67400 Illkirch
[2] CDS, Observatoire de Strasbourg, UMR-CNRS 7550, 11, Rue de l'Université F-67000 Strasbourg
[3] Queen's University Belfast, IRL
[4] CEA, Saclay, France

The Image archive model described here has been developed within the scope of the IDHA (Images Distribuées Hétérogènes pour l'Astronomie) project. The IDHA project led by CDS, associates image processing scientists of LSIIT (Strasbourg), and astronomers of several French Institutes including TERAPIX from IAP, and Queen's University (Belfast). IDHA goal is to describe the information content of image archives and datasets. The metadata are designed as objects interacting with each other, using the standard UML notation. To describe the user logical approach to such image archives we studied different scenarii for the queries of images and their scientific interpretation. We emphasize the need for scalability in the metadata, from description of one single astronomical image to a complete information about an image collection such as a survey or a telescope archive.

The Class Diagram in Fig. 1 shows the objects and their relationship. The content of each class can be browsed at `http://alinda.u-strasbg.fr/IDHA/lastmodel`. The main classes of the model are **RawObservation** and **ProcessedObservation**. The **Processing** classes and **Reduction** classes contains mostly mission specific information about the reduction pipe-line and describe the processus leading from **RawObservations** to **ProcessedObservations**. At a lower level the **Observations** have links to their **AstrometricSolution** or **PhotometrySolution**, their **ObservingConditions**, and their related **StoredImages** in the database archive. The **View** class offers the user a representation of the region he requested and is created on the fly by the client application. At an upper level **Observations** belong to **ObservingProgram**. Selecting a subset of images according to spectral, positional or data quality criterion is supported by the **ObservingGroup** object. **Filter**, **Telescope**, **Instrument**, etc... describe in detail the various elements of this **ObservingConfiguration**.

The XML syntax which is hierarchical cannot support the graph structure of our model. But the choice of some user scenarii helps in defining subtrees in the graph. We start from a root node and explore the different connections avoiding backward links and circularity. For example the user scenario "getting a simple description of processed observations in a given cone, sorted out by spectral bands, in a given **ObservingProgram**", can be seen as a tree with **ObservingProgram** at the root, followed by the various **ObservingGroups**

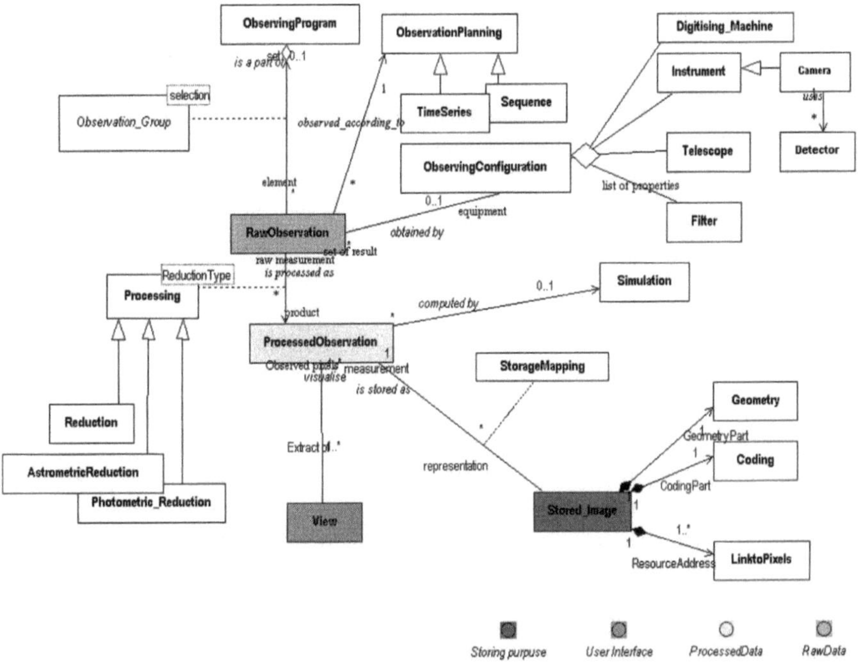

Fig. 1. UML description of basic objects and their connections

corresponding to the selection properties and then by the list of **ProcessedObservations**. If we restrict to such subtrees, it is possible to offer partial XML views of the model that match the different granularity level of description needed. Each node (that is each class of the model) starting from the root can be described as a RESOURCE table element, with attributes arranged (stored) in a table and with its child nodes described recursively by including new adequate RESOURCE elements.

A parallel effort is underway at the Chandra data center (McDowell, 2002...) The goal is to modelize data for development of processing software. Close contacts have been maintained with this group in order to define common classes and attributes names (ie a common understanding of the basic elements) as well as to express the specificities of Xray data versus optical data images.

The IDHA model is able to describe a large range of astronomical image archives. It's now time to use the model to generate XML VOTABLE metadata documents adapted to different user scenarii.

We acknowledge the support of the French Research Ministry program ACI-GRID 2001 for founding the IDHA project.

References

1. MacDowell J. 2002, this conference

Building Interoperable NASA Archives

Thomas McGlynn[1], Alberto Accomazzi[7], Bruce Berriman[3], Kirk Borne[2], Gunther Eichhorn[7], John Good[3], Timothy Kimball[6], Joseph Mazzerella[4], Arnold Rots[5], and Brian Thomas[2]

[1] NASA HEASARC, Code 660.1, Greenbelt, MD 20771
[2] NASA ADC, Code 630, Greenbelt, MD 20771
[3] IRSA IPAC, Pasadena, CA 91125
[4] NED IPAC, Pasadena, CA 91125
[5] CXC SAO, Cambridge, MA 02138
[6] ST ScI, Baltimore, MD 21218
[7] ADS SAO, Cambridge, MA 02138

Abstract. Over the past decades NASA has developed an extensive set of archives for space astronomy data. These archives are organized along mission, wavelength regime and information domains. Interlinking of these archives is essential to enable the fullest use of NASA's data resources. These archives will form key elements of the VO and include many advanced archive systems. Understanding the issues that arise in linking these systems together will be important in understanding how to build the full Virtual Observatory.

1 Current Status and Plans

NASA's current archive infrastructure include three distinct kinds of archives: Domain archives such as the Infrared Science Archive (IRSA), the Multi-mission Archive at Space Telescope (MAST) and the High Energy Astrophysics Science Archive Research Center (HEASARC) provide long-term archive for data for an entire wavelength domain. Mission archives like that at the Chandra Science Center and nascent archives for the SIRTF and SOFIA missions support major missions during their lifetime. Smaller missions may use the appropriate domain archives. Information archives are organized across mission and domain boundaries but provide a coherent view of some subset of astronomy data: The Astrophysics Data System (ADS), NASA Extragalactic Database (NED), and the Astronomy Data Center are organized around bibliographic, extragalactic and catalog information respectively.

Since astronomer's scientific questions over cross the boundaries of NASA's archives, NASA archives have begun to build systems that link these resources. These efforts have been formalized as part of an interoperability technical working group (ITWG) of NASA's Astrophysics Data Centers Executive Committee (ADEC). Figure 1 shows some of the links that are already in place among these centers. The first step towards fully interoperable systems is to enable users to easily transition from one resource to another. Discussions in the ITWG suggested that the main linking concepts include position, bibliography information, target name, observation ID and author. The figure indicates that these are indeed the elements used, and that no one concept is dominant.

Link from \ To	ADC	ADS	CXC	HEASARC	IRSA	MAST	NED
ADC		**Obs. Row-Abstract**		IMPReSS-Archive VizieR-Browse		IMPReSS-Archive	CatsEye-Objects Name Resolver
ADS	*ADS-Catalogs*		*ADS-archive*	*ADS-archive*	*ADS-archive*	*ADS-archive*	*ADS-Objects*
CXC		**Obs.-Abstracts**					Name Resolver
HEASARC	Browse-VizieR	BibServer-Refs	**Browse-archive**		SkyView-2MASS	**Browse-EUVE QL** S'View/Morph-DSS2	Name Resolver Browse-Position serv. BibServer
IRSA	OASIS-VizieR			OASIS-SkyView OASIS-archive		OASIS-DSS	Name Resolver OASIS-objects
MAST	MAST-Catalogs	**Obs/Target-Abstract**		MAST-ROSAT **MAST-EUVE**			Name Resolver
NED	NED-IMPReSS NED-VizieR	*Reference-Abstract*		NED-Browse	NED-2MASS/IRAS	NED-archive NED-FIRST	

Figure 1: Existing Links Between NASA Archives
Font indicates linking concept:
Position *Bibcode* Target name **Observation**

The existing links between services do not use any standard mechanism for publicizing the availability or syntax of services. Links are often fragile,depending up very specific details of the service remote URLs Upgrades to services can easily break remote sites that depend on them. As more and more links are developed there is substantial of developing *weblock* where the need to continue to support existing linkages stifles further development. Building a framework these links can be standardized and evolve is a major goal of the ITWG and the VO. We anticipate that some archives may need to provide summary services that provide a precis of the data available at a site for a given position or time, or which indicate which particular catalogs or archives meet user specified criteria. Such services will enable other sites to decide when it is appropriate to provide links to the remote service. In the general VO context one could imagine a hierarchy of discovery queries, but in the more restricted venue we address here, a single level of resource discovery may be sufficient. Since such services need to have clearly defined output formats, standards such as WSDL are particularly applicable to them.

NASA's archives are a key resource for the development of the Virtual Observatory. The efforts to build interoperable systems within NASA are necessarily complement to our ongoing efforts to make NASA data available within the broader Virtual Observatory community.

*i*Astro: Computational and Information Infrastructure in the Astronomical DataGrid

F. Murtagh[1], D. Egret[2], G. Longo[3], V. Di Gesu[4], G. Allen[5], J. Nunez[6], R. Molina[7], I. Llorente[8], A. Shearer[9], A. Golden[9], R. Butler[9], A. Holl[10], M. Tsvetkov[11], D. Boyd[12], L. Sastry[12], V. Golev[13], P. Jetzer[14], and A. Csillaghy[15]

[1] Queen's University Belfast, UK
[2] Observatoire astronomique Strasbourg, F
[3] Univ. Federico II, I
[4] Univ. Palermo, I
[5] Albert-Einstein-Institute, Potsdam/Golm, D
[6] University of Barcelona, E
[7] University of Granada, E
[8] Universidad Complutense, E
[9] National Univ. of Ireland, Galway, IRL
[10] Konkoly Observatory, H
[11] Space Research Institute, Sofia, BG
[12] Rutherford Appleton Laboratory, UK
[13] Department of Astronomy, University of Sofia, BG
[14] Inst. Theoretische Physik, Zurich, CH
[15] Univ. Applied Sciences, Windisch, CH

Abstract. The *i*Astro network is a European action supported by the COST program, to foster coordination in the domain of computational and information infrastructure for the astronomical data grid.

1 Institutional Context

The COST program is a European program designed to promote European co-operation in the field of scientific and technical research. It is one of the best mechanisms for integrating research in Eastern and Western Europe. COST helps maintain research in many member states.

*i*AStro is one of the 290 actions currently supported by the COST program.

2 Mission Statement

The Grid is the infrastructure of the virtual organization of the future, providing high performance and high added value services relating to computational, data, information and knowledge processing requirements. Those involved in *i*Astro aim at ensuring best application of new theory and tools in the astronomy application domain, and simultaneously Grid-enabling the most appropriate areas of the application domain. The means applied by *i*Astro to achieve these ends are, respectively, further developing and bridging the many ongoing projects (i.e., disseminating exciting theories and good practice in the direction of astronomers), and through selection and bringing Grid-appropriate areas of the application domain into focus (i.e., expressing and focusing application requirements in the direction of computer scientists, and data and information analysts). An additional objective is the spinning out of new national and international projects, where appropriate and needed.

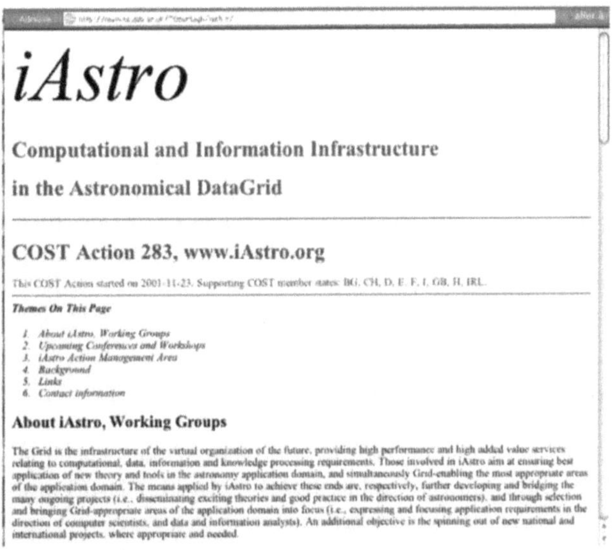

Fig. 1. The *i*Astro home page at `http://www.iAstro.org`

3 Working Groups

The *i*Astro action is organized along the following four working groups:

- WG1 Interoperability, data correlation and federation (D Egret) – Interoperability (D Egret) – Data quality/correlation/fusion (V Di Gesu)
- WG2 Visualization (G Allen) – Advanced visual user interfaces for data mining (V Di Gesu)
- WG3 Heterogeneous, multimedia data (R Molina) – Image/signal restoration (R Molina) – Data mining (V Di Gesu)
- WG4 Surveys (J Nunez, G Longo and M Tsvetkov) – Wide field imaging (G Longo, M Tsvetkov) – Robotic observatories (J Nunez).

PRIME: A Deep Near-Infrared Survey Project

A. Omont[1], on behalf of The PRIME Team[2], W. Zheng, H.C. Ford, J.W. Kruk, Z.I. Tsvetanov, A.S. Szalay, P.K. Shu, M.A. Greenhouse, G. Hartig, M. Postman, H.S. Stockman, G.M. Voit, R. Lenzen, H.-W. Rix, S. Kent, C. Stoughton, and Y. Mellier

[1] Institut d'Astrophysique de Paris, France
[2] Johns Hopkins University, GFSC, STScI, MPIA, FNAL, IAP

Abstract. PRIME is a proposed space mission for a deep survey of 25% to 100% of the sky in four near-IR bands (0.9-3.5 μm), to $M_{AB} \approx 24$ (K $\approx$ 22). The PRIME survey should be $\sim$ 600 times deeper than 2MASS. Deeper surveys in selected regions are planned. Its main goal is to find the first massive objects formed: quasars up to z=15-25 if any, galaxies up to z $\sim$20, clusters of galaxies up to z $\sim$ 3-5. PRIME should also be a major step in the study of brown dwarfs and nearby free floating warm planets, and serve many other goals, in particular as a pilot mission for NGST. It should be a central piece of The Virtual Observatory, complementary to other surveys.

1 Sky Surveys and PRIME

Sky surveys have been among the main drivers of the development of astronomy, e.g.: POSS, SDSS in the visible, IRAS, 2MASS, DENIS in the IR, various radio surveys, ROSAT, etc. They are unique in detecting very rare objects and powerful in revealing structures in the Universe. They are essential in providing target selections for major facilities.

In the recent years near-IR surveys have been extraordinarily boosted by major developments in panoramic detectors. Even before the completion of 2MASS and DENIS, new surveys have been decided to gain one or two orders of magnitude in a significant fraction of the sky, by 2007-2015: UKIDSS, WIRCAM/CFHT, and especially VISTA. Infrared surveys will thus catch up visible surveys, such as SDSS and CFHTLS, in studying the high z Universe up to z $\sim$ 1-2. However, even near-IR observations are much better made from space, where the thermal background is much lower than from the ground. This is essential to study the very highly redshifted Universe, as with NGST.

PRIME will observe in the main spectral range of NGST. It has the same driving goal of exploring the Universe at z $\geq$ 5, just after the end of the dark age when the first luminous objects are formed. The advantage of space will make PRIME, with a 0.8m telescope and 0.6$''$ pixels, $\sim$ 600 times deeper than 2MASS and maybe two orders of magnitude faster than VISTA. It will be the first survey between 2.5-3.5μm. Deeper surveys by up to three magnitudes will be performed in $\sim$10-100 deg^2. The combination of multi-wavelength data will be powerful in picking out rare, important objects among millions and providing approximate photometric redshifts. Technically, PRIME is a very simple mission, with passive cooling on a near polar, sun-synchronous orbit. It was originally designed as a

SMEX mission, but its science performance could be substantially enhanced under a MIDEX concept with a possible extension to the entire sky.

2 Main Scientific Goals

- *Formation of the first massive objects (AGN or galaxies) at $z \geq 5$-15*, in the highest peaks of density fluctuations. Their luminosity and redshift distribution will have strong implications for cosmology and for the relationship between galaxies and black holes. This includes the reionization epoch, when these first lighthouses ionize the intergalactic medium. PRIME will also find the most strongly lensed galaxies at highest z, allowing NGST to increase its actual sensitivity.

- *Cluster Formation and Galaxy Evolution at $z \sim$ 1-3.* Measuring the rate of growth of galaxy clustering, from first clusters to z=1, will check a key prediction of cosmological theories. PRIME will also test the very high z evolution and clustering of various classes of star forming galaxies. It will trace dark matter by the effect of cosmic shear on galaxy count density. It will detect very rare and outstanding cases of galaxy and cluster lensing, also tracing dark clusters. It could extend the detection of supernovae to $z \sim 2$.

- *Substellar Populations and other goals.* PRIME will have by far the best power of detection of L and T brown dwarfs up to several hundred parsec, and of young objects of planetary mass down to 1 $M_{\rm Jup}$. It could detect planet transits in front of brown dwarfs, as well as hundreds/thousands of asteroids and Kuiper Belt Objects. It will have many other goals, especially in the frame of multi-λ AVO, in particular for the discovery of rare extraordinary objects.

Thus, PRIME will be unique for surveying the sky for the earliest galaxies, QSOs and clusters. It will also considerably extend our knowledge of substellar populations, and serve many other purposes, including serendipity.

3 PRIME and The Virtual Observatory

The amount of pixel data to process will be ~ 10–30 Tera-Bytes. Previous experience with similar surveys – SDSS at FNAL, CFHTLS at IAP/Terapix – will be used to provide the basis for methodology model: image processing and source extraction will use Terapix (+ SDSS) software tools. The operation of the data pipeline will be similar to SDSS at Fermi National Accelerator Laboratory.

IAP will develop the pipeline software and image simulations, with the help of JHU for calibration. FNAL will operate the production system and data processing. The data will be archived at STScI and delivered via STScI MAST. The SDSS /GALEX/PRIME multiband archive at STScI will thus become a central element of The Virtual Observatory, to be complemented by other surveys:

- Visible surveys deeper than SDSS (e.g. CFHTLS) will be required to fully match PRIME's sensitivity, in particular for the determination of photometric redshifts in the range z=1-3.

- X-ray (XMM-*Newton, Chandra*); medium-far-IR, submm (SIRTF, ASTRO-F, *Herschel*, *Planck*, ALMA); radio (FIRST, NVSS); etc.

A Practical Approach to Catalogued-Data Visualisation

Patricio F. Ortiz[1,2] and Joachim Köppen[3,4,5]

[1] **AstroGrid** Department of Physics and Astronomy, University of Leicester, LE1 7RH, UK
[2] INAF: Osservatorio Astronomico di Capodimonte, Naples, Italy
[3] Observatoire de Strasbourg, 11 Rue de l'Université, Strasbourg, France, F67000
[4] International Space University, Illkirch, France
[5] Inst.Theoret.Phys.u.Astrophysik, Univ. Kiel, Germany

Abstract. Catalogued data is one of the ways information will be collected from VO. The selected material will contain many physical quantities (columns) and in an increasing number of cases, a large number of sources (rows). The larger the catalog, the more interesting it becomes from an statistical point of view. However, describing the parameter space coverage of any catalogue is not a trivial task and neither it is to envision relationships. Besides the need for developing automated methods of data exploration, visualisation techniques contribute enormously to get a good idea of the global properties of a catalogue and to select regions of interest in the parameter space without actually moving large data chunks. A method which associates a database system (datoz, Ortiz, 1998), and a Java graphics interface (Köppen, 2001) is presented as one approach to solve this problem.

1 Introduction

Within VO, catalogues will be as important as other components: images, spectra, etc, for which visualisation tools already exist. The size of catalogues of the future is expected to increase considerably, eg, GAIA, collections of catalogues from multi-wavelength wide image surveys, etc. Large catalogues are attractive for getting better statistics, for giving us the chance to find those *one in a million* rare objects which do not show up in smaller samples.

Although we have been creating plots of 2 or 3 dimensions for years once we have a table in our hands, this approach crashes if a user requests a colour-magnitude diagram based on USNO ($\sim$ 600 million objects). Painting points one by one would create a monster plot-file and consume a lot of CPU cycles to create it. CPU gets cheaper, but bandwidth doesn't. On the other hand, even if the catalogue is not so large, tools to interactively inspect its content are required; tools which could exploit available technology.

Why visualisation and not automatic methods? A large percentage of our perception of the world is visual. Our brain is trained and wired to interpret information and to recognize features when presented with images rather than with arrays of numbers, hence the importance of taking advantage of visualisation in the context of scientific applications.

We present a working model which incorporates visualization by merging a database system (datoz2k) with Java applications to provide a wide range of options including 3D animations.

Because of the limited printing space available in this paper, we have created a web page presenting all the poster material plus the chance to directly interact with the system.
URL: `http://barbara.star.le.ac.uk/datoz/visualisation.html`

2 Visualisation in datoz2k-Catplot in a Nutshell

The first level of visualisation is to provide a histogram for each numeric variable in a table plus data like mean value, standard deviation and extrema.

This approach works well when dealing with independent variables, but in cases like sky position, where both RA and Dec are involved are treated differently: ranges and distributions of RA or Dec (for separate) do not account for features like patchiness in the sky, either real or induced by observational strategies. datoz2k presents an Aitoff projection showing the density of sources across the sky. The characteristics of a catalogue's coverage are quickly recognizable in this way.

Interactivity is vital. Users can select a set of objects based on arbitrary criteria and produce 1-D (histograms), 2-D (scatter, all-sky, etc), and 3+D PostScript plots of any valid expressions involving numerical variables. Figures are camera ready and animated GIF techniques are used to create **movies** in 3-D space, yet much more should be expected.

Java provides an excellent platform for real interactions. JK has developed a number of scientific applications which when applied to a database system real interaction can occur. Two levels of exploration have been tested: **a) Catview** on the fly creation of 2-D plots with the chance to pick points, zoom in and out, transform scatter plots into density plots, regions marked in one plane are mapped into any other plane. This method is currently limited to view up to 5000 objects. **b) GraphEx** Exploration of 2-D spaces by using density plots. Users can start with the whole catalogue and select regions of interest to work with them. The amount of information transmitted is independent of catalogue size, making this method particularly useful for use with large catalogues. When fully implemented, frequently requested 2-D spaces may be pre-stored without needing to interact with a large database until the very last stages of selection.

References

1. Ortiz, Patricio F.: 'Accessing astronomical Data over the WWW using datOZ'. In *Astronomical Data Analysis Software and Systems VII, A.S.P. Conference Series, Vol. 145, 1998, R. Albrecht, R.N. Hook and H.A. Bushouse, eds.,p.371*
2. Köppen, Joachim and Acker, Agnes: 'An interactive visualization of the Catalogue of Galactic Planetary Nebulae'. In: *IAU Sympos. 209, Planetary Nebulae: Their Evolution and Role in the Universe, Canberra, Australia, 19-23 November 2001. (in press)*

Multi-Object Spectroscopy: Automatic Extraction and Previewing of Spectra

Norbert Pirzkal and Florian Kerber

ST-ECF, Karl-Schwarzschild-Str. 2, D-85748 Garching bei München, Germany

Abstract. With multi-object spectroscopy becoming an integral part of modern astronomical instrumentation, the amount of such data in archives is bound to increase manyfold in the very near future. One challenge is to make individual spectra easily accessible for preview by the archive users. We have used aXe, a software package developed to perform slitless spectral extraction for the Advanced Camera for Surveys (ACS) on the HST, to extract spectra from FORS MXU data. One key element of routine spectral extraction is that the extraction process must be robust, take place in a reasonable amount of time, and be general enough so that it works equally well independently of the type of spectrum. We believe that the current approach to the extraction of spectra from FORS MXU satisfy these conditions while being more sophisticated than a simple quick and dirty box extraction method. The extraction process follows the general principle of ACS spectral extraction which is outlined in [1] and [2]. This allows us to perform tilted spectral extraction and to produce wavelength calibrated spectra which can be readily examined using current spectral analysis tools which are part of IRAF.

For the future we foresee the possibility to add further functionality facilitating scientific research in an AVO context. For a rough classification of the objects the extracted spectra could be cross-correlated with templates from a library of digital spectra helping the archive user to judge the relevance of the spectra for his research interest.

1 Experience Generating Preview Spectra

Our experience extracting spectra from FORS2 MXU data has shown us that simply archiving these images will likely not lead to many users investing the considerable amount of time required before one can start looking at the spectra of objects. A considerable amount of knowledge is required before one can even start to extract information from these data. At the very least, a wavelength calibration of the data is required to identify spectral feature in the data. In this respect, multi object spectroscopic data is significantly more difficult to use right away than imaging data and requires a large amount of buy-in time in order to be used even in the most basic form (e.g. to simply select interesting spectra from a large pool of available data). In our case, we wanted to extract all of our spectra in an automatic manner because we had several tens of images, each containing 60 to 70 spectra. But in order to even start selecting interesting objects, we had to combine information obtained from various sources so that we could first obtain wavelength and flux calibrated spectra of all the objects. Our goal was not to automatically generate optimally and perfectly calibrated spectra right away but rather to use a relatively straight forward extraction method and apply calibration to the extracted spectra which would be good to within 5 to 10 percent. This, allowed us to examine the content of our images in a much more efficient way.

The following steps were performed:

- Dark and flat-fielding the science and wavelength calibration data
- Finding the positions of the slits in the field of view using the slit acquisition images
- Finding the exact position of the objects in the slits
- Extracting the lamp wavelength calibration spectra
- Extracting the science spectra, using tilted extraction along the direction of the slit
- Wavelength and flux calibrating the extracted spectra

If performed manually, for one spectrum at a time, these steps took a significant amount of time, on the order of several minutes per object. On the other hand, once enough had been learned about these data, mainly in terms of the wavelength and flux calibration of this observing mode, it was relatively easy to set things up so that we could generate hundreds of preview spectra in the course of a few minutes. Scripting various aXe tasks together and adding some Python scripts to pre-compute the location of the slits in our data, we were able to extract and calibrate our data in a manner which is robust (all spectra were extracted successfully), fast (on the order of 1 to 2 minutes per data sets), and produces calibrated spectra in a FITS format which can be readily analyzed using IRAF tasks.

The same methodology, also based on an automatic processing using aXe, was recently used to analyze new HST/ACS grism observations of the HDF-N. Using this method, 1700 objects were extracted in a matter of minutes, and each extracted spectra could then be quickly examined. This resulted in two supernova being discovered within a matter of hours from when the data was first archived. This, together with our experience with FORS MXU data illustrates the strength of being able to archive spectra which have been generated in an automatic manner using relatively robust extraction parameters. Additional information about aXe can be found in

References

1. Pirzkal et al., ST-ECF Newsletter, July 2001, p.5.
2. `http://www.stecf.prg/software/aXe/`

The *INTEGRAL* Archive at the ISDC

Katja Pottschmidt[1,2] on behalf of the ISDC team

[1] INTEGRAL Science Data Centre, Ch. d'Écogia 16, 1290 Versoix, Switzerland
[2] Max-Planck-Institut für extraterrestrische Physik, Giessenbachstr. 1, 85748 Garching, Germany

1 The *INTEGRAL* Mission

A wide variety of scientific objectives will be addressed by ESA's International Gamma-Ray Astrophysical Laboratory *INTEGRAL* – launch date 2002 October 17 – which will primarily observe γ-ray bursts (GRBs), the diffuse emission of our own galaxy, binaries containing an accreting black hole or neutron star, and Active Galactic Nuclei (AGN).

The *INTEGRAL* payload consists of the two main γ-ray instruments, the spectrometer SPI and the imager IBIS, as well as of the X-ray monitor JEM-X and the optical monitor OMC. All instruments are co-aligned and have comparatively large fields of view. The high energy instruments are based on the coded mask principle, therefore requiring complex and advanced data processing. A detailed description of the *INTEGRAL* mission is given in the *INTEGRAL* AO1 manual
(`ftp://astro.estec.esa.nl/pub/integral/AO/AODocB.pdf`).

INTEGRAL is an observatory which is open to the world community. The distribution of the data to the observers is performed by the *INTEGRAL* Science Data Centre (ISDC) near Geneva. After receiving the telemetry from the Mission Operations Center in Darmstadt, the ISDC is responsible for processing the data, evaluating their scientific content (e.g., alerts are issued in case of GRB detections), and archiving the raw and processed data as well as results from a standard scientific analysis. For a recent overview of the tasks performed by the ISDC see Beckmann et al. (2002, astro-ph/0206506).

2 The *INTEGRAL* Archive

A new grouping approach was introduced by the ISDC system to provide easier and more effective access to logically connected data: Any data element can belong to many different data groups simultaneously, thus allowing data objects to share a given collection of data. The hierarchical nature of *INTEGRAL* data (which are, e.g., multi-instrument, multi-time scale, multi-processing level, and multi-pointing) makes such an approach very rewarding.

The association to file formats is made via the FITS Hierarchical Group Convention: a data group corresponds to a FITS group file and a base element corresponds to a FITS Header Data Unit. If the members of a group are of the same type, this format becomes even more efficient: in addition to the contents

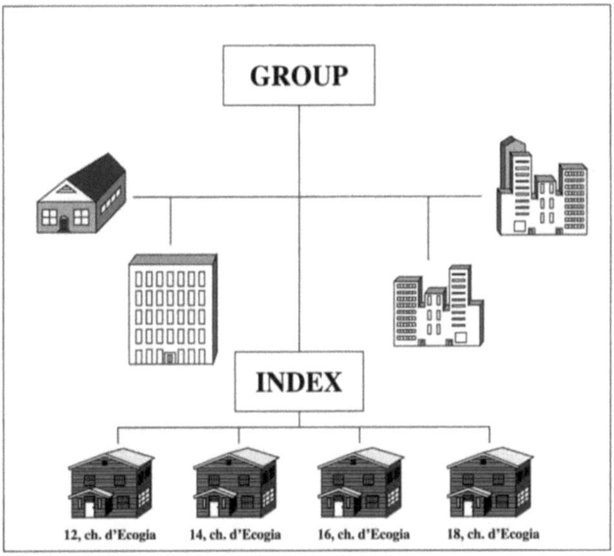

Fig. 1. Schematic of a data group containing an index group

of a normal FITS group file, the corresponding files may also contain columns which list the values of keywords of interest for all members. Those FITS group files, first introduced by the ISDC, are called index groups. For example, consider that due to the dithering strategy employed by the *INTEGRAL* spacecraft, each astronomically meaningful observation usually contains a set of several "science window" pointings as well as additional files. The whole ensemble is accessed as an observation *group*. However, one of the members of the observation group is the associated science window group *index* (Fig. 1).

The grouping concept has been implemented by the means of using the Data Access Layer library (DAL), an ISDC development which was built on top of the CFITSIO library. The DAL library is based on standard FITS files templates.

The archive system consists of the following main components:

- Ingestion: During the ingestion process a given data set is copied to the correct place within a pre-determined UNIX directory structure, meta-data for the database are extracted, and an index is created. In the ISDC system, the archive repository can be accessed directly, where proprietary rights are enforced via setting the appropriate UNIX access permissions.
- Database: The Oracle-based archive database stores administrative data related to the observations – e.g., proposal data and observation planning parameters – and archive repository meta-data – i.e., descriptions of the files in the archive repository and their locations.
- Browse: For external users the ISDC provides access to the *INTEGRAL* archive by the Browse facility, originally developed by HEASARC. For the *INTEGRAL* mission several new features were added by the ISDC, e.g., the selection by multiple object coordinates. If the data sets associated with a search are public, they can be directly downloaded through the browser.

- Distribution: The system distributes private and public data to the different kinds of astronomical community users. Management and control are provided by the OPUS environment. Depending on the distribution request, the files can then be transferred via FTP or on hard media.

ESA is developing an interoperable access (JAVA/XML) to its mission archives. An adaption for the *INTEGRAL* archive is foreseen as a long term project.

HyperLeda: Virtual Instruments for Extragalactic Astronomy

Philippe Prugniel[1], Georges Paturel[1], Vladimir Georgiev[2], Igor Chilingarian[3], and Chantal Petit[1]

[1] CRAL-Observatoire de Lyon, F-69561 Saint Genis Laval, France
[2] Astronomical Observatory, University of Sofia, P.O. Box 36, BG-1504 Sofia, Bulgaria
[3] Sternberg Astronomical Institute, MSU, Moscow 119899, Russia

Abstract. HyperLeda offers a series of tools of interest for Galactic and extra-galactic astronomy. The most prominent service is the LEDA catalogue which presents 70 homogenized parameters collected for over 3 million galaxies.

It is build on compilations from the literature plus specific measurements from the available sky surveys. HyperLeda also offers on-line tools for analysing stellar populations and contains an archive of reference data (in FITS).

1 Introduction

The HyperLeda project [4] is driven by two astronomical key-programmes. The first is interested in the study of the local Universe (distance scale, structure and kinematics, luminosity function) and the second is concerned with the study of the stellar populations in nearby galaxies.

A strong common point between the two programmes is the need for homogeneous multiparametric data on large-area samples, or possibly on whole-sky samples. Assembling such homogeneous datasets rises specific problems: (1) many different sources have to be combined and first of all transformed to a common system (2) the cross-identification between the different sources of data (survey or small datasets from the literature) is not trivial. The tools offered by HyperLeda handle these issues.

HyperLeda is a collaboration between astronomers from Observatoire de Lyon, Observatoire de Paris (GEPI), Sofia Observatory and Moscow University.

2 Description of the Public Access

The data distributed by HyperLeda are compilations of measurements and homogeneous parameters computed from these catalogues. The compilations concern global parameters of galaxies (as total magnitude, central velocity dispersion, ...) or 1D spatially resolved data (aperture photometry, kinematical profiles, ...).

We also distribute reference data (in FITS format) images and spectra: for example a remarkable dataset is the library of high resolution stellar spectra

(R=10000) [3] used for spectroscopic synthesis of galaxies [1]. These reference data can be processed on the fly by a pipeline which can perform various analysis functions. The spectroscopic synthesis package PEGASE allows to create a customized stellar population whose spectrum can be downloaded or analysed by the pipeline.

The users can connect to HyperLeda at: `http://leda.univ-lyon1.fr/` or to any of the other mirrors linked from the above address. The database can be queried for a single object specified by a designation (any available name used in the literature), for objects in a circle around a position (cone search), for pre-defined or user-defined samples of objects or, in the case of the LEDA catalogue, full SQL queries on the whole set of available parameters can be executed.

Beside the html interface, the HyperLeda servers provide other output formats, such as Astrores used by Aladin (CDS) or by the ESO *Virtual Observatory* prototype, Querator.

3 Internal Organization & Future

The idea of opening the database to the public [2] raised, two decades ago, from the will of making re-usable the enormous work of compiling data from the literature, and now from the Virtual Observatory (VO). Though the methods have changed, the task is still much time consuming, and to keep the problem scalable, HyperLeda share the catalogue production between different astronomers in their own field of expertise. The catalogues are maintained in the different laboratories of the collaboration and the software automatically and asynchronously gathers the new versions, checks integrity and cross-identifications, applies the homogenization recipes and re-constructs the database. The whole system, software and data is then available for mirroring (automatic replication) on the various HyperLeda sites or private mirrors.

The basic steps of catalogue production consists in locating relevant references in the literature, extracting the data and uniformizing them (i.e. convert them in a standard format with standard units), cross-identify the sources and homogenize the parameters. The first steps become easier thanks to the effort of the datacenters: extracting uniform data from the VO will become transparent in a few years. Two major points, at least, justify a strengthening of our efforts: (1) The cross-identification cannot be fully automatic (positional or multi-parametric identification occasionally fail, hiding the interesting rare objects in a crowd of miss-identifications). (2) The rules for deriving homogeneous parameters from multiple data sources require strong scientific expertise. We hope that HyperLeda will continue in the frame of the International VO Alliance.

References

1. Leborgne, D. et al. 2002, *SF2A 2002*, 325
2. G. Paturel et al., 1997 A&AS 124, 109
3. Prugniel, Ph. and Soubiran, C. 2001 A&A 369, 1048
4. Ph. Prugniel et al. 2001, *Mining the Sky*, ed. AJ. Banday et al. (Springer-Verlag), 683

A Galactic Model as a Useful Tool for Virtual Observatories

Annie C. Robin, Céline Reylé, Sébastien Picaud, and Bernard Debray

CNRS UMR6091, Observatoire de Besançon, BP1615, F-25010 Besançon Cedex, France

Abstract. Many observational data will be available through Virtual Observatories, but at some wavelengths and in some directions, no data yet exist. Concerning star counts, photometry and classifications, galactic models can be usefully included in VO for statistical inputs. The Besançon model of population synthesis well calibrated at visible and NIR wavelengths can be useful to complement observational data: (i) to extrapolate the stellar statistics to fainter magnitudes, at new wavelengths or interpolate at directions where no data are available yet; (ii) to compute bayesian classification probabilities. The revised version of the model as well as the classification scheme are briefly described here.

1 The Besançon Model of the Galaxy

The model is based on a semi-empirical approach, where physical constraints and current knowledge of the formation and evolution scenario of the Galaxy are used as a first approximation for the population synthesis. The model involves 4 populations (disc, thick disc, spheroid and bulge) each deserving a specific treatment.

Previous versions were described in Bienaymé et al. (1987ab) and Haywood et al. (1997ab). Many new features have been implemented and the parameters have been re-constrained with new available data. The revised version is fully described in Robin et al (2002). Main modifications include:

- New parameters for the thin disc, due to significant improvements after Hipparcos (potential, velocity ellipsoid and local luminosity function).
- A new fit of the overall parameters of the spheroid (density law, local density and Initial Mass Function (Robin et al., 2000).
- New constraints on the thick disc population: kinematics (Ojha et al, 1996), density law, local density, metallicity (Robin et al., 1996), and the Initial Mass Function at low masses (Reylé & Robin, 2001).
- The disc is warped and flared in the external part following new constraints from near-infrared star counts (Derrière & Robin, 2000).
- A triaxial outer bulge has been included (Picaud et al, in preparation).

2 Model Predictions and Applications

The model produce reliable predictions in different wide band photometric systems, such as the Johnson-Cousins (UBVRIJHKL), GAIA G band, RGU system.

The Sloan system is being studied. Samples of model predictions are given in Robin et al. (to be submitted to A&A). The revised model will be soon available on the web in place of the old version[1].

Applications run from simple star count predictions in many directions, to photometric and astrometric statistics (proper motion, radial velocities distributions, colours, etc.). However predictions may not be reliable at very low latitudes ($|b| < 5^o$) where the spiral structure and the local extinction features are not modeled in detail.

3 A Bayesian Classification Tool for a Virtual Observatory

We propose to use simulated catalogues to help classifying objects from their observed properties. The simultaneous use of different kind of observables (photometry, astrometry, spectroscopy) to classify the objects will significantly improve the efficiency of the classification. Its adds, to the standard classification method, informations from current knowledge of the stellar distribution and stellar kinematics in the Galaxy.

The model allows to compute easily Bayesian probabilities for a star having measured observables to belong to a given population, to a given type, or simply the probability to have been found in this direction, assuming that the model is correct. The model may also help to determine which observable is more suitable for getting a better efficiency in the classification. These informations may be usefully added to virtual observatories to complement the observational data when the observables are too few to be conclusive.

References

1. Bienaymé, O., Robin, A. C., Crézé, M., 1987a, A&A, 180, 94
2. Bienaymé, O., Robin, A. C., Crézé, M., 1987b, A&A, 186, 359
3. S. Derrière, A.C. Robin, 2001. In *The new era of wide-field astronomy* ASP Conference series, vol 232, Edited by Roger Clowes, Andrew Adamson, and Gordon Bromage, p.229.
4. Haywood, M., Robin, A. C., Crézé, M., 1997, A&A, 320, 440
5. Reylé, C., Robin, A. C., 2001, A&A, 373, 886
6. Ojha, D. K., Bienaymé, O., Robin, A. C., Crézé, M., Mohan, V., 1996 A&A, 311, 456
7. Robin, A. C., Haywood, M., Crézé, M., Ojha, D. K., Bienaymé, O., 1996, A&A, 305,1 125
8. Robin, A. C., Reylé, C., Crézé, M., 2000, A&A, 359, 103
9. Robin, A. C., Reylé, C., Derrière, S., Picaud, S., 2002, to be submitted to A&A

[1] **http://www.obs-besancon.fr/modele/**

The SolarWeb Semantic Data Model for Describing Relationships Between Heterogeneous Solar Databases

Isabelle F. Scholl, Romain Linsolas, and Eric Legay

Institut d'Astrophysique Spatiale, CNRS Université de Paris-Sud, Bâtiment 121, F-91405 Orsay Cedex, France

1 Introduction

The number of archives of solar observations is continuously growing and the location of their storage is getting more and more scattered. Consequently, the number of tools or websites existing for consulting these observation catalogs (ie. metadata) is equally increasing. Therefore the interest of a single program capable of accessing multiple and heterogeneous archives is obvious. The Solar Web Project, developed by the MEDOC/IAS team, is clearly designed to come up to these expectations. In this paper, we are presenting the underlying data model of this Virtual Solar Observatory software.

2 The Model

The main objective is to design a generic tool that implements technical aspects of the problem and not specific business logic. Therefore, adding new data sets is a role for the Data Manager and no longer for the software developer. This explains why all parts related to metadata manipulation are kept external to the main source code.

The architecture of SolarWeb creates a homogeneous semantic level above remote site catalogs or archive access systems. The key component consists of a dictionary which role is two-fold:

∘ during the pre-processing query stage, it maps input fields into specific semantics for each archive system.

∘ in return, it translates results coming from heterogeneous sources in a high level dynamic view.

As shown on Fig. 1, the model is built from 3 components corresponding to 3 angles under which the model can be viewed.

First, the *System Manager* view is the description of all sites included in the virtual observatory in terms of access types and parameters representing their specific access methods and archive content and structure.

Second, the *Data Manager* view is a data dictionary describing high level definitions of all the solar observation concepts. In this dictionary, a cross-link between all archives content is established by the System Manager.

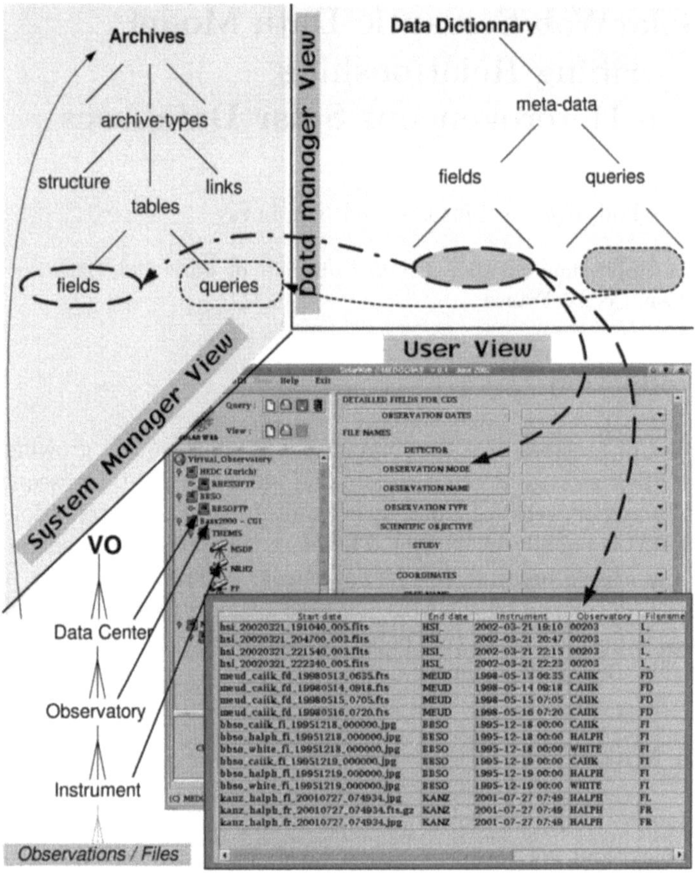

Fig. 1. The model and its views

Third, the *User view* is a higher level above the 2 other views. It shows the user components of a virtual observatory in terms of data center, observatories or missions and instruments. At this stage, the user visualizes information about solar observations only through the data dictionary definitions: these are homogeneous concepts and no longer instrument specific FITS keywords.

The dictionary is also used to build the graphical user interface in terms of window content and user profiles.

3 Future Works

The SolarWeb programs is currently under testing. Two SQL databases, two CGI archive access systems, and several FTP archives are already available through the server. The next step is to provide Data Managers with a graphical interface to specify dictionary's rules.

References

1. I. Scholl, E. Legay, R. Linsolas: "An XML-based generic tool for information retrieval in solar databases", in ASP Conf. Ser. Vol 281, ADASS XI, Conference on Astronomical Data and Analysis Software and Systems (Victoria, CA., 2001).

Development of a Multi-Mission Data Server at LAEFF

E. Solano, R. Gutiérrez, B. Montesinos, C. Morales, J. García

LAEFF, P.O. Box 50727, 28080 Madrid, Spain

Abstract. Astronomical archives have been built independently of each other showing a variety of access levels and policies. This has caused that their remote interoperability is neither easy nor efficient as demonstrated by the fact that the simultaneous query to multiple databases is presently done, most of the times, by hand.

A system providing a common search interface and delivering data in a common format and in absolute wavelength and energy units is, therefore, an essential requirement for an efficient multi-wavelength astrophysical research. In this paper, the main functionalities and contents of the system that is being developed at LAEFF are presented. In the long-term, the system is aimed at being incorporated at the Astrophysical Virtual Observatory (AVO) federation consolidating LAEFF as the reference centre for astronomical databases in Spain.

Functionalities

In a first phase, the system focused on the integration of ultraviolet archives (EURD (`http://www.laeff.esa.es`) and INES (`http://ines.laeff.esa.es`)) with flux-calibrated Strömgren photometry. At the time of writing we are working in the accessibility to the XMM and ISO data as well as other missions in the UV and spectrophotometric catalogues in the optical.

- **Query:** The query to the access catalogues is made by means of an HTML fill-in form permitting different types of queries. Particularly interesting is the possibility of making queries by object type. A built-in name resolver has been implemented allowing queries using any of the object names provided by SIMBAD.
- **Result from search:** The HTML output format includes several utilities like the data preview that allows to visualise either individual spectra or a combined average spectrum of different missions that can be further saved as a GIF file (MultiPlot option, see Fig. 1). Spectra can be retrieved using different compression and packed formats. Retrieval of individual spectra or grouped by mission and/or instrument is also implemented.
- **Future implementations:** The efficiency of a multi-mission data server can be greatly enhanced if tools for data evaluation and analysis are developed. We are presently working in the implementation of calibrations to determine the stellar physical parameters as well as the possibility of fitting synthetic models atmospheres. Also, the possibility of using neural networks for spectral classification and characterisation is also contemplated.

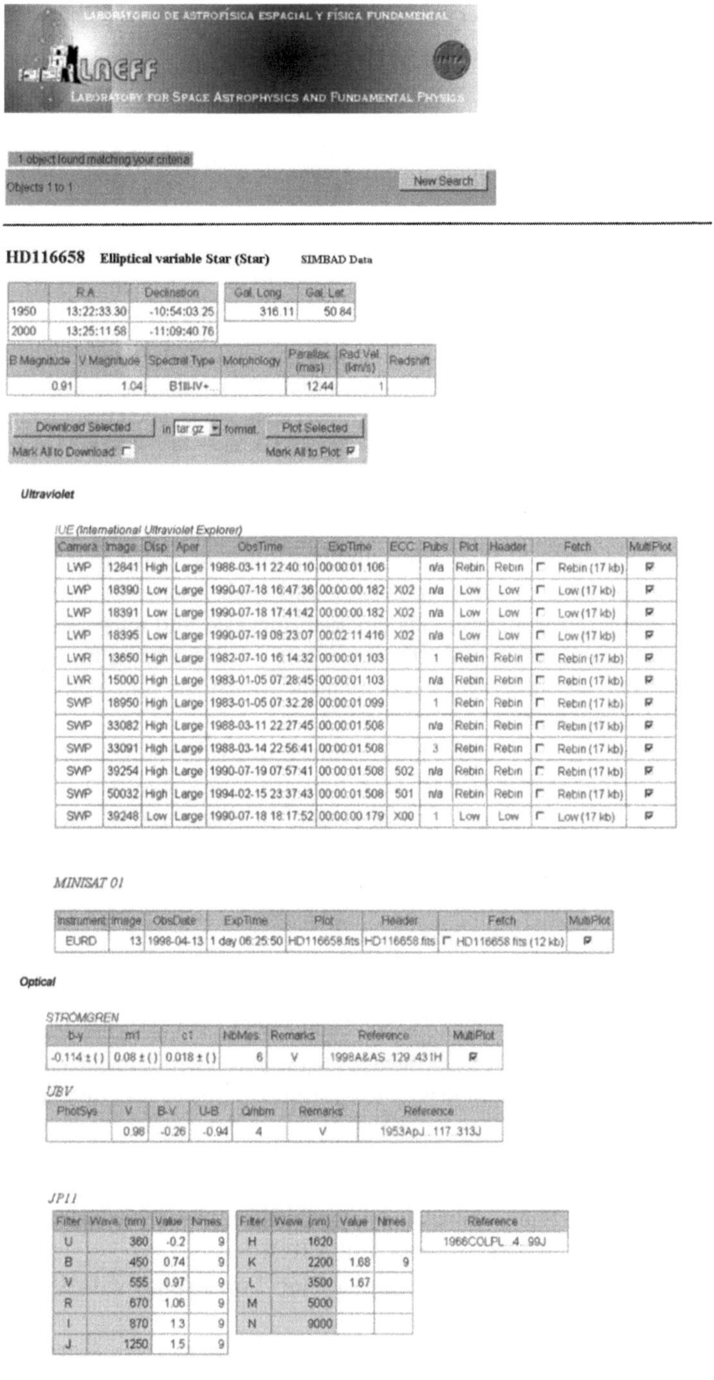

LABORATORIO DE ASTROFÍSICA ESPACIAL Y FÍSICA FUNDAMENTAL

LAEFF

LABORATORY FOR SPACE ASTROPHYSICS AND FUNDAMENTAL PHYSICS

1 object found matching your criteria

Objects 1 to 1 — New Search

HD116658 Elliptical variable Star (Star) SIMBAD Data

	R.A.	Declination
1950	13:22:33.30	-10:54:03.25
2000	13:25:11.58	-11:09:40.76

Gal. Long	Gal. Lat
316.11	50.84

B Magnitude	V Magnitude	Spectral Type	Morphology	Parallax (mas)	Rad Vel (km/s)	Redshift
0.91	1.04	B1III-IV+...		12.44	1	

Download Selected in tar.gz format. Plot Selected

Mark All to Download: ☐ Mark All to Plot: ☑

Ultraviolet

IUE (International Ultraviolet Explorer)

Camera	Image	Disp	Aper	ObsTime	ExpTime	ECC	Pubs	Plot	Header	Fetch	MultiPlot
LWP	12841	High	Large	1988-03-11 22:40:10	00:00:01.106		n/a	Rebin	Rebin	☐ Rebin (17 kb)	☑
LWP	18390	Low	Large	1990-07-18 16:47:36	00:00:00.182	X02	n/a	Low	Low	☐ Low (17 kb)	☑
LWP	18391	Low	Large	1990-07-18 17:41:42	00:00:00.182	X02	n/a	Low	Low	☐ Low (17 kb)	☑
LWP	18395	Low	Large	1990-07-19 08:23:07	00:02:11.416	X02	n/a	Low	Low	☐ Low (17 kb)	☑
LWR	13650	High	Large	1982-07-10 16:14:32	00:00:01.103		1	Rebin	Rebin	☐ Rebin (17 kb)	☑
LWR	15000	High	Large	1983-01-05 07:28:45	00:00:01.103		n/a	Rebin	Rebin	☐ Rebin (17 kb)	☑
SWP	18950	High	Large	1983-01-05 07:32:28	00:00:01.099		1	Rebin	Rebin	☐ Rebin (17 kb)	☑
SWP	33082	High	Large	1988-03-11 22:27:45	00:00:01.508		n/a	Rebin	Rebin	☐ Rebin (17 kb)	☑
SWP	33091	High	Large	1988-03-14 22:56:41	00:00:01.508		3	Rebin	Rebin	☐ Rebin (17 kb)	☑
SWP	39254	High	Large	1990-07-19 07:57:41	00:00:01.508	502	n/a	Rebin	Rebin	☐ Rebin (17 kb)	☑
SWP	50032	High	Large	1994-02-15 23:37:43	00:00:01.508	501	n/a	Rebin	Rebin	☐ Rebin (17 kb)	☑
SWP	39248	Low	Large	1990-07-18 18:17:52	00:00:00.179	X00	1	Low	Low	☐ Low (17 kb)	☑

MINISAT 01

Instrument	Image	ObsDate	ExpTime	Plot	Header	Fetch	MultiPlot
EURD	13	1998-04-13	1 day 06:25:50	HD116658.fits	HD116658.fits	☐ HD116658.fits (12 kb)	☑

Optical

STROMGREN

b-y	m1	c1	NbMes	Remarks	Reference	MultiPlot
-0.114 ± ( )	0.08 ± ( )	0.018 ± ( )	6	V	1998A&AS..129..431H	☑

UBV

PhotSys	V	B-V	U-B	Q/nbm	Remarks	Reference
	0.98	-0.26	-0.94	4	V	1953ApJ...117..313J

JP11

Filter	Wave. (nm)	Value	Nmes
U	360	-0.2	9
B	450	0.74	9
V	555	0.97	9
R	670	1.06	9
I	870	1.3	9
J	1250	1.5	9

Filter	Wave. (nm)	Value	Nmes
H	1620		
K	2200	1.68	9
L	3500	1.67	
M	5000		
N	9000		

Reference
1966COLPL...4...99J

Fig. 1. Example of a result from search

The CATS Database as a Tool to Study Radio Sources

Oleg V. Verkhodanov[1,2], Sergei A. Trushkin[2], Vladimir N. Chernenkov[2], and Heinz Andernach[3]

[1] Theoretical Astrophysics Center, Julian Maries Vej 30, Copenhagen, Denmark
[2] Special Astrophysical observatory RAS, Nizhnij Arkhyz, Russia
[3] Depto. de Astronomía, Univ. Guanajuato, Mexico

Abstract. The CATS database unifies more than 350 catalogues mostly radio astronomical. It allows a user to execute over these data some operations including the search in a given area and cross-identification with a user list. The task requests can be sent via Web-page or e-mail. FTP access the CATS data is opened too.

The large majority of over three million different radio sources cataloged by now have not been optically identified and their nature is their nature can only be extrapolated from previous, sparser surveys, and surprises are sure to happen...: the classic example is the source 4C+04.66 known from 60ies. Only in 1978 this variable radio source was identified with the remarkable X-ray and optical close binary, SS 433.

On the other hand the most radio sources (outside the Galactic plane) are extragalactic, while nearby sources ($z < 0.2$) are rare amongst these sources.

In 1995, based on their radio astronomical interests authors decided to create the CATS *Astrophysical* **CAT***alogues* **S***upport System* database firstly as a system for support of the RATAN-600 radio telescope observations.

CATS Structure

The CATS database operates under the Linux OS (1200 MHz) at the PC-server without involving any commercial database systems. Therefore, we have designed our database using the simple principles, permitting us to add new data, tables, procedures:

- every new catalogue of objects with the descriptions of the catalogue (.dat, readme, .doc, .txt) are contained in a itself subdirectory named after the catalogue; a few of Russian texts are in KOI-8 codes.
- the binary programs or shell scripts for local operations with the catalogue are also in the same directory;
- main catalogues parameters, references and names of the search programs are stored in the special file cats_descr for reading by manager program;
- the common search program is created for simple ASCII tables, the option "*pattern*" are used for descriptions and reading of fields.

The CATS homepage (`http://cats.sao.ru`) has entries to the list of catalogs, descriptions of catalogues and forms of search procedures.

Most of the CATS catalogues are source lists from radio surveys or other tables of radio measurements (NVSS, WENSS, FIRST, GB, TEXAS, PMN, PKS, MRC, 6C, 7C...).

Main Functions

Now the CATS database foresees the following procedures.

- There are the documentation of two kind for each catalogue: short **readme** and detailed **catalog.html**, accumulated in **CATS_list.html**. The *context search* (based on *Glimpse system*) around HTML-docs, Author's index (JavaScript) and references to NASA ADS system are available.
- Selection of objects from one to all catalogues matching user–specified criteria: equatorial and Galactic coordinates, flux densities, observing frequency or band etc. It works as CGI-script from Web HTML 3.0 forms, as e-mail query and as local network procedures for RATAN users.
- Cross–identification of different catalogues and matching for one or for a list of sources. This procedure works as CGI-binary from Web forms or as e-mail query.
- In the first time the spectra fitting and plotting in GIF format are introduced **on-fly** from the multi–frequency catalogues: Kuhr'79/81, PKSCAT90, RC(COLD), SNR, ZGT, PSR, POL_Tabara. The Java-applets and CGI-bin work for spectra plotting.
- The e-mail interface of the CATS is designed like using keywords and parameters from the e-mail sent to the address **'cats@cats.sao.ru'**. Send a letter with a single word in body: *'help'* to obtain instruction.
- Anonymous ftp–access is possible to CATS from Web pages, recommended from CATS_list or via ftp–server (`ftp://cats.sao.ru/`).

CATS is being expanded and updated continuously. Now it comprises over 350 radio, IR, optical and X-ray catalogs including most of RATAN–600 catalogues and **all** published radio source catalogues with more than 1000 entries; a lot of smaller but unique tables of radio data; hypertext documentation; coordinate transformation, software for spectral data processing. It occupies near 10 Gb.

Among the nearest plans of the CATS development, we suggest to incorporate 2MASS and ISO catalogues, design the CATS source name resolver, include HTML cross-references to the other database in the output files, i.e. make a real step the virtual radio observatory.

Acknowledgements

This work was partly supported by the Russian Foundation of Basic Researches (grant No 02-07-90038). OVV thanks the LOC for a support during a visit at the Conference.

To Creation of Virtual Radio Observatory

Oleg V. Verkhodanov[1,2], Vladimir K. Kononov[1], Sergei A. Trushkin[1], Alexander I. Kopylov[1], Natalia V. Verkhodanova[1], Alexander G. Gubanov[3], Olga P. Zhelenkova[1], and Vladimir N. Chernenkov[1]

[1] Special Astrophysical observatory RAS, Nizhnij Arkhyz, Russia
[2] Theoretical Astrophysics Center, Julian Maries Vej 30, Copenhagen, Denmark
[3] Astronomical Institute of St.Petersburg University, Russia

Abstract. The approach to the creation of the virtual radio observatory in Russia is described. The proposed system is based on the system operating in the present time. They are the astrophysical CATalogues Supporting system, the system "Evolution of radio galaxies" RATAN-600 Data Bank ODA and database of clusters of galaxies. The problems of these systems interacting and developing are discussed.

General Background

Under the Virtual Observatory (VO) we understand a system (sample of tools) of access to presenting data of different types: catalogues, spectra, one- and multi-dimensional experimental data and images. VO in the specific sense should be an analogue of the real observatory. The access includes not only network packages, but the systems of selection, cross-identification and processing of information (e.g. calibration). The virtual observatory can closely interact with the virtual library, but it is not a topic this discussion. The problem of the VO creation appeared in the development of the electronic tools of processing, storage and access to the astronomical information both observational and simulational (modeled). Let us note 2 important moments which we use in our work:

1. We consider the presence of real data as the main factor for VO (even more important the operational tools).
2. The system providing access to the distributed resources exists almost in each astronomical institute.

Necessity of the VO creation is discussed now at various workshops. We just note why we started to work in this problem:

1. VO is necessary for the complex study of astronomical data and search for hidden dependences (e.g. in data mining);
2. Is is needed for fast search for information about studied object via cross-references;
3. It should realize a simple and universal mechanism of complex data visualization;
4. It should provide a possibility to process huge arrays by external astronomers;
5. It should give a tool for critical analysis of results to escape false discoveries.

Problems of Realization

Is it possible to realize VO in local observatory, e.g. in SAO (Russia)?
The question is not monosemantic, especially taking into account comparatively low access to the observatory electronic resources and and only few organized archives. However, there are some examples of systems, operating in SAO and servicing the users. They are

1. Database of astrophysical catalogues CATS (`http://cats.sao.ru`) (Verkhodanov et al., 1997);
2. Bank of observational data of RATAN-600 (*oda.ratan.sao.ru, rain.sao.ru*) (Kononov, Mingaliev, 1998; Kononov et al., 1999);
3. Database of clusters of galaxies (`http://www.astro.spbu.ru/CLUSTERS/`) (Gubanov, Titov, 1997);
4. Server "Evolution of radio galaxies", containing galaxy SEDs and archive of observations of radio sources at 6 m telescope (`http://sed.sao.ru`) (Verkhodanov et al., 2000);
5. As an additional item we plan to organize a close interaction with the French *HyperLeda* database being a tool for studying galaxies (see P. Prugniel contribution) which will allow one to study separate characteristics of objects.

Besides the systems mentioned above, SAO has a sample of local data processing systems with the designed algorithmic and graphic interface parts for data obtained both in observations and in simulation. The software allows users to process initial data (the FADPS and *fgr* systems, Verkhodanov, 1997) and continuum radiospectra including of infrared and optical data (the graphic system *spg*, Verkhodanov, 1997). The problems of unification and organization of the joint systems discussed in SAO in the previous century (e.g. Kononov, Lipovetsky, 1994; Trushkin et al., 1996). In that time we discussed a problem of global unification of data formats under FITS ideology (Kononov, 1995; Verkhodanov et al., 1997). However, just now we have a real chance to finish organization of these systems interaction. We define our niche in the radio wavelength range and identifications (optical, infrared, X–ray) connected with radio sources.

We mention the following problems to be solved to create radio VO in SAO:

- The presence of the real data for VO (the most important problem) [`partly solved`];
- Intrusion of the FITS format ("FITSation") for all the data types and subsystems [`partly solved`];
- Organization of a standard graphic interface understanding data recorded in various FITS families [`partly solved`];
- Organization of standard request systems for search and calculations [`solved`];
- Organization of cross-identification system for all catalogues [`solved`];
- Use of the cross-references to studied objects in the results of VO processing;
- Organization of standard requests to the servers containing maps with the brightness distribution [`partly solved`];

- Design of the sources name resolver [`partly solved`];
- In a case of slow access, one has to organize *e-mail* access in a batch-mode [`solved`].

As we see the most of the problems are solved. After the elaboration of global interface and visualization systems will be finished, one may say about a protoplast of the virtual radio observatory.

Acknowledgements

This work was supported by the Russian Foundation of Basic Researches (grants 02–07–90038, 02–07–90247, 01–07–90274). NVV and OVV thank the LOC for a support during a visit at the Conference.

System to Study Evolution of Radio Galaxies as a Part of a Virtual Radio Observatory

Natalia V. Verkhodanova[1], Oleg V. Verkhodanov[1,2], Alexander I. Kopylov[1], Olga P. Zhelenkova[1], and Vladimir N. Chernenkov[1]

[1] Special Astrophysical observatory RAS, Nizhnij Arkhyz, Russia
[2] Theoretical Astrophysics Center, Julian Maries Vej 30, Copenhagen, Denmark

Abstract. The system for study of radio galaxies evolution using photometry data has been created at the site `http://sed.sao.ru`. It consists of several components and includes the data base of spectral energy distributions (SED), software tools for SED fitting, infrared maps for extinction calculation. It also contains an image archive of radio galaxies observations at the 6 m telescope of SAO RAS and papers by close topics.

Introduction

The fast grow of the photometric observations and slow spectroscopic data collecting (Parijskij et al., 1996) have forced us to create a system for the estimation of a redshift and age of radio galaxies.

This system (called "Evolution of radio galaxies") is developed at the site `http://sed.sao.ru` and solves the following tasks:

- estimation of ages when fixed z;
- estimation of ages and z;
- development of the archive of optical observations of RC radio galaxies in the FITS, JPEG, PS formats with text comments;
- development of the archive of publications;
- organizing a standard access by HTTP, FTP and e-mail protocols;
- organizing local SED operations.

The SED fitting is based on the two main models known as PEGASE (Fioc, Rocca-Volmerange, 1996; Le Borgne, Rocca-Volmerange, 2002). and GISSEL (Charlot, Bruzual, 1993; Bolzonella et al., 2000).

We use a standard sample of filters U, B, V,Rc,Ic,J,H,K, g,r,i et al. plus the published SDSS and Hubble Space Telescope bands.

To calculate extinction in the corresponding bands, the FITS-maps from (Schlegel et al., 1998) are used.

Algorithms

The two main algorithms are developed to produce the best fitting of the photometric observational data to the SED curves.

The first one is a smoothing algorithms applied to the SED curves to simulate observations in the current band:

$$S_{ik} = \frac{\sum_{j=0}^{n} s_{i-n/2+j} f_{jk}(z)}{\sum_{j=0} f_{jk}(z)}$$

s_i – the initial synthetic SED, S_{ik} – the smoothed SED by the k-th filter, $f_k(z)$ – the transmission in the k-th filter, "compressed" in $(1+z)$ times when "moving" along the SED, $j = 1, n$ - pixel index in a curve of filter transmission.

The second algorithm is the standard SED fitting through χ^2 minimization procedure:

$$\chi^2 = \sum_{k=1}^{Nfilters} \left(\frac{F_{obs,k} - pSED_k(z)}{\sigma_k} \right)$$

Access Modes and Plans

To include this system to the astronomical software we developed the standard access modes as FTP, HTTP and e-mail. These modes are close to ones applied in the CATS (`http://cats.sao.ru`) database. An example *e-mail* form is shown below:

```
seds start
object 3C65; model PEGASE1,  type=E
z_limits: 0 6,    age_limits: 200 16000
B=23.73+0.21 V=23.57+0.2 R=22.36 I=20.81
extinction off
seds end
```

This system unified with the CATS data base is suggested to be an important part of the virtual radio observatory. It is planned to designed a close interaction in the part of data access and exchange between this server and the CATS.

Acknowledgements

This work was partly supported by the Russian Foundation of Basic Researches (grant No 02-07-90038). NVV and OVV thank the LOC for a support during a visit at the Conference.

References

1. G. Bruzual, S. Charlot: Astrophys. J., **405**, 538 (1993)
2. M. Bolzonella, J.-M. Miralles, R. Pelló: Astron. Astroph., **363**, 476 (2000), astro-ph/0003380
3. M. Fioc, B. Rocca-Volmerange: Astron. Astroph., **326**, 950 (1997)
4. D. Le Borgne, B. Rocca-Volmerange: Astron. Astroph., **386**, 446 (2002)
5. Yu.N. Parijskij, W.M. Goss, A.I. Kopylov et a.: Bull. SAO, **40**, 5 (1996)
6. D.J. Schlegel, D.P. Finkbeiner, M. Davis: Astrophys. J., **500**, 525(1998)

The Russian Virtual Observatory Project Insight

Vladimir V. Vitkovskij, Olga Zhelenkova, Nadejda Kalinina,
Vladimir Chernenkov, and Vladimir Shergin

Special Astrophysical Observatory of RAS, Nizhnij Arkhyz, Russia

Abstract. The main principles and approaches of the Russian Virtual Observatory (RVO) Project are described.

The more wide view on the problem of Virtual Observatory (VO) creation makes it clear that the VO concept mast be expanded. There are no principal differences between data stored in archive and data receiving from telescope or data computing from numerical experiment. There are no big differences between the real science experiment and students training that. There are small differences, in the collaboration sense, between the live association of investigators and virtual one. Therefore the VO would be include not only data collections and software, but the experimental complexes, the education/training features and the virtual intercommunities. As the base principles of RVO creation it is offered:

- uniting the information archives and the information sources (telescopes) in the one distributed system;
- association in that system of experimental complexes, archives and data centers of RVO by means of network access and standard Internet interfaces;
- creation for the requests for deriving of an information uniform source portal that try to satisfy request in existing information basis, and for want of impossibilities it, statement in turn on providing of experiment on some RVO telescope that adequate to the request (see Fig. 1);
- "network transportation" of the contributor to data sources, instead the data transportation to the contributor, implying under it extended access of the user to information, computing and tool resources including a possibility of use on basis RVO own programs and hardware;
- Creation on the RVO basis of an Open Virtual Medium (OVM) of the astronomical education performing function of tutoring, reference (encyclopedic) system and means of communion.

Principal components RVO will be:

- Centers of astronomical data.
- Automated telescopes and acquisition systems.
- Computing facilities and/or a system of access to supercomputer centers.
- System of remote access to information and technical resources.
- System of the resources schedule and control.

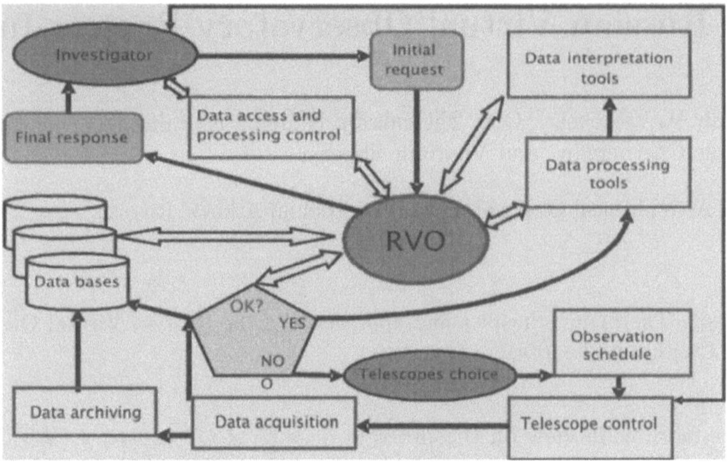

Fig. 1. Scheme of the RVO data request

- The self teaching consulting model.
- Internet portal "The Russian Virtual Observatory".
- Open Virtual Medium of the astronomical education.

The general properties of a Remote Access System (RAS) (supported by the MNTP projects 22/04 and 25/01; the RFBR grant 00-07-90014) is supposed:

- The territorial distributed system.
- Multi-mode quality: the simultaneous experiments are provided on the several instrumental units
- Connection to the system can be realized by any standard interfaces as ATM, Ethernet, etc.
- The interactive Internet access possibility by means of a standard web-browser.
- The resources scheduling/controlling contains planning and experiment fulfilling.
- The data exchange band with in a local net is up to 655MBps (Internet exchange speed is defined by outer channels band with). It is admitted the remote tasks input and on-line results getting.
- The system realizes simultaneous transmit video images and service audio, video conference connection during the experiment that allow to execute all together resources.
- It is provided for multi level authorization access.

One of the RVO goals is the training for the students of the main Russian universities, with the possibility for them to take part of the scientific investigations. At the present time a remote access system for BTA and RATAN-600 implies the possibility for research and experiments for students and post-graduate students.

Possibilities for a Convenient Desktop Access to the GRID and to Virtual Observatories

Hans Weghorn

University of Cooperative Education, Rotebühlplatz 41, 70178 Stuttgart, Germany

1 Introduction

Whether the upcoming virtual observatories (VO) will be operated as independent isolated systems or as integrated global solution, will have big impact on the usability of these systems. An astronomer working on a certain scientific question, or on particular objects will desire to collect all available and relevant information in a convenient manner. Today, there are certainly many useful sources for observational data and related contents on the WEB, but all these services appear extremely different and require individual handling. A similar situation may be avoided for the planned VOs, if the below discussed aspects for the user interfacing and for service integration of the VO sites are regarded.

2 User Access Point

Nowadays desktop computers at office working places typically are equipped with operating systems like Microsoft Windows^{tm} and all the related products for text processing, spread sheet calculation, graphical tools, and many other useful utilities. In contradiction to that, scientific computing mainly bases on UNIX systems. Both worlds can be integrated as single point of access for the user by different system approaches, because in the recent years desktop PCs have reached or even exceeded the power of workstations, which are required as user interface for scientific applications. For a seamless merge of these access technologies from the user point of view, different Linux-based projects – namely VM ware^{tm} (`www.vmware.com`), CYGWIN (*cygwin.com*), WINE (`www.winehq.com`) – can be employed. In principal, in all these concepts a host operating system (OS) supplies an embedded or simulated environment for a second OS. This can be achieved by providing a complete layer of OS functions (CygWin, WINE), or by simulating an entire virtual computer (VM ware). Which construction realizes the best solution depends on the local IT infrastructure, but these concepts enable the user to benefit from all the tools of different OS worlds with only a few technical restrictions, while handling only one physical computer.

Another related point of discussion is the WEB access to astronomical services. Here a mismatch can be identified between the used WEB technologies and the high-performance requirements of astronomical applications. XML and JAVA-based systems certainly are capable of exchanging textual data with a certain efficiency [2], but these technologies are insufficient for handling huge astronomical contents, e. g., exchanging observational raw data. In consequence WEB-based services mainly can be used for querying and searching mechanisms, while for real-time control or interactive processing direct communication (e.g., UNIX login services) controlling an in-place data processing will yield much higher working throughput.

3 Agent-Based GRID Mining

While the goals of GRID-based concepts can be displayed quite easily [1], it appears much more difficult to implement a truly-working system with an intelligent task split, because of the many incoherent technical details on the different sites of service. As sample for distributed computing, the SETI project (`www.seti.org`) sometimes may be mismatched with GRID computing, but it is far away from the GRID idea, because it only shares CPU time in a hierarchal setup, which is controlled centrally through individual point-to-point communication links. Between the clients in SETI there is no interaction. Agent technology, which is one of the most recent research topics in computer science [3], can solve the requirements of an automated interoperation of VOs. For enabling an agent or multi agent mining mechanism in cooperating VOs, a common data representation and appropriate processes would be required, which enable the necessary module interactions [4, section 16.1.3]. The collaborative use of VOs will more require to solve questions of information management – like it was done already for other areas [4, section 6.4.2] – rather than to supply a sharing model for CPU time. Presumably mobile agents, which are sent out for a return journey to autonomously collect the desired contents from the distributed heterogeneous sources, will realize the best technical solution.

4 Concluding Remarks

Agent technology is capable of coupling the individual VOs in a manner that the distributed astronomical contents can be used as unique source. For this, appropriate services have to be installed at all the contributing sites, and a similar internal site structure certainly would be advantageous. With an adequate local interfacing, the user will be able to take best advantage of the new valuable services.

References

1. M. Kunze: 'e-Science and the GRID'. In: *Toward and international virtual observatory, Garching, March 2002*, this book issue
2. H. Weghorn: 'Reducing Data Bandwidth in Web Transfers'. In: *WWW/Internet 2002, Lisbon, November 2002*, ed. by P. Isaías (IADIS Proceedings, 2002)
3. N.R. Jennings, et al.: 'A roadmap for agent research and development', Journal of Autonomous Agents and MultiAgent Systems, 1, **1**, (1998), pp. 7–36
4. G.F. Luger, and W. Stubblefield: *Artificial Intelligence: Structures and Strategies for Complex Problem Solving*, 4th edn. (Addison-Wesley, New York 2002)

Applying Data Compression Methods in Astronomical Archives

Hans Weghorn

University of Cooperative Education, Rotebühlplatz 41, 70178 Stuttgart, Germany

1 Introduction

The operation of virtual observatories will require extremely huge data archives. Capacities of hundreds of TB up to several PB will exceed the capabilities of reasonable-sized IT systems. This fundamental problem may be moderated by introducing compression techniques for the archived files. Today FITS appears as common storage format, but it stores data in native number representation like its generated by detectors and processing software, instead of using a convenient compressed coding. A compacted storage would also help to increase download speed on the network (especially on the Internet), and it would also reduce the required disk space during local astronomical processing.

2 General vs. Astronomy-Specific Compression Methods

How storage space can be saved, early was investigated in computer science. Algorithms like Huffman [1] and Lempel-Ziv [2] coding were developed for minimizing the size of text coded documents (ASCII), but these are also working on astronomical contents to a certain extend. Unfortunately, high-performance image compression methods like JPEG / MPEG are not feasible for scientific applications, because with these the original data is not preserved fully [3].

On base of the specific properties of its data contents, astronomy should define own compression techniques for achieving best performance in respect to size and processing speed. For the particular field of interferometric imaging this was already investigated [3], but a more general concept is required now for the VOs. Usable methods can be invented straight forward in different difficulty, effort, and efficiency levels. A very simple approach is to measure the truly required number range and map this to a coding bit length starting from a certain bias value (linear transformation). Instead of the fixed bit width coding, Huffman or Lempel-Ziv adopted to the truly-required bit width instead of 8 bits (ASCII length) will further reduce file size. An alternative could be the use of a look-up-table because rarely all possible sample values will be used in a data set. One very specific property of astronomical data is that high signal often covers only a small fraction of the whole data field (consider, e. g., 2D images with spectral scans or compact bright sources). Hence, further space can be saved by dividing the data field into subregions with high and low dynamic range. In general, the entire data field can be reconstructed from a set of several or many subregions, each of which appropriately coded with minimal size.

3 Experiments on FITS Data Sets

FITS – the flexible image transport system – is an open standard file format applicable to astronomy [4]. Concerning compression up to now only the reservation of a header keyword is known [5], but a few FITS header extensions can already enable size compaction. The simplest method can be defined by allowing any bit length of data instead of only 8, 16, and 32 bits, while storing the debias value in the header. The more elaborated compression approaches should be added by some kind of plug-in definition rather than a deep coding rule within the FITS standard. This would make FITS even more flexible, and it would avoid specification upgrades for each improvement of the compression methods.

The sample data coming with the MIDAS processing package can be used to demonstrate the benefit of compression: The data files were used to measure the achievable compaction size (two samples in below table). The classical methods obtain values between 22% and 81% for the data content, and values down to 17% for the FITS header (in terms of the original size). The simple astronomy-specific methods also reach the order of these ratios for data. Further experiments and investigations will presumably show that the described elaborated techniques can exceed the performance of the conventional methods.

File		Huffman	Lempel-Ziv	Bit depths	Bit depths + LUT	Entropy
casth	header	40%	19%	87,5%	88%	34%
	data 32 bit	37%	27%	75%	35%	18%
ngc2997	header	57%	44%	87,5%	88%	26%
	data 32 bit	56%	45%	75%	42%	31%

4 Concluding Remarks

As seen, there are good reasons to apply data compression to astronomical storage formats and archives. Using native formats like in the current standard, and applying de/compression manually is not a convenient approach, because everything that can be automated should be automated. By overcoming manual handling, astronomers win more space for the relevant working contents.

References

1. R. G. Gallager: 'Variations on a Theme of Huffman', IEEE Transactions on Information Theory, vol. IT-24, **6**, (1978), pp. 668–674
2. T. A. Welch: 'A Technique for High Performance Data Compression', IEEE Computer, vol. 17, **6**, (1984), pp. 8–19
3. H. Weghorn, et al.: 'Real-Time Image Compression Method for Astronomical Inferograms'. In: *Science with the VLT Interferometer, Garching, June 1996*, ed. by F. Paresce (Springer, Heidelberg 1997) pp. 397–398
4. D. Wells, et al.: 'FITS: A Flexible Image Transport System', Astron. Astrophys. Suppl. **44**, 363–370 (1981)
5. NOST specification: *Definition of the Flexible Image Transport System (FITS)*, March 1999 (NASA, source: `www.fits.org`)

Multiwavelength Studies of Microquasar GRS 1915+105

J.S. Yadav

Tata Institute of Fundamental Research, Homi Bhabha Road, Mumbai 400 005, India

Abstract. The microquasar GRS 1915+105 has shown spectacular variability and has been extensively studied in multiwavelengths during last few years. The β class is supposed to be the most complex out of 12 separate X-ray classes observed in this source. This X-ray class is always accompanied by IR and radio flares and has been extensively studied in multiwavelengths which have greatly enhanced our understanding of accretion disk-jet connection.

1 Introduction

GRS 1915+105 has shown spectacular X-ray variability since its discovery in 1992 [1]. The superluminal relativistic radio jets were first seen in 1994 [2]. Recently, Belloni et al. (2000) have classified the X-ray activity of this source in 12 separate classes on the basis of their light curves and the color-color diagrams. Out of these, the β class is described as the most complex and always accompanied by IR and radio flares. The presence of "spike" in the X-ray light curve which separates the dips with hard and soft spectra clearly distinguishes this class from others. This X-ray class has been extensively studied in multiwavelengths.

2 Results and Discussion

In Table 1 we list details of some of the observations which have been studied in multiwavelengths. The source was in high/flaring state during the course of these observations. The first observation of 1996 October 7 belongs to class λ which is a "radio-quiet" state. All the other observations belong to class β which is clearly shown by the presence of "spike" and the half hour IR and radio flares (Yadav, 2001).

Eikenberry et al. (1998) have studied X-ray and IR profiles during 1997 August 14 observations and have found that the spikes in X-ray coincide with the beginning of IR flares. The IR intensity reaches its peak flux little after the X-ray peak. The IR flares have decaying phases very similar in their smoothness and time scale to the rising phases. On the other hand, the X-ray enters the fast oscillation mode after reaching the peak as mentioned earlier. During another simultaneous observation of β class in IR and radio on 1997 September 9, Mirabel et al. (1998) have confirmed the role of spike in initiating IR flares and have found that the radio flares follow the IR flares with some time delay. The profiles of

Table 1. SUMMARY OF THE SELECTED OBSERVATIONS OF GRS 1915+105.

Date	Exposure	ASM^a	Radio	Period^b	Ampl.(mJy)^b
1996 Oct 07	7000	98.8	3	-	-
1997 Aug 14	10000	44.5	29	$\sim$ 30	$\sim$ 12 †
1997 Sep 09	10000	81.8	47	$\sim$ 30	$\sim$ 60 §
1997 Oct 30	4000	86.7	200	$\sim$ 25	$\sim$ 200 ‡
1997 Oct 31	10000	48.1	170	$\sim$ 30	$\sim$ 150 ‡

^{a} Mean ASM flux (c/s) for a day, †IR flares (2.2 μm), § Radio flares (8.3 GHz), ‡ Radio flares (15 GHz), ^{b} IR or radio flare period (min) and flare peak amplitude

IR and radio flares are quite similar. Fender et al. (1999) have observed four relativistic superluminal radio jets during 1997 October/November and have seen radio flares of 20-30 min. period on 1997 October 30-31 in 15 GHz band.

Yadav (2001) has studied temporal and spectral properties of both the classes. From observed temporal properties it is suggested that classes λ and β of Belloni et al. (2000) are very similar except for (1) the presence of long soft dip in the beginning of the burst phase in case of class β while it is short and shallow in case of class λ, and (2) the short but frequent soft dips during the fast oscillations are relatively deeper and for longer times in case of class β than that in class λ. On the basis of calculated spectral properties and the accretion disk parameters it is suggested that the "spike" which separates the dips with hard and soft spectra of β class marks the beginning of the burst phase when the luminosity of the soft X-rays (5−15 keV) increases by a large factor [7]. This produces a major ejection episode of the synchrotron - emitting plasma termed as "baby jets" which are associated with infrared (IR) and radio flares of about half an hour period. This explicitly bring out the role of "spike" in initiating the IR and radio flares [8]. The radio emission during "baby jets" is consistent with synchrotron emission from an adiabatically expanding small cloud with flat radio spectrum and it can satisfactorily explain the delay observed between different frequencies [9].

References

1. Castro-Tirado, A.J., Brandt, S., & Lund, N. 1992, IAU Circ., 5590
2. Mirabel I. F., Rodriguez L. F., 1994, Nature, **371**, 46
3. Belloni T. et al., 2000, A&A, **355**, 271
4. Eikenberry S. S. et al., 1998, ApJ, **494**, L61
5. Mirabel I. F. et al., 1998, A&A, **330**, L9
6. Fender R. P. et al., 1999, MNRAS, **304**, 865
7. Yadav J. S. et al., 1999, ApJ, **517**, 935
8. Yadav, J. S., 2001, ApJ, **548**, 876.
9. Ishwara-Chandra, C. H., Yadav, J. S., Pramesh Rao, A., 2002, A&A, **388**, L33.

Automated Data Reduction and Analysis Tools for the VIMOS Integral Field Unit

Alessandra Zanichelli[1], Bianca Garilli[2], Marco Scodeggio[2], Davide Rizzo[3], Giampaolo Vettolani[1], and Olivier Le Fevre[4]

[1] Istituto di Radioastronomia, CNR Bologna, Italy
[2] Istituto di Astrofisica Spaziale e Fisica Cosmica, CNR Milano, Italy
[3] Osservatorio Astronomico di Brera – Merate, Italy
[4] Laboratoire d'Astrophysique de Marseille, Marseille, France

1 The VIMOS Spectrograph and the Integral Field Unit

The VIsible Multi Object Spectrograph (VIMOS, [1]) is a high multiplexing spectrograph with imaging capabilities installed on the third unit of the Very Large Telescope. The VIMOS field of view is split into 4 quadrants, each one covering $7' \times 8'$ of sky and is by all means a full spectrograph on its own.

VIMOS main characteristics are the capability to simultaneously obtain up to 800 object spectra in MOS mode, and the availability of a microlens-fiber unit designed to perform integral spectroscopy of a whole region of the sky: the Integral Field Unit (IFU). The IFU consists of 6400 microlenses organized in an 80×80 array (IFU Head) covering a maximum field size of $54'' \times 54''$. Microlenses are coupled with optical fibers that carry the light to a second set of microlenses on the 4 IFU masks. Output microlenses are arranged in four "rows" on each mask, playing the role of the slits for this kind of spectrograph. A focal elongator makes it possible to have 2 spatial samplings of $0.33''$ and $0.67''$ per fiber.

2 Handling VIMOS Data

When in MOS mode, VIMOS acquires up to 800 spectra in each exposure and, when observing with the Integral Field Unit, each exposure produces 6400 spectra. To successfully exploit the capabilities of an instrument like VIMOS it is thus required a rather new approach to the data reduction, based on the implementation of an almost completely automated pipeline for data processing.

The VIMOS Data Reduction Software (DRS, [2]) can be used as a stand-alone package, and is also incorporated in the VIMOS pipeline developed at the European Southern Observatory. A complete automation is achieved by means of auxiliary tables, storing all the information needed by the DRS procedures (e.g. optical distortion models) and the output information they produce. Fundamental is the IFU Table, which for each fiber lists parameters like relative transmission, fiber spatial profile on the detector, and the correspondence between fiber position on the IFU Head and spectra on the CCD. The main VIMOS DRS products consist of a set of 1D extracted, sky subtracted, fully calibrated spectra, plus a 3D datacube and a 2D reconstructed image for IFU data.

3 The IFU Data Reduction Software

The DRS for Integral Field Unit data has been designed to be as similar to MOS reduction as possible. However, there are some aspects peculiar to IFU spectra data reduction that need to be addressed by dedicated software procedures.

First, the high density of IFU spectra on the detectors causes some degree of cross-talk, i.e. flux contamination from neighbouring spectra. The amount of cross-talk in a spectrum depends on both the flux in the adjacent spectra and the fiber spatial profile. Correction of the cross-talk is done using the information on the fiber profile stored in the IFU Table.

A further step is the calibration of differences in fiber transmission efficiencies. Relative transmission coefficients are periodically computed by a calibration procedure which measures the differences in sky line flux on 1D extracted spectra from a twilight sky image or a set of shift-and-stare images. These coefficients are applied to science exposure spectra before sky background determination.

Sky subtraction for IFU data is different than for MOS ones, since on fiber spectra no regions of pure sky can be identified. For shift and stare observations a sky spectrum for each fiber is built by stacking all the 1D spectra for that fiber. When a single exposure is acquired, pure sky spectra are identified from the distribution of total intensities for all the 1D extracted spectra in a slit, and combined to give a sky spectrum for the given slit.

3.1 IFU DRS Products and Analysis Tools

Once obtained the 1D extracted, fully calibrated spectra from the 4 quadrants of each exposure, the final products of the IFU DRS are the 3D datacube and the 2D reconstructed image. The datacube is a 3D FITS image with the (x,y) axes sampling the sky region seen by the IFU Head and the z axis sampling the wavelength. The knowledge of the correspondence between the position of a fiber in the IFU Head and the position of its spectrum on the CCD allows for a complete and spatially coherent reconstruction of the given dataset.

The reconstructed 2D image is obtained by collapsing in wavelength the spectra in the 3D cube. The 2D reconstructed image is extremely useful in presence of crowded fields and/or faint objects, when the automatic DRS cannot guarantee optimal results. As an example, when dealing with crowded fields the 2D image is analyzed by an interactive tool in order to identify fibers in object-free regions to perform accurate sky subtraction and, at a later stage, to sum spectra coming from microlenses that fall on the same object.

References

1. O. Le Fevre, M. Saisse et al. In: Proc. SPIE Vol. 4008, *Optical and IR Telescope Instrumentation and Detectors*, ed. by M. Iye, A.F. Moorwood (2000), p.546
2. M. Scodeggio, A. Zanichelli, B. Garilli et al. In: ASP Conf. Ser., Vol. 238 *Astronomical Data Analysis Software and Systems X*, ed. by F.R. Harnden, Jr., F.A. Primini, & H.E. Payne (2001), p. 451

Classification of Active Objects in the Multiwavelength Parametric Space

Yongheng Zhao, Yanxia Zhang, and Chenzhou Cui

National Astronomical Observatories, Chinese Academy of Sciences, Beijing 100012, P.R. China.

Abstract. In this paper, we apply two algorithms for clustering analysis: support vector machines (SVMs) and learning vector quantization (LVQ) to classify AGNs from stars and normal galaxies. Our results indicate SVMs shows better performance than LVQ when classifying multi-wavelength data. Furthermore, these methods will be part of the toolkits of the International Virtual Observatory (IVO).

1 Method Used

LVQ is a supervised variation of the Kohonen self-organizing map (SOM) and shares the same network architecture as SOM. There are several versions of the LVQ algorithm for which the learning rules differ in some details. See Kohonen [1] for an explanation of the differences between these algorithms.

SVMs was originally developed by Vapnik [2], who is now leading this research. Now SVMs is widely used in prediction and classification. For more information on SVMs see Stitson et al.[3] and Gunn [4].

2 Chosen Sample and Parameters

We positionally cross-identify the catalogue of quasars and active nuclei [5] with the ROSAT Bright Source Catalog [6] and Faint Source Catalog X-ray sources, and cross-match the result with the USNO A-2.0 catalog. Then cross-identifying 2MASS released data, we get 1656 AGNs. Likewise, we apply stars from SIMBAD and galaxies from Third Reference Catalogue of Bright Galaxies [7] to get 3718 stars and 173 normal galaxies. Hereafter, active objects represent AGNs while non-active objects contain stars and normal galaxies. The chosen parameters from different bands are B-R (optical index), B+2.5log (ct), ct (source count-rate in the broad energy band), HR1 (hardness ratio 1), HR2 (hardness ratio 2), ext (source extent), extl (likelihood of source extent), J-H (infrared index), H-K (infrared index), J+2.5log (ct).

3 Results

Using the total sample as training set and test set, we apply LVQ and SVMs to classify the sample. The results are given in Table 1. Then dividing the sample to 2774 for training and 2773 for testing, we use LVQ and SVMs to classify the sets. The results are summarized in Table 2.

Table 1. Classified Result by LVQ/SVMs with the total sample for training and testing

classified↓known→	active objects	non-active objects
active objects	1617/1633	112/25
non-active objects	39/23	3778/3866
accuracy	97.6%/98.6%	97.1%/99.4%

Table 2. Classified Result by LVQ/SVMs with two parts from the sample

classified↓known→	active objects	non-active objects
active objects	795/767	76/24
non-active objects	33/61	1869/1921
accuracy	96.0%/92.6%	96.1%/98.8%

4 Conclusion

The results show that LVQ and SVMs are effective methods to classify active objects from non-active objects with multi-wavelength data, moreover, SVMs shows much more excellent performance than LVQ. We believe the performance will increase if the data are complete or the quality and quantity of data improves. Moreover, these methods can be used to preselect AGNs from large numbers of sources in large surveys avoiding wasting time and energy, when studying AGNs or cosmology. Our results will be applicable to other digitized and digital sky surveys that will be part of IVO.

References

1. T. Kohonen: Self-Organization Maps (Springer, Berlin 1995)
2. V.N. Vapnik: The Nature of Statistical Learning Theory (Springer, New York 1995)
3. M.O. Stitson, J.A.E. Weston, A. Gammerman, et al.: Theory of Support Vector Machines, Technical Report CSD-TR-96-17, Department of Computer Science, Royal Holloway College, University of London (1996)
4. S.R. Gunn: Support Vector Machines for Classification and Regression, ISIS Technical Report ISIS-1-98, Image Speech & Intelligent Systems Research Group, University of Southampton (1998)
5. M.P. Véron-Cetty, P. Véron: ESO Scientific Report 19 (2000)
6. W. Voges, B. Aschenbach, Th. Boller, et al.:A&A **349**, 389 (1999)
7. G. de Vaucouleurs, A. de Vaucouleurs, H.G. Corwin, et al.: Third Reference Catalogue of Bright Galaxies (RC3) (Springer-Verlag, New York 1991)

Science Data Archives: Feedback Relations

Olga Zhelenkova, Vladimir Vitkovskij, Nadejda Kalinina, Galina Malkova, Vladimir Chernenkov, and Vladimir Shergin

Special Astrophysical Observatory of RAS, Nizhnij Arkhyz, Russia

Abstract. SAO RAS archive consists from local archives that to receive observational data from diverse acquisition systems. To get access to observational data we work out a search information system. A server and web-based technology Oracle are expected to use for the realization of the system. Besides access to observation data the information system can use as a tool for investigation of the science archive.

Sets of the astronomical data: the catalogues, different wavelength range sky surveys, observation archives can under condition of organization of interoperable interactions give unique opportunities for scientific investigations. The virtual observatory concept forces in a new fashion the consideration of questions got in touch with science archives [1,2].

SAO RAS archive consists from local archives that to receive observational data from diverse acquisition systems and the one is a good test example for development of similar systems. The history of the digital observation data archive of the observatory begins with 80-th by realization of the sky strip survey in a radio wave range. For 20 years of its existence the general archive of SAO RAS has passed a few of development stages in data performance and storage, granting of service functions. The distinctive feature of the archive is a heterogeneity of data formats, data quality and methods of data processing. In archive there are data of about 40 observation methods, which are grouped in 15 archives.

We have put a task to analyze the available data and to organize archival system for transparent access to diverse local archives, giving attention to data quality, automation of data re-processing and integration with the astronomical catalogues and surveys. It is possible to consider development stages of the SAO RAS archival system the following chain: (1) *local archive*; (2) *basic archive*; (3) *general archive*; (4) *warehouse*; (5) *"VO-ready" archive.*

Each acquisition system has the local archive. Feature of these archives is the heterogeneity of the data: multi wave range, internal structure and format, acquisition methods and data processing. The basic archive is a collection of the local archives. It does not change formats and parameters of the stored data: in what format the data have obtained on an input of an archive, in the same format them have received at request. The general archive is the basic archive with organization of a network/web-data access. At this level the information system is added[3]. At this stage we decided to be limited to several standard types of queries: by observation date, observation program, device, coordinates and name of a source. In the information system the following storage architecture is used:

SIB (Service Information Block), FITS-header, file with observation data. The observation schedule database and the telescope parameter database are used for necessary correction of missing file parameters. The possible mistakes (date, name of the observation program, meteorological parameters) are corrected in parameters describing each observation file. The control and dynamic updating of the keyword dictionary after viewing file headers is carried out.

One of interest questions is an evaluation of science output of an observation science archive. Link between a data archive and a bibliographical database will provide a count of bibliographical references relate to certain observation data. Similar worldwide unification archive and bibliographical resources will allow evaluating a quality of archive and survey data and frequency of requests to them. Each archive user will improve an observation archive by means of a procedure of "qualifying" data. The procedure will consist from a confirmation of right or wrong parameters, data quality evaluation and so on. On the base of that information we could correct and range available data.

As an example of our practical result in feedback relations we bring an information about computer and astronomical resources which we used in the development of the web-archive (`http://sed.sao.ru/rc-archive.html`) of optical identification of SSRG sample from the RATAN-600 RC-catalog. The site contains superpositions of the VLA radio maps and the 6-m R band CCD images for 75 sample sources. We prepared more then 2600 files for web-presentation of the archive. Up to 40 files were used for each source. We worked with the next observation archives: SAO RAS archive, VLA archive; sky surveys: DSS1, DSS2, FIRST; USNO-A2 catalog; astronomical data processing system and packages: MIDAS, STARLINK (GAIA), WCSTools. For each object we used 2-5 arcmin sky area. The amount of computer storage was increased from 100 bytes (one row of the RC catalog) up to 10 MB data. The necessary storage volume is up more the 100000 times.

The work is carried out with support of the Federal Program "Scientific instrument development", project 22/04, and RFBR, the grant 99-07-90335 and 99-07-90334.

References

1. The National Virtual Observatory, Alexander S., Szalay, Astronomical Data Analysis Software and System X, ASP Conf. Ser. 2001. v.238, pp 3-12
2. The Russian Virtual Observatory. Principles and a Way of Realization, V. Vitkovskij et al., Astronomical Data Analysis Software and System XI, in press
3. The Project of Distributed Information System OASIS, V.Vitkovskij et al. 2000. Baltic Astronomy, v.9, N4, pp 578-582

The Hamburg/RASS Catalogue of Optical Identifications of ROSAT Bright Source X-Ray Sources

F.-J. Zickgraf[1], D. Engels[1], H.-J. Hagen[1], D. Reimers[1], and W. Voges[2]

[1] Hamburger Sternwarte, Gojenbergsweg 112, 21029 Hamburg, Germany
[2] Max-Planck-Institut für extraterrestrische Forschung, Postfach 1312, 85741 Garching

Abstract. We used digitized Schmidt direct and prism plates taken for the northern hemisphere Hamburg Quasar Survey (HQS) to obtain optical identifications for all high galactic latitude X-ray sources in the ROSAT Bright Source Catalogue (RASS-BSC) at $|b| \geq 30$ deg and $\delta \geq 0$ deg. In this part of the sky the RASS-BSC contains 5341 X-ray sources. We found 51% extragalactic counterparts (AGN, galaxies, clusters of galaxies), and 31% stellar sources. 3% are empty fields and for further 15% no unique identification could be given. The digitized database of Schmidt plates and tools to use this database for identification work will be implemented in the German Virtual Observatory (GAVO).

1 RASS Sample

The ROSAT Bright Source Catalogue (1RXS-B) contains a total of 18 811 ROSAT all-sky survey sources with a count-rate limit of $0.05\,\mathrm{cts\,s^{-1}}$ in the 0.1-2.4 keV energy range and with a typical positional accuracy of $\sim 30''$ [3]. The goal of our project was to obtain optical identifications for all 1RXS-B sources in the northern hemisphere with $\delta \geq 0$ deg and galactic latitudes $|\, b \,| \geq 30$ deg. In this part of the sky the 1RXS-B catalogue contains 5341 sources. The previously published first version of the Hamburg/RASS Catalogue [1] contained ~ 4100 identifications. The final *Hamburg/RASS catalogue of optical identifications* (version 3.0) containing all 5341 RASS-BSC sources is available on

`http://www.hs.uni-hamburg.de/hrc.html`

2 Method of Identification

For the identification of the optical counterparts of the remaining ~ 1200 sources we made use of the digitized objective prism and direct Schmidt plates of the Hamburg Quasar Survey [2]. A description of the digital data base is given by Engels et al. (this conference). The identification procedure was as follows. We first searched on the digitized direct plates for all possible counterparts in the X-ray error circle. A search radius of $60''$ was adopted. The limiting magnitude of the direct plates is $B \approx 20$. Then spectra were extracted from the data base of the digitized spectral plates at the positions of the objects. The limiting magnitude of the spectral plates is $B \approx 18.5$. Using these spectra the objects were classified interactively. Finally, all possible counterparts were entered into a catalogue and the most likely optical counterpart was identified by a flag.

3 Statistics

The statistics of the source classes is summarized in Table 1. The largest group of objects in the catalogue are the Active Galactic Nuclei comprising $\sim$ 42% of the RASS-BSC sources, followed by the stellar counterparts which represent $\sim$ 31% of the X-ray sources. Note that most F/G stars appear in the group of "bright stars" comprising $\sim$ 23% of stellar counterparts. Due to their brightness the spectra of these objects are saturated and therefore a detailed classification was not possible. In about 3% of the sources no optical counterpart was visible within $40''$ around the X-ray source on our blue direct plates ("blank field"). $\sim$ 15% of the sources could not be identified, partly due to overlapping spectra and low S/N ratio.

Table 1. Statistics of the various object classes identified in the catalogue.

class	number	%
AGN, QSO	2215	41.5
galaxies	238	4.5
galaxy clusters	262	4.9
M stars	197	3.7
white dwarfs	45	0.8
K stars	141	2.6
F/G stars	45	0.8
cataclysmic variables	26	0.5
bright stars	1219	22.8
blank fields	155	2.9
unidentified	798	14.9

References

1. Bade N., Engels D., Voges W., et al., 1998, A&AS 127, 145
2. Hagen H.-J., Groote D., Engels D., Reimers D.,1995, A&AS 111, 195
3. Voges W., Aschenbach B., Boller Th., et al. 1999, A&A, 349, 389

Author Index

ESO ASTROPHYSICS SYMPOSIA
European Southern Observatory

Series Editor: Bruno Leibundgut

G. Meylan (Ed.), **QSO Absorption Lines**
Proceedings, 1994. XXIII, 471 pages. 1995.

D. Minniti, H.-W. Rix (Eds.),
Spiral Galaxies in the Near-IR
Proceedings, 1995. X, 350 pages. 1996.

H. U. Käufl, R. Siebenmorgen (Eds.),
The Role of Dust in the Formation of Stars
Proceedings, 1995. XXII, 461 pages. 1996.

P. A. Shaver (Ed.),
Science with Large Millimetre Arrays
Proceedings, 1995. XVII, 408 pages. 1996.

J. Bergeron (Ed.),
The Early Universe with the VLT
Proceedings, 1996. XXII, 438 pages. 1997.

F. Paresce (Ed.),
Science with the VLT Interferometer
Proceedings, 1996. XXII, 406 pages. 1997.

D. L. Clements, I. Pérez-Fournon (Eds.),
Quasar Hosts
Proceedings, 1996. XVII, 336 pages. 1997.

L. N. da Costa, A. Renzini (Eds.),
Galaxy Scaling Relations: Origins, Evolution and Applications
Proceedings, 1996. XX, 404 pages. 1997.

L. Kaper, A. W. Fullerton (Eds.),
Cyclical Variability in Stellar Winds
Proceedings, 1997. XXII, 415 pages. 1998.

R. Morganti, W. J. Couch (Eds.),
Looking Deep in the Southern Sky
Proceedings, 1997. XXIII, 336 pages. 1999.

J. R. Walsh, M. R. Rosa (Eds.), **Chemical Evolution from Zero to High Redshift**
Proceedings, 1998. XVIII, 312 pages. 1999.

J. Bergeron, A. Renzini (Eds.),
From Extrasolar Planets to Cosmology: The VLT Opening Symposium
Proceedings, 1999. XXVIII, 575 pages. 2000.

A. Weiss, T. G. Abel, V. Hill (Eds.),
The First Stars
Proceedings, 1999. XIII, 355 pages. 2000.

A. Fitzsimmons, D. Jewitt, R. M. West (Eds.),
Minor Bodies in the Outer Solar System
Proceedings, 1998. XV, 192 pages. 2000.

L. Kaper, E. P. J. van den Heuvel, P. A. Woudt (Eds.), **Black Holes in Binaries and Galactic Nuclei: Diagnostics, Demography and Formation**
Proceedings, 1999. XXIII, 378 pages. 2001.

G. Setti, J.-P. Swings (Eds.), **Quasars, AGNs and Related Research Across 2000**
Proceedings, 2000. XVII, 220 pages. 2001.

A. J. Banday, S. Zaroubi, M. Bartelmann (Eds.),
Mining the Sky
Proceedings, 2000. XV, 705 pages. 2001.

E. Costa, F. Frontera, J. Hjorth (Eds.),
Gamma-Ray Bursts in the Afterglow Era
Proceedings, 2000. XIX, 459 pages. 2001.

S. Cristiani, A. Renzini, R. E. Williams (Eds.),
Deep Fields
Proceedings, 2000. XXVI, 379 pages. 2001.

J. F. Alves, M. J. McCaughrean (Eds.),
The Origins of Stars and Planets: The VLT View
Proceedings, 2001. XXVII, 515 pages. 2002.

J. Bergeron, G. Monnet (Eds.),
Scientific Drivers for ESO Future VLT/VLTI Instrumentation
Proceedings, 2001. XVII, 356 pages. 2002.

M. Gilfanov, R. Sunyaev, E. Churazov (Eds.),
Lighthouses of the Universe: The Most Luminous Celestial Objects and Their Use for Cosmology
Proceedings, 2001. XIV, 618 pages. 2002.

R. Bender, A. Renzini (Eds.), **The Mass of Galaxies at Low and High Redshift**
Proceedings, 2001. XXII, 363 pages. 2003.

ESO ASTROPHYSICS SYMPOSIA
European Southern Observatory

Series Editor: Bruno Leibundgut

W. Hillebrandt, B. Leibundgut (Eds.),
From Twilight to Highlight:
The Physics of Supernovae
Proceedings, 2002. XVII, 414 pages. 2003.

P. A. Shaver, L. DiLella, A. Giménez (Eds.),
Astronomy, Cosmology
and Fundamental Physics
Proceedings, 2002. XXI, 501 pages. 2003.

M. Kissler-Patig (Ed.),
Extragalactic Globular Cluster Systems
Proceedings, 2002. XVI, 356 pages. 2003.

P.J. Quinn, K.M. Górski (Eds.),
Toward an International Virtual Observatory
Proceedings, 2002. XXII, 341 pages. 2004.

MIX
Papier aus verantwortungsvollen Quellen
Paper from responsible sources
FSC® C105338

If you have any concerns about our products,
you can contact us on
ProductSafety@springernature.com

In case Publisher is established outside the EU,
the EU authorized representative is:
Springer Nature Customer Service Center GmbH
Europaplatz 3, 69115 Heidelberg, Germany

Printed by Libri Plureos GmbH
in Hamburg, Germany